Basic Statistics: An Introduction

George W. Summers
University of Arizona

William S. Peters
University of New Mexico

Charles P. Armstrong
University of Rhode Island

Wadsworth Publishing Company, Inc.
Belmont, California

Students: There is extra help available . . .

This text has been written to make your study of statistics as simple and rewarding as possible. But even so there may be concepts or techniques that will be difficult for you. Or you may want a little "insurance" for success as mid-terms and finals come up. If so, you will want to buy a copy of the *Student Supplement for Basic Statistics: An Introduction*. It's described in the preface of this book. Your campus bookstore either has it in stock or will order it for you.

Production Editor: Joanne Cuthbertson
Designers: Nancy Benedict, Gary Head
Copy Editor: Don Yoder
Technical Illustrator: John Foster

Printed in the United States of America

1 2 3 4 5 6 7 8 9 10—81 80 79 78 77

Library of Congress Cataloging in Publication Data

Summers, George W.
Basic statistics.

1. Mathematical statistics. I. Peters, William S. II. Armstrong, Charles P.
III. Title.
QA276.S893 519.5 76-18201
ISBN 0-534-00471-7

Contents

Preface

One objective of *Basic Statistics: An Introduction* is to illustrate the many uses of statistics in making decisions, in guiding operations, and in isolating important influences controlling desired results. Another objective is to explain statistical concepts and procedures in ways that foster full comprehension of the underlying concepts and logical structure. Verbal and graphic description with minimal mathematical formalization is the approach taken. Hopefully, accomplishment of these two objectives will incite a persistent interest in and appreciation for statistics that will lead to its use as an active instrument in the pursuit of trustworthy information.

Examples have been selected to emphasize a wide variety of applications. It is intended that some of these examples be so close to actual experience that almost everyone can see important uses for statistics in their lives. Readability is also an important means to this end, but without sacrificing understanding of the concept structure. This structure is essential to support more advanced work on the one hand and to enhance appreciation on the other.

Basic Statistics: An Introduction is designed for courses encompassing less than an entire academic year. The material is divided into four parts. In order, the first three are descriptive statistics, background for inference, and basic statistical inference. These three parts constitute an integrated logical sequence of topics leading up to drawing conclusions from incomplete quantitative information, the central objective of statistical analysis.

The final part consists of a collection of additional topics. Here there is no required sequence within the section. Instead this part offers a choice of topics that occupy important places in applied statistics. They are meant to serve as optional material to suit the purposes of different courses.

Chapters are divided into two parts of approximately the same length. Each part has its own exercise set. These usually make convenient study units. They also permit some additional flexibility in selection of topics and their order of presentation. At the end of each chapter, the set of overview exercises requires an integration of topics from the chapter and an occasional extension of concepts. A glossary of the most important equations introduced in the chapter follows the overview exercises. Answers to selected exercises appear at the end of the book.

In statistics, practice material is vital to the learning process. Comprehension is established when students know they can correctly choose and apply statistical procedures or explain basic concepts. At the same time, well-chosen contexts for practice material convey appreciation of the broad applicability of statistics.

While the text has its own practice materials and can be used alone, an additional learning aid has been provided. The student supplement furnishes students with individualized learning materials that are based on frequent detailed feedback of correct responses. Statistics is a subject which builds new concepts largely by synthesizing concepts presented earlier. Most students learn best by making certain they fully understand each new concept when it is introduced. Feedback helps to provide this assurance. The supplement consists of self-correcting exercises, programmed learning modules, and practice examinations. There is an exercise set for each half chapter in the text. The exercises themselves appear in the earlier portion of the supplement and detailed solutions appear in the latter portion. Programmed learning modules are given only for the first three sections of the text. This is the content that will be common to most courses and it should furnish an adequate foundation for the additional topics without the aid of programmed modules. Multiple-choice practice examinations are given at the ends of related chapters. Percentiles are presented along with the answers to the examinations and references to the place in the text from which the item came. The percentiles represent averaged experience in similar examinations by several hundred students.

References to applicable self-correcting exercises and programmed learning modules appear in the text. The self-correcting exercises are referred to at the end of each half chapter. References to the modules appear as marginal notations. A bar over the number ($\overline{2.4}$) indicates the beginning of text material for which programming is provided. A bar under the number ($\underline{2.4}$) correspondingly indicates the end. In all practice material one goal was to keep the numbers as simple as is consistent with the context of the exercise. Hence students can gain command of the procedure with a minimum of sheer arithmetic.

We want to express our sincere appreciation to Susan Potter, whose deep concern for learning guided her in preparation of the text exercises. We also want to thank the editorial staff of Wadsworth Publishing Company, Inc. Specifically, we want to mention Jon Thompson and commend his skill in organizing this project and his insightful suggestions. We also want to mention Joanne Cuthbertson, whose production skills are largely responsible for the book's style and appearance. Acknowledgment must be made of the help rendered by reviewers. They are: William Bray, SUNY Albany; Steve Bilkins, Schenectady Community College; Frank Landram, West Texas State University; William P. Lovell, Cayuga County Community College; Daniel Strong, SUNY Genesco; John Van Druff, Fort Seilacom Community College. Many of our colleagues and students have made excellent suggestions which we very much appreciate. Finally, we wish to thank the authors and publishers who granted us permission to reprint tables. Their permission is specifically acknowledged in each case.

George W. Summers
William S. Peters
Charles P. Armstrong

Introduction

A person about to invest considerable time and effort in learning something about a new field usually wants the answers to at least three questions:

1. What is the general nature of the field?
2. What contribution does the field make to human affairs?
3. What motivates people to devote their careers to such a field?

This introduction is intended to provide brief answers to these questions.

The Nature of Statistics

In popular usage, the term *statistics* refers to collections of numerical facts. Publications of government bureaus and chambers of commerce, sport handbooks, and financial pages of newspapers come to mind as cases in point. When such collections satisfy certain specifications with regard to the way they were collected, professional statisticians call them collections of data.

As a professional field, statistics refers to study and development of effective techniques for collecting, summarizing, and drawing conclusions from numerical data. The field can be divided into two branches: descriptive statistics and statistical inference.

Descriptive statistics is concerned with the use of tables, graphs, charts, and summary measures that bring out important features and patterns in collections of data. As an example, the U.S. Census Bureau gathers the ages of persons. Imagine the list of data representing these ages for just one city. Some summarizing technique must be used to make the list intelligible. A glance at the table of contents makes it apparent that descriptive statistics is covered in the first three chapters. But examples of those techniques appear throughout the rest of the book because almost every statistical study uses descriptive techniques to communicate the results.

Statistical inference is the second branch of the field of statistics and constitutes the major portion. This branch is concerned with how to use a sample to draw valid and reliable conclusions about the whole. For instance, in the development of expensive new chemicals for the treatment of disease, sampling is used to study the effects. It would be ethically questionable to put a chemical into widespread use before checking for possible dangerous side effects on a sample of the types of people who will be exposed. It would be economically foolish to bring the product to full-scale production and usage before demonstrating intended beneficial effects on enough properly selected patients to be convincing. Such considerations usually make the complete enumeration of a population prohibitive.

Careful study has established that high-quality samples, gathered in accordance with laws of probability, produce sound estimates of characteristics of the populations from which they come. Furthermore, such estimates are accompanied by information about the expected sampling error they contain. The theory underlying such sampling procedures enables a public opinion pollster to say, for example, that 42.4 ± 2.3 percent of registered voters in the United States prefer presidential candidate Jones on the basis of a properly drawn sample of a few thousand voters. We are practically certain that between 40.1 and 44.7 percent of all registered voters preferred Jones at the time the survey was made. An even larger sample would narrow the interval, but at a price which may be unwarranted.

Careful study has also established that properly gathered samples can be used to make reliable decisions about alternatives. For instance, an electrical appliance manufacturer must decide which of two newspaper advertisements will have the greater readership. A carefully chosen sample from the potential audience can indicate which has the greater appeal. Furthermore, the decision can be made with a preselected probability of error. For example, the finding might state that the second ad's readership is greater than the first's and that the odds are 99 to 1 that the statement just made is correct for the entire audience.

We have given only brief examples to illustrate statistical description and statistical inference. But when we consider that a great part of our everyday lives is concerned with making decisions in the absence of complete information, it becomes evident that there is a large field for application of statistical techniques. Which brand of a product will provide the best service per dollar of cost? Which activity will furnish the greatest satisfaction? How much will a contemplated investment earn? Which method of communication will provide the greatest comprehension? These are just a few examples of problems that confront large numbers of us.

No claim is made that statistics is a universal guide to action in all cases concerned with decision making without complete information. Sometimes the evidence at hand is judged conclusive without sophisticated sampling studies. At other times, the expected cost of error is too small to justify the expense of a statistical study. Then again, time constraints and lack of resources may present prohibitive obstacles. More and more often, however, statistical techniques are being called on to furnish additional guides to action with regard to important decisions in private and public affairs and in practically every field of the social and natural sciences. As dependable knowledge becomes more necessary, a field that concerns itself with efficiently designing the gathering of information is almost certain to gain in importance.

The ways in which statistics gathers information can perhaps be suggested, at least in summary, by describing reasonably representative steps in a statistical project. As with almost any systematic effort, a project normally starts with a statement of objectives, which is almost always somewhat tentative to begin with. Efforts are then made to render the statement more precise so that it will be easier to define success in pursuit of the objectives.

Given a refined statement of objectives, those responsible for the project construct a preliminary research design. The design answers questions such as: What is to be observed? How? When? Where? How will observations be analyzed—that is, what types of statistical evidence will support the alternative conclusions possible? The content of such procedures constitutes a major portion of the subject matter in statistics and in this book.

Most important studies include a pilot test of the preliminary design. The suppositions on which the design is based are checked. Any discrepancies are used to revise the preliminary design.

Once the research design is set, full-scale collection of data is undertaken. Usually a number of inspections are carried on during this process to protect the final quality of the data. No matter how sophisticated the subsequent analysis, it cannot compensate for unknown mistakes in data collection.

Some pitfalls in data collection can be illustrated by a brief discussion of public opinion polling using questionnaires. In the first place, care must be exercised to make sure that the proper audience is being polled. A list of registered voters for a certain county is not an adequate source for a sample of persons who can be expected to actually vote in an election within this county, for example. In addition, the questions on the questionnaire must be well stated. In surveying opinions about a political candidate, a question such as "You don't favor Gilligan, do you?" is obviously biased. But biases are not always so apparent.

Selection of appropriate language and vocabulary for the intended audience is another necessity. The form of questions is also of concern. Multiple-choice format has a great deal to recommend it provided that the choices are mutually exclusive and exhaustive, that is, provided that the choices cover all possible alternatives once and only once. Finally, respondents must have the information being sought. Requesting the heat of fusion for a particular chemical element would almost certainly not be an appropriate question for the general public. To bring many such features to light, a pilot test of a questionnaire in which respondents are asked to restate questions in their own words can be run.

When data collection has been completed, the final step in the study is taken. The data are processed and analyzed. Conclusions are drawn. The results are presented and interpreted in ways meant to disseminate correct impressions clearly and easily.

Nothing in the entire process is especially mysterious. On the other hand, failure to be guided by the experience of those who have worked in the field almost always leads to wasted effort.

Contributions of Statistics

Some areas in which statistics has made contributions of major importance are politics, public policy at national, state, and local levels, education, psychology, sociology, agriculture, medicine, business, industry, and consumer affairs. This list is suggestive rather than exhaustive.

Nearly everyone is aware of the impact of the public opinion polling organizations on the course of major political campaigns. These organizations also provide information which influences legislation and public policy. Additional information important for public policy comes from federal, state, and local agencies that survey and sample the labor force, the physical volume of production, the time pattern of prices, the size and composition of the population, automobile traffic and accident factors important to the design of thoroughfares, fire and crime occurrences, migration, weights and measures, and myriad other facets of a complex society.

In educational research, statistics provides a cornerstone in the foundation of measurement and testing theory. This theory, in turn, provides measures of capacity and achievement. Experimental designs from statistical theory are crucial in studies of learning effectiveness. Selection tests for college and university study in many fields are the product of elaborate statistical projects.

Psychology and sociology use statistical techniques in their search for meaningful dimensions of individual and group behavior. Where prior theory exists, statistical experiments are designed to test the predictive or explanatory adequacy of such theory.

Modern agricultural crop productivity owes a large debt to statistically guided experiments on the relative effectiveness of different types and combinations of fertilizers, moisture, soil chemicals, climate, seed stock, and similar factors. Animal husbandry has been similarly improved by statistical comparisons of different genetic strains, feeding regimens, disease treatments, and so forth.

In medicine, carefully designed national statistical studies verified the effectiveness of the Salk polio vaccine. Such studies provided the evidence leading to the surgeon general's report on the effects of cigarette smoking on health. Statistical principles are used continually in designing laboratory studies to evaluate the effectiveness of pharmaceuticals and all manner of treatments for abnormal conditions.

In manufacturing, quality control of output has become nearly synonymous with the branch of statistics that goes by the same name. Statistical analyses of time and motion studies are sometimes used in job design. Estimating demand for products, ascertaining buyer characteristics, evaluating effects of product features and advertising—all are based on statistical techniques. In purchasing, acceptance sampling is the statistical process for deciding whether a supplier's wares meet specifications. Statistically analyzed application blanks guide personnel selection in employment offices and selection of credit risks in financial institutions. Statistical aspects of mortality are the foundation on which the life insurance industry is built. Auditing is now often carried out in accordance with statistical sampling procedures. Consumer research groups use sophisticated sample surveys to gather many sorts of information on product effectiveness and consumer attitudes on important issues.

This compilation only begins to suggest the extent to which statistical techniques have permeated human affairs in the last 75 years. The more people work with current knowledge in the field, the greater the number of applications and extensions that are developed.

The Challenge of Statistical Work

It would be presumptuous to speak for professional statisticians as a whole in assessing motivations. Certainly, however, some gain intrinsic satisfaction from helping to search out controllable factors that influence other factors which affect us vitally but are not subject to direct control. Finding that the Salk vaccine was dramatically effective in preventing crippling poliomyelitis surely must have given the statisticians responsible for survey aspects a worthy challenge in prospect and deep satisfaction in retrospect.

The task of creating designs for extracting that last increment of information, the cost of which is just balanced by reduction in the expected cost of error, contains incipient elements of artistry. When the results of a designed statistical experiment show that enough data were collected to establish an anticipated difference without an expensive excess of observations, the statisticians involved can take satisfaction in having been efficient and thoroughly professional. Husbanding scarce resources constitutes a highly worthwhile endeavor for most people.

So long as humankind regards knowledge as crucial in their efforts to live in harmony with themselves and their environment, statistics is almost certain to occupy a major position among techniques for producing knowledge. A central objective of statistics is to provide the dependability which converts opinion into knowledge. Conducting statistical projects which determine that it is safe to predict or decide and which show how to make the prediction or decision provides a very high type of motivation. And now it is time to begin learning how to conduct such projects.

Part One
Statistical Description

1

Observations and Basic Operations

Numbers are the raw materials for statistics. Since the term **statistics** identifies an entire branch of knowledge, the numbers themselves are called **data**. The first part of this chapter describes how data arise from observations of people or other subjects in a study. This description involves, among other things, the types of measuring scales that produce data. Then some important basic operations with data are discussed.

Just as baseball managers place more emphasis on the number of hits a batter gets in an entire season than on his hits in a particular game, we in statistics usually are more interested in an entire collection of data than in just a few numbers in the collection. The second half of this chapter describes some of the most useful characteristics of collections of data.

Data

Early in a typical statistical study, a **unit of observation** must be selected. For instance, in studies of consumer purchasing patterns, the household is often treated as a single purchasing unit. For example, for purchases such as refrigerators the entire household is treated as a unit. In other studies, events such as traffic accidents or students taking a final examination are the units to be observed.

The particular property to be observed in each unit is the second choice that must be made. The beef consumed by a household in a week may be such a property. Other examples are the relative success experienced on an examination by each student in a class and a light bulb's ability to illuminate when proper voltage is applied.

After the unit of observation and the property to be observed have been selected, a decision must be made on how to measure the varying amounts of the property possessed by the units–that is, the measurement scale to be used. Beef consumption can be measured in dollars per week or pounds per week, for instance. A student's relative success on a test can be measured by her rank on number of items correct or by her rank on number of items correct less some fraction of the number incorrect. A light bulb's ability to illuminate can be measured as 1 if it lights and 0 if it doesn't, or the intensity of the light emitted can be metered.

Usually several choices for the unit of observation, the property, and the measurement scale are available. The decisions made almost always affect the time and money used for the study as well as the type of conclusion and the level of precision that can be achieved.

Having explained the basic terms used in gathering data, we are now ready to define data:

Definition | **Data** are the numbers that result when a designated measuring scale is applied to a specified property of each unit of observation.

Data are numerical values associated with a property possessed by every unit observed.

Measurement

Scales

Four types of measurement scale can be used for assigning numerical values to varying amounts of a property. They are, in order of increasing precision, nominal, ordinal, interval, and ratio scales. The characteristics of a scale that determine its level of sophistication are **order** (of numbers), **distance** (between numbers), and **origin** (a zero point). As we will see, the most primitive scale, the nominal scale, does not use any of these properties. The most sophisticated, the ratio scale, uses all three.

The results of applying a selected scale to the amounts of a property possessed by given units of observation are numbers. We have already defined such numbers as data. Other terms used are **measurement** and **observations**. Unless we state otherwise, you are to treat measurements, observations, and data as synonyms.

Nominal Scales When we use numbers merely to classify an observational unit with respect to a property, we are employing a **nominal** scale. If, for example, we assign the number 1 to the red pencils in a drawer, 2 to black pencils, and so forth, we are using a nominal scale. Note that the numbers can be chosen arbitrarily. We could just as well use any other numbers, making no use at all of the principle of *order* in the real-number system. We simply substitute numbers for class names in this case.

A typical situation in which nominal scales are useful is the two-state, or **binary,** situation. Consider, as an example, a light bulb to which the number 1 or 0 is assigned, according to its ability to illuminate or not. As bulbs pass an inspection station, they produce a stream of numbers that might look like this:

0,1,1,0,1,1,1,1,0,0,1, . . .

Many statistical techniques are designed to deal with data of this sort.

Classification systems having more than two categories also may use the nominal scale. Educational status and colors of automobiles are examples. Once again, the numbers chosen to identify the classes make no use of the *order* property of the real-number system.

Ordinal Scales A scale that uses the order principle in numbers but does not use the distance and origin principles is an **ordinal** scale. When a judge uses numbers to rank contestants in order of spelling skill, for instance, he applies an ordinal scale. Any number can be chosen to represent the amount of skill possessed by either the best or the worst speller, although, in practice, most investigators start with number 1. The natural numbers that follow the initial number can be used to rank the others. Successively larger numbers can represent either increasingly competent or increasingly incompetent spellers. Grades of meat, military ranks, and morale are examples of other properties to which ordinal scales can be applied.

In using an ordinal scale to rank spelling contestants, the judge makes no use of the distance principle in the sequence of natural numbers. Assume that contestants have been ranked from best to worst, and consider those who rank 2, 5, and 8, for instance. The distance from 2 to 5 is 3, and the distance from 5 to 8 is also 3. Yet the ordinal scale does *not* imply that the contestant who ranked second is as much better than the one who ranked fifth as the contestant who ranked fifth is better than the one who ranked eighth.

Interval Scales An **interval** scale makes use of both the order and the distance properties of numbers but does not use the origin property. The origins of interval scales are arbitrary. For instance, the zero point on the Celsius temperature scale is arbitrarily set at the freezing point of water. Other interval scales are Fahrenheit temperature, clock time, longitude, and test scores. The origins for clock time, for example, are the middle of the night and the middle of the day. Since these points in the daily cycle have no particular advantage over any other points, they represent arbitrary or conventional origins.

When two interval scales are used to measure the amount of change in the same property, the proportionality of differences (intervals) is preserved from one scale to the other. For example, suppose three pans of oil register 50°, 100°, and 120° on the Celsius scale. The oil which registers 50° on the Celsius scale registers 122° on the Fahrenheit scale. The other two Fahrenheit readings are 212° and 248°, respectively.

As an example, for the Celsius readings the proportion of differences between the intervals 100 to 120 and 50 to 120 is

$$\frac{120 - 100}{120 - 50} = \frac{2}{7}.$$

On the Fahrenheit scale the proportion of the respective differences is

$$\frac{248 - 212}{248 - 122} = \frac{2}{7}.$$

Thus the proportionality of differences is preserved as we change from a Celsius to a Fahrenheit scale or vice versa.

The origin or zero point on interval scales is arbitrary. For example, 0° Celsius obviously does not represent the condition of having no temperature. Instead, the zero point is simply an arbitrarily chosen reference temperature (the freezing point of water) with which we can compare other temperatures. For some purposes it may be more convenient to define another temperature scale for which the zero point is different. For example, the freezing point of alcohol could be chosen.

Ratio Scales **Ratio** scales are similar to interval scales in that they also use the order and distance properties of numbers. In addition, ratio scales have meaningful origins. Some examples of properties that can be measured on ratio scales are weight, length, sales, and number of children per family. In each case the origins (zero points) on such ratio scales signify that none of the property is present.

Application of Scales There are many reasons why it is important to understand which type of measurement scale is being used to produce a given collection of data. One reason is that most statistical techniques are not applicable to data arising from all four types of measurement scales. The majority of techniques discussed in this text are for use with data from interval or ratio scales. We will also examine some techniques for analyzing nominal and ordinal data, however.

Variables

Whatever the measuring scale being used in a study, many different values can be expected to result when the selected scale is applied to all the units of observation. To facilitate reference to such values, we will use the term **variable**. A variable is a symbol that can designate any value of the data that is produced when the selected measurement scale is applied to any unit of observation under consideration. Usually the symbols used for variables are either lowercase or uppercase letters near the end of the alphabet. More succinctly:

Definition | A **variable** is a letter that represents any numerical value a given measuring scale can produce in a given application.

We can, for instance, define the variable X to be the profit earned by any firm in the drug industry last year. Or X could be the rank of any contestant in a typing contest.

A variable may be either continuous or discrete. It is said to be **continuous** if, theoretically, it can assume any value between any two values possible on the measuring scale used. A variable which can assume values at only countable points is **discrete**. Weight in pounds and portions of pounds is a continuous variable. Theoretically, a person can weigh 112 pounds, 112.3 pounds, 112.3278 pounds, and so forth, to any desired degree of precision. On the other hand, the number of coins in a cash drawer on each of several days is a discrete variable.

In practice, of course, there is a limit to the accuracy with which a continuous variable can be measured. Length can be measured only to the nearest eighth of an inch with most yardsticks, for example. When a piece of cloth is measured by such a yardstick as being 18 inches long, it may, in fact, have any length from $17\frac{15}{16}$ to $18\frac{1}{16}$ inches.

In many instances, there is no need to carry measurements of continuous variables to the limit of accuracy possible with the measuring device. As an example, measuring weights to the nearest tenth of a pound may be sufficient for a given purpose. Actual measurements are *rounded* to this level of accuracy. A package that weighs 10.247 pounds is recorded as 10.2 pounds, and one weighing 10.281 pounds is recorded as 10.3 pounds. For measurements that fall exactly halfway between values on the scale, we will use the *round-even* rule:

Rule | Original measurements that lie exactly halfway between values on the rounded scale must end in an even number after rounding.

According to this rule, if we round to tenths of a pound, weights measured as 10.250 and 10.350 pounds are recorded as 10.2 and 10.4 pounds, respectively. (By rounding the first number down and the second up, we cause both to end in even numbers.) The rule tends to average out errors produced by rounding. Incidentally, we can say that these two observations have been rounded to three *significant digits.* The original observations were made to five significant digits; note that the final zero is relevant in both cases.

Operations with Statistical Data

1.1 In statistics, summation is almost certainly the operation most often used with data. This section reviews this operation and discusses the summation symbol along with two related concepts, indexes and parentheses.

Summation

In accordance with general practice among statisticians, we will let $\sum$(uppercase Greek sigma) represent the summing operation. Suppose that we are given the ordered collection

$$x = 7,4,10,2,1,8$$

with

$$i = 1,2,3,4,5,6,$$

where i is the **index** that specifies the ordinal number for any value of x in the collection. For example, in the foregoing, $x_1 = 7$ and $x_2 = 4$, because 7 is the first and 4 is the second value. With i as the index we can indicate that certain numbers in the collection are to be summed; thus

$$\sum_{i=1}^{6} x_i$$

is interpreted as follows:*

$$\blacksquare \quad \sum_{i=1}^{6} x_i = x_1 + x_2 + x_3 + x_4 + x_5 + x_6 = 7 + 4 + 10 + 2 + 1 + 8 = 32. \quad \blacksquare \qquad 1.1$$

Similarly, for the same collection,

$$\sum_{i=1}^{6} x_i^2 = 7^2 + 4^2 + 10^2 + 2^2 + 1^2 + 8^2 = 49 + 16 + 100 + 4 + 1 + 64 = 234.$$

*In this book the most important equations are set off by boxes and numbered. These equations are grouped for convenience at the end of each chapter in the Glossary of Equations.

When an entire collection of data is to be summed (as in the first example above), we typically substitute N for the upper number used with the summation sign. If we were to write $\sum_{i=1}^{N} x_i$ for the collection just defined, the interpretation would be exactly the same as that for $\sum_{i=1}^{6} x_i$.

It is apparent that the use of index notation permits us to describe summation for any collection of numbers, so long as we consider the members of the collection to be ordered.

Parentheses

When required for clarity, parentheses are used to designate the portion of a statement to which an operation such as summation applies. For example, we can define the ordered collection

$$t = 4,11,3.$$

Then

$$\sum_{i=1}^{N} (t_i - 6)^2 = (4 - 6)^2 + (11 - 6)^2 + (3 - 6)^2$$
$$= 4 + 25 + 9, \text{ or } 38.$$

But

$$\left(\sum_{i=1}^{N} t_i\right)^2 - 6 = (4 + 11 + 3)^2 - 6 = 324 - 6, \text{ or } 318.$$

Furthermore,

$$\sum_{i=1}^{N} (t_i)^2 - 6 = 4^2 + 11^2 + 3^2 - 6$$
$$= 16 + 121 + 9 - 6, \text{ or } 140$$

and

$$\sum_{i=1}^{N} t_i^2 - 6 = \sum_{i=1}^{N} (t_i)^2 - 6, \text{ or } 140, \text{ also.}$$ 1.1

Summary

The raw materials of statistical analysis are data—that is, observations, or measurements, stated numerically. These are synonyms for the numbers that describe the varying amounts of some property of interest possessed by each element in a set of units of observation. The amounts are expressed as values on an appropriate measuring scale, which may be of the nominal, ordinal, interval, or ratio type. Characteristics of order, distance, and origin in the real-number system form the basis for distinguishing among the four scales.

It is often convenient to use a variable in reference to a collection of data and to identify the variable with a letter from the end of the alphabet. Operations commonly encountered in statistics can be described succinctly and accurately in terms of such letters. Summation of the values which a variable may assume is needed so frequently in statistics that a special symbol, Σ, is used to signify this operation. This symbol is an uppercase sigma from the Greek alphabet. Indexes and parentheses sometimes appear in conjunction with the summation symbol.

See Self-Correcting Exercises 1A.

Exercise Set A

1. A quality control inspector wishes to classify shipments of replacement automobile bumpers. If a shipment is found to have no scratched bumpers, he records a 2. If any bumper is scratched, he records a 1.
   (a) Is the variable discrete or continuous?
   (b) What constitutes the unit of observation?
   (c) What type of measurement scale is being used?

2. A personnel worker in a certain plant is collecting data on employee attitudes. She presents the following statement to employees and asks them to choose the appropriate response: "I feel that I am performing a valuable service for society when I do my job well."

1	2	3	4	5
strongly agree	agree	no opinion	disagree	strongly disagree

   (a) Name the property under study.
   (b) What type of measurement scale is being used?
   (c) What constitutes the unit of observation?

3. What is the scale having the greatest precision that can be applied in collecting data on the following properties?
   (a) Human eye color
   (b) Human babies' lengths at birth
   (c) Number of accidents occurring at each intersection in a certain community during a 1-year period
   (d) Television viewers' choices among the different network offerings at a particular time
   (e) Concentrations of sodium chloride in samples of seawater
   (f) Human cranial circumference
   (g) Whether or not consumers prefer Sudso beer to all other brands
   (h) Number of minors per household in a community
   (i) Noon wind velocities at a weather station over a certain period of time
   (j) Academic rank in class at graduation
   (k) Year of birth
   (l) Percentages of persons receiving unemployment benefits in each state at a certain time
   (m) Stock prices at closing time on the New York Stock Exchange

4. Referring to Exercise 3, consider the values which the variable may assume in each of the instances mentioned.
   (a) In which of these is the variable discrete?
   (b) In which is it continuous?

5. Round each of the following numbers to three significant digits and then to four significant digits, using the round-even rule where necessary.
   (a) 3.2465
   (b) 0.004444
   (c) 1.02006
   (d) 1.300548
   (e) 3.555038
   (f) 11.2521

6. Consider this ordered list of values: 4,7,3,−1,0,1,4,−2.
   (a) To index this collection of values, we would start with $i = 1$. What value for the index i would we end with?
   (b) Give the value for: (i) x_1; (ii) x_4; (iii) x_5.
   (c) For the ordered collection given, evaluate each of the following:

(i) $\sum_{i=1}^{8} x_i$

(ii) $\sum_{i=1}^{4} x_i$

(iii) $\sum_{i=3}^{7} x_i$

(iv) $\sum_{i=1}^{6} (x_i - 2)$

(v) $\sum_{i=1}^{6} x_i - 2$

(vi) $\sum_{i=1}^{6} (x_i - 2)^2$

(vii) $\sum_{i=1}^{6} x^2 - 2$

(viii) $\sum_{i=1}^{4} i$

(ix) $\sum_{i=1}^{4} 2i$

(x) $\sum_{i=5}^{6} i$

7. Express each of the following using sigma notation, indexes, and parentheses as needed:
   (a) $x_1 + x_2 + x_3 + x_4 + x_5$
   (b) $x_5 + x_6 + x_7$
   (c) $x_2 + x_4 + x_6 + x_8$
   (d) $(x_1 - 1)^2 + (x_2 - 1)^2 + (x_3 - 1)^2$
   (e) $(x_1 - 1)^2 + (x_2 - 2)^2 + (x_3 - 3)^2$
   (f) $3 + 4 + 5 + 6$
   (g) $5x_1 + 5x_2 + 5x_3$
   (h) $(x_3 + a) + (x_4 + a) + (x_5 + a) + \cdots + (x_N + a)$

Properties of Collections of Data

Populations and Samples

In a statistical study, the collected data may come from all units of observation under consideration or from only a portion of those units. A credit manager may be interested in the balances owed by all credit customers. She may, however, have immediately available only the accounts of every tenth customer in the file.

Definition The complete collection of data of interest in a given study is called the **statistical population** or the **statistical universe**. Any portion of the statistical universe is called a **sample.**

In the foregoing example, balances owed by all credit customers constitute the statistical population, or universe, and balances owed by every tenth credit customer in the file constitute a sample from that population.

The observational units in the credit-customer example are the customer account cards or similar unit records of account. The data which constitute both the statistical population and the sample mentioned are the *numerical credit balances* on the cards, but not the cards themselves nor the people whose names appear on them. Statistical populations and samples always are *collections of measurements.* Hence they are always collections of numbers and never collections of units of observation.

Other examples of statistical populations and samples can be mentioned. Household incomes in the United States for a certain year make up an important statistical population for many studies. Household incomes in any one state would constitute a sample, though not a very representative one, from that statistical population. The number of children that each living adult female in New York City has borne could be a population under investigation. The number of children borne by every fourth such female selected from a complete list would be a sample.

The concepts of *statistical population* and *sample* are fundamental. Perhaps the single most important objective of statistical techniques is to find ways to infer the major characteristics of a statistical population from the information contained in a sample from that population.

We have seen that collections of data can be classified as populations or samples. Statistical populations also may be classified as **finite** or **infinite**. The lengths of a certain number of copies of a particular part produced by a lathe on a given day will serve as an example. If this collection of lengths is viewed as the entire collection of interest, the statistical population is finite. On the other hand, this collection of measurements can be viewed as a portion of all measurements that might conceivably be made on all output produced by all possible lathes of this type. For all practical purposes this latter statistical population has an infinite number of elements in it. In the first part of this book we will be concerned with finite populations of measurements. Later we will consider infinite populations.

The primary interest of investigators seldom centers on the individual numbers in a collection of data. Once the accuracy and relevance of the individual numbers have been established, the main concern becomes the entire collection considered as an entity in itself. The credit manager referred to earlier probably is not interested in each credit balance. Rather, she might want to know how the average amount currently owed compares with the average amount owed a year ago. She would, therefore, be interested in a *property* of the entire statistical population (the average) rather than in the individual numbers which make up that population.

To illustrate our discussion of properties of statistical populations, we can suppose that we have two statistical populations of x values:

$$x = 3,3,4,4,4,6 \qquad \text{and} \qquad x = 5,7,9,11,11,12,12,12,14.$$

The first population might be the number of automobiles passing a certain point on a street between 3:00 and 3:15 P.M. on six successive workdays. The second might be the data for the hours 5:00 to 5:15 P.M. on nine successive workdays. We can plot both on the same x axis. In Figure 1.1, numbers in the first population are represented by dots. Crosses are used for the second.

The first visual impression that emerges from Figure 1.1 concerns the relative locations of the two populations. The general location of the first is to the left of the

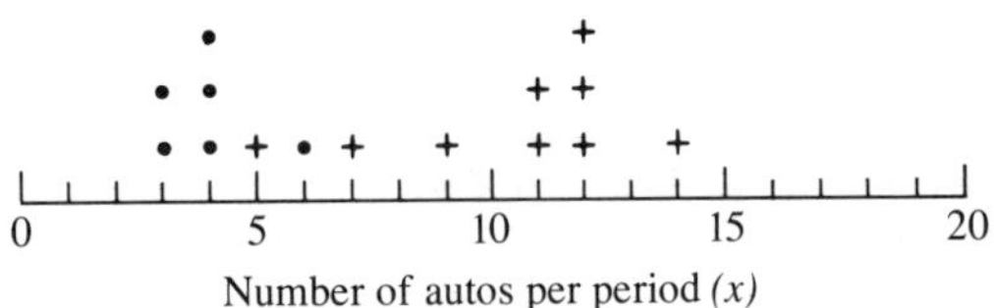

Figure 1.1 *Two statistical populations*

second. Generally, there seems to be more traffic in late afternoon than there is earlier in the afternoon. This first impression gives rise to the concept of **central location**.

The second impression gained from Figure 1.1 is that the crosses spread over a greater portion of the x scale than do the dots. In other words, there seems to be more variability in the amount of traffic during the late afternoon. This second impression gives rise to the concept of **dispersion**.

Central location and dispersion are two of the most useful basic properties of statistical populations and samples. They are of prime importance throughout the entire field of statistics. Later on, in Chapter 3, we will discuss skewness, a third important property. The remainder of this chapter is concerned with measures of central location and dispersion.

Measuring Central Location

1.2 **The Mean** This is the measure familiarly known as the "average." Because there are several measures of central location, or "average," however, each must have its own name. The full name of the measure we are about to describe is the **arithmetic mean**, but we usually refer to it simply as the mean. This is the most important measure of central location. The symbol commonly used to represent the mean of a statistical population is the Greek letter μ (lowercase mu, pronounced "mew").

Definition

In a statistical population of N measurements, the **arithmetic mean** is the sum of the measurements in the population divided by the number of such measurements. In symbols,

$$\mu = \frac{\sum_{i=1}^{N} x_i}{N},$$

which we usually state more simply as

$$\blacksquare \quad \mu = \frac{\sum x}{N}. \quad \blacksquare \qquad 1.2$$

Consider the two statistical populations in Figure 1.1. If μ_1 and μ_2 are the means of these populations, then

$$\mu_1 = \frac{3 + 3 + 4 + 4 + 4 + 6}{6} = 4$$

and

$$\mu_2 = \frac{5 + 7 + 9 + 11 + 11 + 12 + 12 + 12 + 14}{9} = 10\tfrac{1}{3}.$$

The two means are at the centers of "mass" of their respective populations and satisfactorily describe the impression that the central location of the first population is well below that of the second. The mean of the population illustrated by dots is only 4 autos per period while the mean of the other population is $10\frac{1}{3}$ autos per period. 1.2

The Weighted Mean Sometimes we need to find the mean of a statistical population that is composed of two or more populations for which we already know the means and the numbers of observations. Rather than adding and counting all the measurements again, we can find the mean of the combined population from the information we already have on the individual means and numbers of observations of the separate populations. This process is called *finding the weighted mean.*

Statement Consider k statistical populations with means $\mu_1, \mu_2, \mu_3, \ldots, \mu_k$. Let the numbers of observations in these populations be $N_1, N_2, N_3, \ldots, N_k$ and let the total number of observations in all these populations be N. For the combined population composed of all N measurements, the mean μ is the weighted sum of the means of the k original populations with weights N_i/N. Symbolically,

$$\blacksquare \quad \mu = \frac{N_1}{N}\mu_1 + \frac{N_2}{N}\mu_2 + \cdots + \frac{N_k}{N}\mu_k. \quad \blacksquare \qquad \textit{1.3}$$

Consider the two populations 2,7,9 and 4,18. The mean of the first population is 6 and the mean of the second is 11; that is, N_1 is 3, N_2 is 2, μ_1 is 6, and μ_2 is 11. If we combine these two populations into a single one, the result is 2,7,9,4,18, which has five observations; that is, N is 5. Since there are three observations in the first population and two in the second, the statement says that we can find the mean of the combined population by adding $\frac{3}{5}$ of the first mean to $\frac{2}{5}$ of the second. The weights are therefore 0.6 and 0.4. The result is $0.6(6) + 0.4(11)$, which is 8. As a check, we can use the basic definition of the mean to find that the mean of the combined population is also 8 by this approach.

The Median For a finite statistical population, the median $\tilde{\mu}$ (the Greek letter μ with a tilde to distinguish it from the mean) is loosely defined as the middle value when the population is arranged in order of magnitude. In the population composed of the three values 7, 4, and 8, the middle value *in order of magnitude* is 7. Therefore, $\tilde{\mu} = 7$.

The foregoing definition of the median is useful when there is an odd number of values, but it is sometimes ambiguous when there is an even number. In four families composed of 2, 3, 3, and 5 persons each, it is clear that the median is 3 persons per family. But for four families composed of 2, 3, 6, and 10 persons each, what is the median family size? Any value between 3 and 6 can be the middle value. By convention, the median in such a situation is arbitrarily defined to be the mean of the two values nearest the middle. For the last example, the median is the mean of 3 and 6, or 4.5 persons per family. In general:

Definition When a collection of data is ordered from smallest to largest, the **median** is the value of the $[(N + 1)/2]$th observation, where N is the number of observations.

For the first example above there are three observations, so $(N + 1)/2$ is 2 and the value of the second observation in order of magnitude is 7. For the last example, N is 4, $(N + 1)/2$ is 2.5, and the median is 4.5 persons per family.

These examples illustrate that the median, as compared with the mean, is sometimes a less precisely defined measure of central location. On the other hand, for a population with a few outliers—that is, numbers widely separated from the majority of values—the median usually is considered a better descriptive measure of central location than the mean. For example, in the population

$$1,2,2,3,3,3,4,4,32$$

the mean μ is 6 and the median $\tilde{\mu}$ is 3. The median is less influenced by the single extreme value and is more representative of the majority of the observations in the population.

Another important measure of central location is the **mode,** the most frequently recurring value in a population. In the foregoing population, the mode is 3 because this value occurs most often. We will defer further discussion of this measure until we have considered grouped data.

Measuring Dispersion

Recall Figure 1.1, in which the larger statistical population is more widely dispersed than the smaller. Having found ways to measure the property of central location, we now want to find a suitable measure for the *dispersion* of data within a collection.

Initially, it may seem reasonable to attempt to measure dispersion by finding the amount by which every value differs from some central reference value and averaging these deviations. Suppose we choose the mean as the reference value. For the population 3,4,4,9, the mean is 5. The *differences* from the mean are

$$-2, -1, -1, 4.$$

In statistics, the differences are called deviations from the mean.

Definition | In general, for the statistical population composed of measurements of the variable X with mean μ, a **deviation** from the mean is $(x - \mu)$.

For our example, given $x = 3,4,4,9$, then the values of $(x - \mu) = -2, -1, -1, 4$ are the deviations from the mean.

We now want to average the deviations from the mean. But when we sum these deviations, we find that $\sum(x - \mu) = (-2) + (-1) + (-1) + 4 = 0$. Indeed, in general,

$$\sum (x - \mu) = 0;$$

the sum of the deviations of the measurements in a statistical population from the mean is *always* zero. Positive deviations are canceled by negative deviations. Whether the values in a population are widely scattered or concentrated, the mean of the deviations will always be zero, because the sum of the deviations is zero. This is a consequence of the mean being the center of mass, or the balance point, of a population. Our initial attempt to develop a measure of dispersion is not a success.

The Variance We can keep negative deviations from canceling positive deviations by squaring all deviations from the mean and then averaging these squares. The result is the most pervasive measure of dispersion in the field of statistics, the variance. **1.3**

Definition | The **variance** σ^2 (lowercase sigma) of a statistical population is the mean of the squared deviations from the mean μ. In symbols,

$$\blacksquare \quad \sigma^2 = \frac{\sum_{i=1}^{N} (x_i - \mu)^2}{N} \quad \blacksquare \qquad 1.4$$

where $x_i - \mu$ is a deviation and N is the number of observations in the population. Usually, the index notation on the summation sign is omitted for simplicity.

Consider the statistical population 4,10,16. The mean μ is 10 and deviations of the observations from the mean are $-6,0,+6$. The squares of these deviations are $+36,0,+36$, and the mean of these squared deviations is 72/3, or 24. Hence the variance of the population is 24. We will, of course, get the same result by direct substitution in Equation 1.4:

$$\sigma^2 = \frac{(4-10)^2 + (10-10)^2 + (16-10)^2}{3}$$

$$= \frac{36+0+36}{3}, \text{ or } 24.$$

In addition to its central role in many other branches of statistics, the variance is a reasonably satisfactory measure of dispersion for descriptive purposes. For instance, consider a statistical population in which all the values are the same. There will be zero dispersion. When we calculate the variance, none of the values will be different from the mean. Consequently, all will have deviations of zero and the variance will also be zero, as we would wish. Furthermore, when the values in one population are more widely scattered than those in another population measured on the same scale, the value of the variance for the first will be larger than the value of the variance for the second. The value of the variance changes in the same direction as does the amount of scatter or dispersion.

One of two techniques for calculating the variance will usually prove to be effective for small populations. The first is used when the deviations from the mean turn out to be whole numbers. In this case we use the equation

$$\sigma^2 = \frac{\sum (x-\mu)^2}{N}$$

to find the variance. To illustrate this procedure we can use the smaller population illustrated in Figure 1.1. This population is listed in the left-hand column of Table 1.1.

Table 1.1 *Calculating Variance from Deviations*

x	$(x-\mu)$	$(x-\mu)^2$
3	-1	1
3	-1	1
4	0	0
4	0	0
4	0	0
6	2	4
$\sum x = 24$	$\sum (x-\mu) = 0$	$\sum (x-\mu)^2 = 6$

$$\mu = \frac{\sum x}{N} = \frac{24}{6} = 4 \qquad \sigma^2 = \frac{\sum (x-\mu)^2}{N} = \frac{6}{6} = 1$$

The first step is to find the mean μ. Having found the mean, we find and record in the middle column of Table 1.1 the deviations of each of the observations from the mean. As we expect the sum of these deviations is zero. We took this step only as a check; we don't need to do it to find the variance. The next step is to square the deviations and write the results in the right-hand column. (Appendix H lists the squares of numbers from 1 to 9.99.) Finally, we sum these squared deviations and average them to find the variance, which is 1.

The second technique for calculating the variance is used when we can see that the deviations from the mean will not be whole numbers. To illustrate this procedure we will use the larger of the two populations in Figure 1.1 (the population represented by crosses). This population appears in the left-hand column of Table 1.2.

Again we begin by finding the mean, which is found to be $10\frac{1}{3}$. This makes it apparent that deviations will not be whole numbers, so we abandon the earlier approach. Instead, we square the original observations, record the squares in the right-hand column, and find the sum of the squares. By doing a little algebra we can show that the variance can be expressed in terms of the sum of the squares of the original observations and the square of the mean. In symbols,

$$\sigma^2 = \frac{\sum x^2}{N} - \left(\frac{\sum x}{N}\right)^2$$

or

$$\blacksquare \quad \sigma^2 = \frac{\sum x^2}{N} - \mu^2. \quad \blacksquare \qquad 1.5$$

For our example in Table 1.2,

$$\sigma^2 = \frac{1025}{9} - (10\tfrac{1}{3})^2, \text{ or } 7.11\ldots.$$

Table 1.2 *Calculating Variance Directly*

x	x^2
5	25
7	49
9	81
11	121
11	121
12	144
12	144
12	144
14	196
$\sum x = 93$	$\sum x^2 = 1025$
$\mu = \frac{93}{9} = 10\frac{1}{3}$	$\sigma^2 = \frac{\sum x^2}{N} - \mu^2 = 7\frac{1}{9}$

The variance of the population represented by crosses in Figure 1.1 is 7.11, to two decimal places, while the variance of the population represented by dots is only 1. This shows that the variance of a widely scattered population is greater than the
1.3 variance of one that is more concentrated about the mean.

1.4 **The Standard Deviation** There are several reasons why another measure of dispersion based on the variance is preferable to the variance. Among them is the fact that the dimension of the variance is awkward. For example, the dimension of the values in the two populations in Figure 1.1 is *automobiles.* Since we *square* the deviations of these numbers from their mean, the dimension of the variance of either population is *automobiles squared.* For descriptive purposes, we would prefer a measure of dispersion stated in the original dimension of the data. Such a measure can be had simply by taking the square root of the variance. The square root of a number is another number which, when multiplied by itself, gives the original number. For example, the square root of 9 is 3, because 3 times 3 is 9.

The positive square root of the variance is the **standard deviation** σ, where

$$\blacksquare \qquad \sigma = +\sqrt{\frac{\sum (x-\mu)^2}{N}}. \qquad \blacksquare \tag{1.6}$$

The standard deviation is probably the most widely used measure of dispersion for descriptive purposes.

To obtain the standard deviation, we first calculate the variance and then find the square root of the variance. Square roots can be found in Appendix H. For the smaller population represented by dots in Figure 1.1 we found that the variance is 1. Hence the standard deviation will be $\sqrt{1}$, which is also 1. But now the dimension will be auto instead of auto squared. For the larger population represented by crosses, we know that σ^2 is 7.11 autos squared. In Appendix H we look under N until we find 7.11. Under $\sqrt{N}$ is a number opposite 7.11 that begins with a 2. Under $\sqrt{10N}$ is another number, which begins with an 8. Since 7.11 lies between 4 and 9 (that is, between 2^2 and 3^2), we know that $\sqrt{7.11}$ lies between 2 and 3. Consequently we choose the digits under $\sqrt{N}$ and fix the decimal point properly to obtain, to two decimal places,

$$\sigma = \sqrt{7.11} = 2.67 \text{ autos}.$$

Complete instructions for using the square-root tables are given in Appendix H.

Some insight into the usefulness of the standard deviation can be gained by recalling the two populations in Figure 1.1. Here is a rule of thumb: In a statistical population you can expect to find from 90 to 100 percent of the observations within three standard deviations of the mean. For example, in the smaller population in Figure 1.1 we know that μ is 4 and σ is 1. By adding and subtracting three standard deviations to and from the mean we find the interval from 1 to 7 automobiles per time period, and this interval contains all the data. Similarly, for the larger population the mean is $10\frac{1}{3}$ and the standard deviation is $2\frac{2}{3}$. Once again, the interval from $2\frac{1}{3}$ to $18\frac{1}{3}$ contains all the data in the population.

Another use for the standard deviation (in conjunction with the mean) is to find standard scores.

Definition The **standard score** z of an observation x in a statistical population is found by subtracting the mean μ from that observation and dividing the difference by the standard deviation σ. In symbols,

$$\blacksquare \qquad z = \frac{x - \mu}{\sigma}. \qquad \blacksquare \qquad \textit{1.7}$$

Suppose that a successful car salesman has taken two psychological tests. One test measures sensitivity to others and the second test measures ability to gain cooperation. On the first test the salesman got 36 answers correct and on the second he got 70 answers correct. For the population of several thousand adults in all types of jobs who have taken the first test, the mean is 20 correct and the standard deviation is 10. For the second test, the mean and the standard deviation are 45 and 15, respectively. Hence the salesman's standard score on the first test is

$$z_1 = \frac{36 - 20}{10}, \text{ or } +1.60,$$

and on the second test the standard score is

$$z_2 = \frac{70 - 45}{15}, \text{ or } +1.67.$$

In the first test the salesman's score was 1.60 standard deviations greater than the population mean and on the second test the score was 1.67 standard deviations greater than the mean. The z scale measures the number of standard deviations any observation in a collection is from the mean of that collection. The salesman's standing relative to the group is roughly the same for both tests. The use of z scores adjusts for differences in both variability and central location for the two tests. His standing is well above the mean in both sensitivity to others and ability to gain cooperation. 1.4

Summary

When considering a finite statistical population, our interest usually is focused on one or more properties of the entire population, such as its central location or its dispersion. Two measures of central location are the arithmetic mean and the median. The mean is the center of mass, or the balance point, of the population. When the data are arranged from smallest to largest, the median is either the middle value or the mean of the two values that bracket the middle. The mode is the most frequently recurring value. Two measures of dispersion are the variance and the standard deviation. The variance is the mean of the squared deviations from the arithmetic mean of the observations. The standard deviation, which is the square root of the variance, has the same dimension as the original observations. A standard score tells us how many standard deviations an observation is from its population mean. Standard scores are an important way to compare observations from different populations.

See Self-Correcting Exercises 1B.

Exercise Set B

1. Find the mean and the median for each of the following populations:
   (a) 7,7,2,9, −1,0,4,3,2,5,2,3, −1
   (b) 3,3,3
   (c) 2,3,4
   (d) 6,4, −2, −2, −4,0,2,8
2. (a) Is it necessary for the mean of a population to be itself one of the values in the population?
   (b) Is it necessary for the median of a population to be itself one of the values in a population?
3. Suppose the observations in a population have been ranked in order of increasing size. The median is the value of which observation if the population is of size (a) 32? (b) 87? (c) 10,000?
4. (a) When population data have been ranked in order of increasing size, which two size-ranked observations must we know the value of to obtain the median of a population of size 32?
   (b) How many values must we know to determine the mean for a population of size 32?
   (c) When the population data have been ranked in order of increasing size, which is the only size-ranked observation we need to know the value of to obtain the median of a population of size 87?
   (d) How many observations must we know to determine the mean of a population of size 87?

   Notice from your results that the mean is affected by all the values in a population while the median is determined only by the middlemost value or values.

5. An investment portfolio consisting of 200 stocks is worth $4000.
   (a) What is the current mean stock price for the portfolio?
   (b) If each stock increased $1 in value, what would the mean stock price for the portfolio be?

6. In a fund-raising drive, two-thirds of all appeals were made by mail and one-third by door-to-door solicitation. If the average income per mail appeal was $0.72 and the average donation per personal solicitation was $1.74, what was the average income resulting per appeal in this fund drive?

7. In a fifth-grade class consisting of 15 boys and 10 girls, the results on a nationally administered achievement test in reading were an average for the entire class of 5.8 years and an average for the boys of 5.6 years. What is the average achievement level for the girls?

8. After reviewing the directions for using Appendix H, determine the square roots of the following numbers:
   (a) 68
   (b) 9.92
   (c) 514
   (d) 0.09
   (e) 121

9. For the populations

   Population 1: 0,0,100,100
   Population 2: 0,50,50,100
   Population 3: 50,50,50,50

   we find that
   (a) $\mu_1 = \mu_2 = \mu_3 =$ _____
   (b) $\tilde{\mu}_1 = \tilde{\mu}_2 = \tilde{\mu}_3 =$ _____

   From the measures of central tendency it would seem that the three populations are identical and yet we know they are not. It is not until we consider measures of dispersion that we find their differences reflected. Find σ_1^2, σ_2^2, and σ_3^2. Which population has the greatest dispersion? Which has the least? Do these results conflict with your initial impressions of the three populations?

10. For the population 3,7,2,4,1,4,3,7,5:
    (a) Determine μ.
    (b) Find the deviation for the mean, $x_i - \mu$, for each x_i.
    (c) Verify that $\Sigma_{i=1}^{9} (x_i - \mu) = 0$, as anticipated from reading the discussion on page 23.
    (d) Find the variance σ^2.
    (e) Find the standard deviation σ.

11. Given that $\mu = 10$ and $\sigma = 2$:
    (a) Convert each of the following into a standard score z:
        (i) $x = 7$
        (ii) $x = 10$
        (iii) $x = 12$
        (iv) $x = 0$
        (v) $x = -4$
        (vi) $x = 3.2$
    (b) Using the rule of thumb that at least 90 percent of the data in any population will fall within three standard deviations of the mean, we may surmise that in this population at least 90 percent of the data fall between what two values?

12. On a kindergarten test for manual dexterity, one child received a standard score of -1.2 and another received a standard score of $+0.2$. On the basis of these scores, what can you say about each child's general level of dexterity relative to other kindergarten pupils?

Overview

1. Use sigma notation, indexing, and parentheses where necessary to express each of the following sums more compactly:

   (a) $(aw_1 + b) + (aw_2 + b) + (aw_3 + b) + \cdots + (aw_N + b)$

   (b) $100 + 200 + 300 + 400$

   (c) $100 + 300 + 500 + 700 + 900$

2. Given that $x_1 = 2$, $x_2 = 8$, $x_3 = 1$, and $x_4 = 4$, evaluate

$$\sum_{i=2}^{4} (ix_j)^2.$$

3. The following measurements were taken by five different researchers. Round each measurement to four significant digits:

   (a) 170.5532

   (b) 0.0032571

   (c) 25,099.36

   (d) 1.234501

   (e) 1.23450

4. (a) Measures of the way data tend to group together are known by what general term? Give two examples of such measures.

   (b) Measures of the way data are spread out are known by what general term? Give two examples of such measures.

5. The daily number of sick employees in a small shop over a 2-week work period is given by 0,2,5,7,4,1,0,4,2,0. Find the values of (a) the mean and (b) the median for this population.

6. An across-the-board 25 percent reduction in fee must be put into effect by a consultant firm consisting of 10 consultants. Before the reduction, six consultants charged \$20/hour and the others charged \$30/hour. What will the mean fee per consultant be after the reduction?

7. Although the 20 management personnel in a purchasing department have a mean vacation period of 4 weeks per year, the mean vacation length per employee for all 120 employees in the department is 1.5 weeks per year. What is the mean vacation period for the non-management personnel in the department?

8. An automobile dealership sold the following numbers of cars each day in a 6-day promotional period: 14,8,8,14,14,8. What is the variance in the cars sold per day?

9. The variance in the weight of steel beams in a certain shipment is 784 pounds. If the mean weight of a beam is 1500 pounds, at least 90 percent of the beams weigh between what two values?

10. From a population of size N:

    (a) What is the smallest sample that may be taken?

    (b) What is the largest sample that may be taken?

11. A population is given as 5,4,7,6,3.
    (a) Convert the 6 and the 4 to z scores.
    (b) Determine the actual percentage of data lying within three standard deviations of the mean.
    (c) Does your answer for part (b) conflict with the rule of thumb concerning the minimum percentage of data one might expect to be within this range?

Glossary of Equations

1.1 $$\sum_{i=1}^{N} x_i = x_1 + \cdots + x_N$$

Uppercase sigma is defined to be the summation operator and i, the index.

1.2 $$\mu = \frac{\sum x}{N}$$

The arithmetic mean of a statistical population is the sum of the observations in the population divided by the number of observations in the population.

1.3 $$\mu = \frac{N_1}{N}\mu_1 + \frac{N_2}{N}\mu_2 + \cdots + \frac{N_k}{N}\mu_k$$

The mean of a composite population composed of k populations is the sum of the weighted means of these populations. The weight for a given population is its proportion of the total number of observations in all k populations.

1.4 $$\sigma^2 = \frac{\sum_{i=1}^{N} (x_i - \mu)^2}{N}$$

The variance of a population is the arithmetic mean of the squared deviations of the data from their arithmetic mean.

1.5 $$\sigma^2 = \frac{\sum x^2}{N} - \mu^2$$

The variance of a statistical population is the mean square of the observations less the square of the mean.

1.6 $$\sigma = +\sqrt{\frac{\sum (x - \mu)^2}{N}}$$

The standard deviation of a statistical population is the positive square root of its variance.

1.7 $$z = \frac{x - \mu}{\sigma}$$

The standard score of an observation is the number of standard deviations the observation lies from the mean.

2

Frequency Distributions

In Chapter 1, concerned with small collections of data, graphic description and measured properties of these collections were based on the individual observations. When the number of observations is small, these techniques for ungrouped data are satisfactory. However, when we are concerned with a collection composed of a large number of observations, the data often must be grouped into classes.

Consider the problem of describing the number of employees per firm in any large city. If you list these observations, the mass of detail will obscure any general pattern. Impressions about the average number of employees per firm and the general pattern of variation are nearly unobtainable. Grouping, however, permits visual display that brings out these characteristics. If you do not have access to a computer, grouping also makes it easier to calculate measures of the desired properties.

Frequency Distributions for Discrete Variables

A major concept in statistics is the notion of frequency. In practice, a large collection of data may have values which occur more than once. For example, in a city containing 10,000 firms there may be 552 firms with four employees each, 637 firms with five employees each, and so forth. In this case the variable is number of employees per firm, and we have noted that the value 4 of the variable occurs 552 times, the value 5 of the variable occurs 637 times, and so forth.

Definition | **Frequency** is the number of times a value of a variable occurs in a collection of data.

In our example, we would say that 4 has a frequency of 552, 5 has a frequency of 637, and so on.

Introducing the notion of frequency allows us to simplify the description of large collections of data because we can state the different values of the variable just once and then note the frequency with which each occurs. In our example, we would note the value 4 and its frequency 552, the value 5 and its frequency 637, and so on until we had accounted for all the different values of the variable. If there were 82 different values of the variable, we could list these 82 together with the frequency of each. Such a list is one form of **frequency table** or **frequency distribution**. This table would have just 82 entries instead of the original 10,000.

Sometimes we summarize even more and make frequency distributions by forming classes with more than one value of the variable in each class. In our employee example, we might put all firms with less than five employees in the same class. Then the frequency recorded for this class would be the count of every firm with zero, one, two, three, or four employees. We will contrast this latter type of frequency distribution with the former by calling the former a frequency distribution with single-value classes and the latter a frequency distribution with many-value classes.

Preparation of a frequency distribution can be described in four steps:

Rules

1. Divide the range of values into convenient classes.
2. Sort the observations into these classes.
3. Count and record the number (frequency) of observations in each class.
4. Summarize the results in a table, a graph, or both.

Now we will turn to the simplest case: single-value classes for discrete variables.

Single-Value Classes 2.1

When values can occur only at isolated points on the scale measuring a variable, the variable is said to be **discrete**. For instance, a quality inspector in a steel plant has counted the number of flaws on each of 50 sheets of galvanized steel. In the results listed below, note that only integral, nonnegative values of this variable can occur. Note also that every value of the variable within the range of observation (0 through 5) occurs several times. For example, there are several 0s, several 1s, and so forth.

1	2	1	0	3	0	2	4	2	1
3	4	2	0	2	1	3	1	3	2
1	0	3	2	2	0	2	0	0	1
5	1	2	1	3	1	0	5	3	1
0	1	2	4	2	1	4	1	0	3

Even for this modest number of observations, however, it is difficult to get any idea of pattern from the list itself.

We can begin constructing a frequency distribution by noting that the variable ranges from 0 through 5, with several instances of each value. These six values will make convenient classes. Formally, we can define the variable X as the number of flaws per sheet and let it assume the integral values 0,1,2,3,4,5.

We now must sort the 50 observations into the six classes and count the number of observations in each class. When sorting is done by hand, tallying is perhaps the easiest approach. The following table illustrates tallying and counting:

x	Tally	Frequency
0	~~||||~~ ~~||||~~	10
1	~~||||~~ ~~||||~~ ||||	14
2	~~||||~~ ~~||||~~ ||	12
3	~~||||~~ |||	8
4	||||	4
5	||	2

The classes are described in the left-hand column. As we come to each observation in the list, a tally mark is made to the right of the appropriate x value as shown. The final column is the result of counting the tally marks in each class.

A frequency distribution table is one form of summary for presentation of grouped data. When the table is intended for general audiences, presumably not familiar with the subject matter, table headings must be fully descriptive. One possible display for our example is shown in Table 2.1.

Table 2.1 *Frequency Distribution of Flaws in Galvanized Steel Sheets*

Number of Flaws per Sheet	Number of Sheets
0	10
1	14
2	12
3	8
4	4
5	2
Total	50

For statisticians in the steel plant the table may be simplified as shown in Table 2.2. Among statisticians, f is customarily used to denote the frequency of

Table 2.2 Frequency Distribution of Flaws per Sheet

x	f
0	10
1	14
2	12
3	8
4	4
5	2
	$N = 50$

occurrence for each value of the variable and N (where $N = \sum f$) is used to represent the total number of observations in the population.

In our current example, every possible value of the variable within the range of observations constitutes a class. For frequency distributions of this type, a **frequency diagram** is suitable for graphic display. As shown in Figure 2.1, the frequencies (f) are plotted on the vertical scale (the ordinate). The number of flaws per sheet (x) are plotted on the abscissa. When the context is clear, the axes can be labeled f and x. Ordinates equal to the proper frequencies are constructed for every relevant value of the variable.

Notice how grouping the data brings out the pattern in the distribution. In Figure 2.1, the average appears to be between one and two flaws per sheet. The major portion of the observations is concentrated in the range zero to three flaws. The distribution is not symmetrical with respect to the average; it tails off toward higher values of the variable. The facility it offers for displaying pattern illustrates one important benefit to be gained from forming a frequency distribution for a large collection of data.

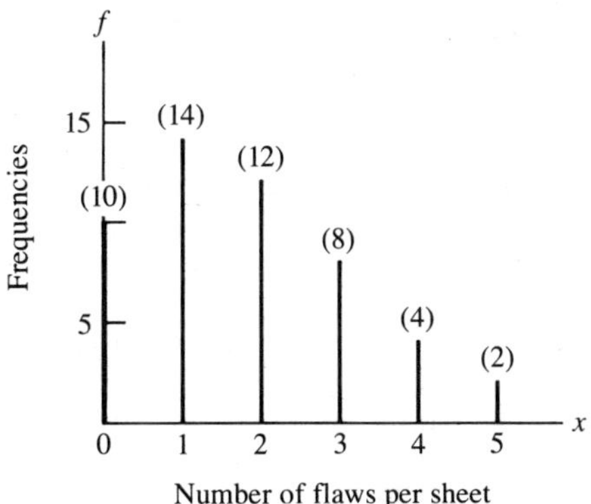

Figure 2.1 *Frequency diagram*

Suppose a new process for galvanizing has been proposed and the result of an initial run of 150 sheets is as shown in Table 2.3. It is difficult to compare the two distributions in Tables 2.2 and 2.3 because different numbers of observations were made. A useful device to make such comparisons is the **relative frequency table**. Instead of listing frequencies, the *proportion* of the total number of observations falling in each class is shown. The symbol f' is often used to designate these proportions.

Table 2.3 *Frequency Distribution of Flaws per Sheet for New Process*

x	f
0	24
1	36
2	33
3	30
4	18
5	9
	150

Table 2.4 shows relative frequency distributions for the two galvanizing processes; Figure 2.2 presents the two frequency diagrams. With the two distributions expressed in relatives it is apparent that the old process produces more sheets with zero, one, and two flaws and fewer with three, four, and five flaws than does the new process. By adding percentages for zero, one, and two flaws, we see that 72 percent of sheets produced by the old process have fewer than three flaws per sheet. Only 62 percent of sheets produced by the new process have fewer than three flaws per sheet. The pattern for the new process shows a heavier concentration of sheets with higher numbers of flaws.

Table 2.4 *Relative Frequency Distributions of Flaws per Sheet*

(a) Old Process		(b) New Process	
x	f'	x	f'
0	0.20	0	0.16
1	0.28	1	0.24
2	0.24	2	0.22
3	0.16	3	0.20
4	0.08	4	0.12
5	0.04	5	0.06
	1.00		1.00

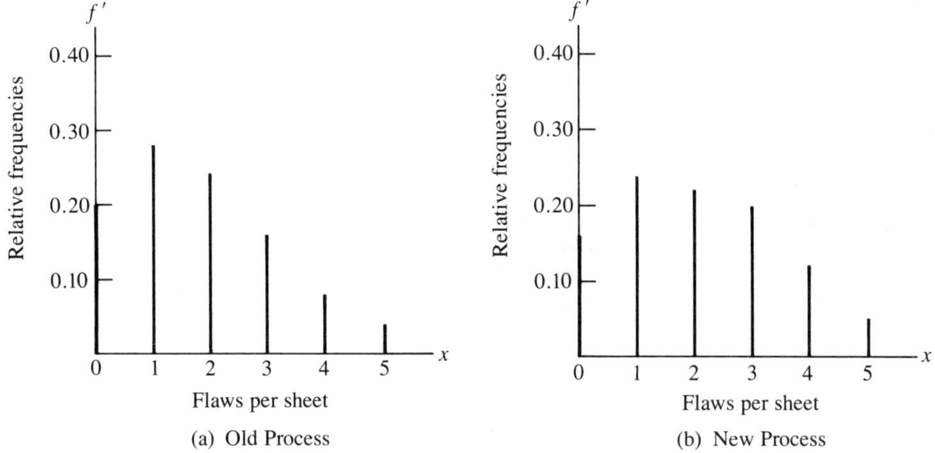

***Figure 2.2** Relative frequency diagrams for two galvanizing processes* 2.1

Many-Value Classes 2.2

In many instances the use of single-value classes is not advisable. For example, when observations fall over a wide range of the variable or when the density of the observations is low, a class should not be defined for every possible value of the variable in the range. Rather a **class interval**, which includes several values of the variable in every class, should be defined. Then the range will include as many classes of this chosen length as this interval makes possible. The objective is to choose enough classes to approximate the smooth underlying pattern in the population but not so many that an irregular sawtooth effect is produced. A balance must be struck between preserving all the accurate detail in the original data and displaying the smooth pattern by grouping the data into classes.

Selection of the number and location of classes in a particular instance requires judgment. It is not an exact procedure. Experience has, however, produced a number of helpful guidelines. The number of classes to use depends on the number of observations in the collection of data. The larger the collection, the greater the number of classes needed. The number of classes shown in Table 2.5 for various numbers of observations usually proves satisfactory, although it is just one of many ways to determine the number of classes.

Another sound practice is to make all class intervals the same length. With this procedure, certain problems, such as preventing erroneous impressions from graphs, do not arise. On the other hand we sometimes encounter collections of data with just a few very widely scattered values toward one end of the range. Using a single open-ended class to contain these observations is sometimes necessary to achieve a practical result.

Table 2.5 *Number of Classes for Frequency Distributions**

Number of Observations	Number of Classes
12–22	5
23–45	6
46–90	7
91–181	8
182–361	9
362–723	10
724–1448	11

**Based on Sturges' formula: $M = 1 + 3.322 \log N$, where M is the number of classes and N is the number of observations.*

A third guideline is to position class intervals so that any clusters of data fall near the centers of intervals rather than toward one end. Later we will see that when this is done values of descriptive measures (such as the mean) calculated from the frequency distribution are almost always more nearly equal to the correct values calculated from the ungrouped data.

We now want to illustrate the process of making a frequency distribution with many-value classes when the variable is discrete and the observations are whole numbers. Suppose that an advertising manager wants a frequency distribution for the number of travel advertisements appearing in the last 60 issues of a certain national weekly magazine. The original data, arranged in numerical order for convenience, appear in Table 2.6.

Table 2.6 *Number of Travel Advertisements in 60 Issues*

14	26	31	35	39	43
15	26	32	35	39	43
18	26	33	35	39	43
19	27	33	35	39	43
19	28	33	36	39	44
21	28	33	36	40	44
22	28	34	37	41	44
23	29	34	38	42	46
25	29	34	38	42	49
26	30	35	38	42	52

For 60 observations, Table 2.5 shows that 7 classes should suffice. The data run from 14 ads to 52, a range of 39. Dividing 39 by 7 classes, we get a class interval length of 5.6 ads, which we round to an integer value of 6 ads per interval to correctly reflect the type of data we have. Seven class intervals of 6 ads per interval is 42. This exceeds the observed range of 39. Hence 7 equal intervals of width 6 ads each will include all the data.

One way to establish the actual classes for our example is to begin by selecting the smallest observation as the lower limit of the first class. This gives us 14 ads. Then we add the class interval, 6, to the lower limit of the first class to get the lower limit of the second class ($14 + 6 = 20$). We continue the process to find the lower limits of successive classes until we run beyond the value of the largest observation. To find the upper limit of a class, we subtract 1 from the lower limit of the next class because the observations are integers ($20 - 1 = 19$). This is the way the class intervals in the left-hand column of Table 2.7 were obtained. The frequencies in the right-hand column were found by counting the observations in each interval in Table 2.6.

Table 2.7 *Frequency Distribution with Many-Value Classes (Discrete Variable)*

Number of Ads per Issue	Frequency
14–19	5
20–25	4
26–31	12
32–37	16
38–43	17
44–49	5
50–55	1
	60

The steps we have taken satisfy the first two guidelines discussed earlier. There are seven classes and all classes have the same interval. Before you accept the arrangement in Table 2.7, however, you should consider the last guideline and the general objective for constructing frequency distributions. Recall that the objective is to produce a smooth pattern. Our initial attempt shows a rather abrupt jump from 0 frequencies to 5. Then there is a drop to 4 and a rise to 12. Toward the other end of the distribution there is another abrupt drop from the peak frequency of 17 to 5 in the following class. The pattern could be smoother. Furthermore, the observations in the first class are 14, 15, 18, 19, and 19. These tend to cluster at the upper end of the class rather than near the center. There is also clustering at the lower end of the class interval 44 through 49.

By starting the first class interval at 13 instead of 14, we may be able to improve the result. The frequency distribution in Table 2.8 reflects this change. Rising rather smoothly to a peak of 16 and dropping smoothly from that peak, the pattern is

Table 2.8 *Alternative Frequency Distribution*

Number of Ads per Issue	Frequency
13–18	3
19–24	5
25–30	12
31–36	16
37–42	14
43–48	8
49–54	2
	60

better. On the other hand, we find when we apply these class intervals to the data in Table 2.6 that there is still clustering at the lower ends of class intervals 19 through 24 and 43 through 48. We could try starting the first class interval at 12 or 11 and we could also try changing the class interval to five or seven ads per interval. Some combination of these might produce better locations for clusters, but that combination could hardly improve the pattern and probably would worsen it. We will stop at this point because the result is satisfactory and marked improvement doesn't seem possible.

In summary, the following guidelines apply when constructing a frequency distribution with many-value classes for discrete data:

Rules

1. Select an appropriate number of classes. (Table 2.5 is one guide.)
2. Find the range of the data. For integers, the range is found by adding one to the difference between the largest and smallest observations.
3. Find the length of the class intervals by dividing the range by the number of classes and rounding the result to the next higher value that has the same number of places as the data.
4. Find the class limits by adding the class interval length to the smallest observation successively until the largest observation is included in the final class.
5. If necessary to center clusters of observations in classes, simultaneously lower all class limits the same amount. This may make it necessary to add a new class to include the largest observations.
6. Count the number of observations in each class and record the result in a table.

A graphic device often used to represent frequency distributions for observations placed in many-value classes is the **histogram**. Figure 2.3(a) presents a histogram illustrating the frequency distributions in Table 2.7, and Figure 2.3(b) presents one illustrating the distribution in Table 2.8. For class intervals of equal length, adjacent

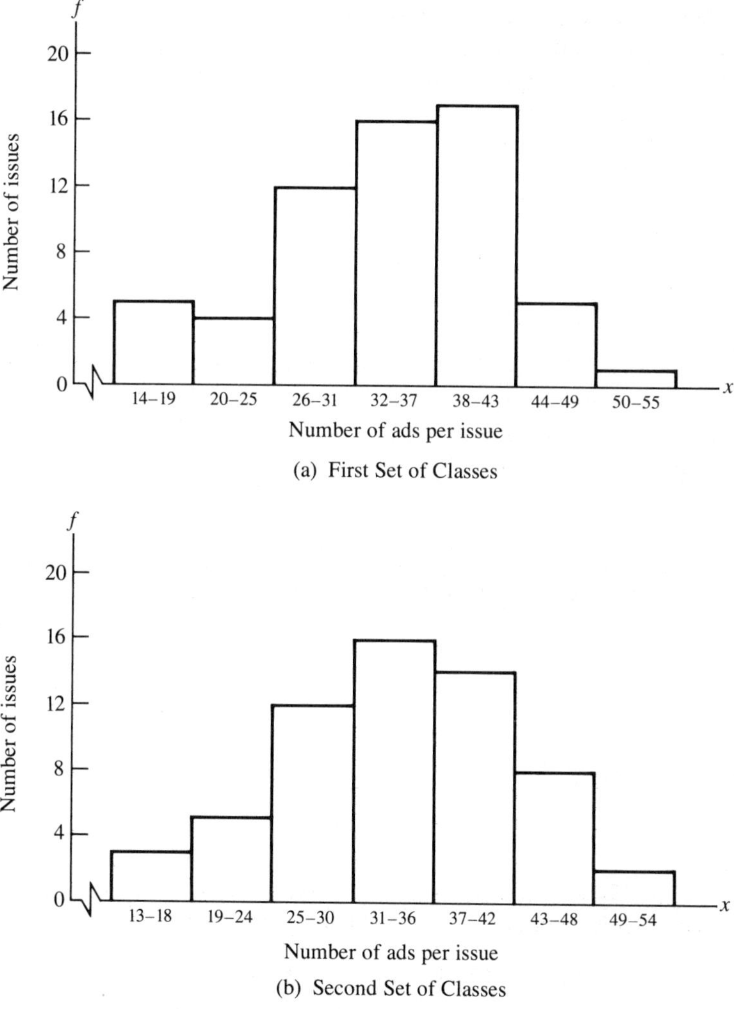

(a) First Set of Classes

(b) Second Set of Classes

Figure 2.3 *Histograms for two frequency distributions*

vertical bars of equal width are created above the horizontal axis. As a result, the range of values is shown on the horizontal axis. Frequencies are shown on the vertical axis. Breaking the scale at the left of the horizontal axis simply makes it possible to avoid a large, empty space. These histograms emphasize the superiority of the pattern with
the second set of class intervals. **2.2**

Summary

A frequency distribution is a device for emphasizing the pattern in a large collection of data. Two types of frequency distributions are appropriate when the variable being observed is discrete. A frequency distribution with single-value classes can be used for a collection of data in which there are several repetitions of every value of the variable throughout the observed range. This type of frequency distribution is represented graphically with a frequency diagram.

When observations occur throughout a wide range of a discrete variable, single-value classes can make the number of classes excessive. This can be avoided by defining classes so that each contains several values of the variable. Where feasible, all classes should have the same length. Moreover, classes should be positioned along the measurement scale to make clusters of observations fall near class midpoints. The histogram is the graphic device used to illustrate frequency distributions characterized by many-value classes.

The column of frequencies in a frequency table can be replaced with relative frequencies. The result is called a relative frequency distribution. Relative frequency distributions are useful for comparing patterns in collections of data with widely varying numbers of observations.

See Self-Correcting Exercises 2A.

Exercise Set A

Exercises 1 through 20 refer to the data below, which represent the quantity of state lottery tickets sold daily at a certain market over a 30-day period.

42 47 46 35 43 39

38 40 50 37 43 37

47 47 44 49 41 34

38 41 36 42 38 38

58 34 32 42 49 52

1. Determine the average number of tickets per day sold during this period.
2. Is the variable under study here discrete or continuous?

3. From the data given, construct a frequency distribution with single-value classes.
4. (a) Give the frequency you obtained in Exercise 3 for each of the following daily totals of ticket sales: (i) 38; (ii) 19; (iii) 48.
   (b) What value has the highest frequency?
5. Construct a frequency diagram from your results in Exercise 3.
6. Give the relative frequency of each of the following daily sales totals for the period: (a) 38; (b) 40; (c) 50.
7. Using your results of Exercise 3, construct a relative frequency distribution with single-value classes for these data.
8. Construct a relative frequency diagram for the daily ticket sales.
9. To find the range of data to be included in a many-value table for these data, subtract the least value from the greatest and add 1 to the difference. What is this range?
10. Using the appropriate table provided in this text section, determine an appropriate number of classes to use in constructing a many-value frequency distribution for these data.
11. Using your results from Exercises 9 and 10, determine the shortest interval length that could be used to set up a frequency distribution having the desired number of classes.
12. Using your results from Exercises 10 and 11, set up a many-value frequency distribution for the daily ticket sales. Let 32 be the lower class limit of the first class.
13. Is it necessary in this case to use an open-ended class for either the least or the greatest values?
14. Construct a histogram for the frequency distribution obtained in Exercise 12. Does your histogram have a sawtooth appearance? How can a sawtooth appearance be avoided or at least mitigated?
15. Construct, for the original data, a frequency distribution having class intervals of the same number and length as those used in Exercise 12, but this time let 30 be the lower class limit of the first class.
16. Would it be possible to construct a frequency distribution of classes of the same number and length as those in Exercise 12 but having 28 as the lower class limit of the first class? Justify your answer.
17. Construct a histogram for Exercise 15.
18. Compare the histograms from Exercises 14 and 17. Does either have a smoother or less sawtooth appearance than the other? Which frequency distribution, if either, would be preferable? Justify your conclusion.
19. From the original data, give a frequency distribution having two equal intervals, and construct a histogram for the distribution you obtain. How effective is this histogram in preserving patterns that may have been present in the original data?
20. From the original data, determine a frequency distribution with 14 equal intervals, and construct a histogram for this distribution. Does this histogram effectively reflect characteristics of the underlying data?
21. Considering the results of Exercises 19 and 20, what is wrong with having too few classes? Too many?
22. When do statisticians use (a) frequency diagrams (b) histograms?

23. Can you think of any visual reason for statisticians to use vertical bars in graphic presentation of single-value frequency distributions and rectangles for many-value distributions?

24. Why do statisticians use such devices as frequency diagrams and histograms to present data that can be stated in frequency distributions with less effort on their part?

Frequency Distributions for Continuous Variables

2.3 As we stated earlier, when a variable theoretically can assume any value between any two given values on a measuring scale, it is a *continuous* variable. For example, consider a bag of sugar which, according to the label, weighs 5 pounds. If the bag were weighed on instruments of increasing sensitivity, weights of 4.9, 4.93, 4.927, 4.9274 pounds, and so forth, to the limit of accuracy for the particular instrument, might be shown.

In practice, observations of continuous variables are recorded to the limit of precision possible with the measuring instrument or they are rounded to some lesser degree of precision. Suppose a bag of sugar is weighed on an instrument capable of weighing to the nearest ten-thousandth of a pound and suppose we read 4.9274 pounds as the weight. For the purpose at hand, weights to the nearest hundredth of a pound are sufficient. We round 4.9274 pounds to 4.93 pounds and record this latter number as our observation. To the nearest hundredth of a pound, Table 2.9 lists the net content weights of 120 bags labeled as containing 5 pounds.

In Table 2.9 the first observation is recorded as 4.84 pounds. Because this value has been rounded to two decimal places, the exact weight of sugar in this sack could be any value from 4.835 to 4.845 pounds. Similarly, every recorded observation must be regarded as having been rounded to two decimal places.

To construct a frequency distribution when observations are values of a continuous variable, we must use many-value classes rather than single-value classes. Single-value classes have class intervals with no length. This indicates that observations have the exact values listed, but such an impression is not correct when we are working with a continuous variable.

As was the case for the travel advertisements in the preceding section, Table 2.5 is the guide for an appropriate number of many-value classes. For the 120 observations being considered, eight classes should prove adequate.

Determining the length of class intervals for continuous variables is somewhat different from doing so for discrete variables. The data in Table 2.9 are ordered from smallest to largest. The smallest observation recorded is 4.84. Since this is rounded, it could represent an exact weight as small as 4.835. The largest value recorded is 5.11 and this could represent an exact weight up to but not including 5.115. The data must be assumed to extend from 4.835 to 5.115 pounds—that is, over a range of

Table 2.9 Content Weights of 120 Bags of Sugar (pounds)

4.84	4.95	4.98	5.00	5.02	5.05
4.89	4.96	4.98	5.00	5.02	5.05
4.90	4.96	4.98	5.00	5.02	5.05
4.90	4.96	4.98	5.00	5.02	5.06
4.91	4.96	4.98	5.00	5.02	5.06
4.92	4.96	4.99	5.00	5.03	5.06
4.92	4.96	4.99	5.01	5.03	5.06
4.92	4.96	4.99	5.01	5.03	5.06
4.92	4.97	4.99	5.01	5.03	5.06
4.93	4.97	4.99	5.01	5.03	5.06
4.93	4.97	4.99	5.01	5.03	5.07
4.93	4.97	5.00	5.01	5.03	5.07
4.94	4.97	5.00	5.01	5.03	5.07
4.94	4.98	5.00	5.01	5.04	5.07
4.95	4.98	5.00	5.01	5.04	5.08
4.95	4.98	5.00	5.01	5.04	5.08
4.95	4.98	5.00	5.01	5.04	5.08
4.95	4.98	5.00	5.01	5.05	5.08
4.95	4.98	5.00	5.02	5.05	5.08
4.95	4.98	5.00	5.02	5.05	5.11

0.280 pound. Dividing 0.28 pound into eight classes of equal length makes each class interval 0.035 pound.

As in the preceding section, we can find the beginning point for the first class interval by referring to the smallest observation in the collection, 4.84. But the variable in this case is continuous, and 4.84 can represent an exact weight as small as 4.835. Hence we establish 4.835 as the lower class boundary.

Definition When classes are established for frequency distributions based on continuous variables, **class boundaries** are the limiting lower and upper exact values that can be included in the class.

Suppose we add the interval we found above, 0.035 pound, to the class boundary we just found, 4.835. The sum is an exact value: 4.870, or 4.87. Then this becomes both the upper boundary for the first class and the lower boundary for the second class. But such a boundary puts us in a quandary with regard to an observation recorded as 4.87. This recorded value can represent any exact value from 4.865 to 4.875. Consequently, we can't tell whether to assign this observation to the first or second class.

One way to avoid this problem is to choose a length for class intervals which makes all class boundaries end in 5 in the next place to the right of that used to record the observations. For our example, a class interval of 0.030 (or 0.03 exactly) in conjunction with a beginning lower class boundary of 4.835 will accomplish this result. When we add 0.03 to 4.835 we get 4.865, for example. Such class boundaries cause no confusion as to where observations should be placed. The number of classes will no longer be eight, however. The range (0.28) divided by 0.03 yields $9\frac{1}{3}$, so 10 classes will be necessary. Alternatively, we can use a class interval of 0.04 pound. This will not only avoid the ambiguity problem but will also include all the data in just seven classes. Since this result is closer to our objective of eight classes, we will select 0.04 as our class interval. To avoid ambiguity, state the class interval to no more decimal places than are the data.

Beginning with 4.835 (the lower class boundary of the first class) and adding successive intervals of 0.04, we find the class boundaries shown in the left-hand column of Table 2.10. The next step is to assign observations from Table 2.9 to these classes. These frequencies appear in the right-hand column of Table 2.10.

Table 2.10 *Class Boundaries and Frequencies*

Class Boundaries	Frequencies
4.835–4.875	1
4.875–4.915	4
4.915–4.955	16
4.955–4.995	30
4.995–5.035	42
5.035–5.075	21
5.075–5.115	6
	120

When we position the class boundaries just found with respect to the observations in Table 2.9, the only observation in the first class is at the very bottom of the class but the remaining classes seem to have no serious clustering problems. Although we might be able to improve the location of this first observation by lowering the class boundaries somewhat, there is only one case involved and the improvement is likely to be slight. So we won't change the boundaries.

Frequency distributions based on observations of continuous variables almost always describe the classes in terms of class limits rather than class boundaries.

Definition

When classes are defined for frequency distributions based on continuous variables, **class limits** are found by rounding class boundaries to the same number of decimal places as the observations.

For our example, we establish the class limits in Table 2.11 by rounding the lower class boundaries in Table 2.10 up and the upper class boundaries down. The resulting class limits are stated to two decimal places as are the observations in Table 2.9, and they are unambiguous. Note that we did not call Table 2.10 a frequency distribution. Table 2.10 utilized class boundaries rather than class limits.

Table 2.11 *Frequency Distribution for Bags of Sugar*

Content Weight in Pounds (x)	Number of Bags (f)
4.84–4.87	1
4.88–4.91	4
4.92–4.95	16
4.96–4.99	30
5.00–5.03	42
5.04–5.07	21
5.08–5.11	6
	120

For data that constitute observations of a continuous variable we can summarize the technique described immediately above for making a frequency distribution:

Rules

1. Select an appropriate number of classes (Table 2.5 is the guide we used).
2. Find the range of the data by subtracting the lowest possible value of the smallest observation from the greatest possible value of the largest.
3. Find the class interval by dividing the range by the number of classes and stating the result to no more places than are used to record the observations.
4. Find the class boundaries by successively adding the class interval to the lowest possible value of the smallest observation until the largest observation is included in the final class.
5. If necessary, simultaneously lower all class boundaries to center

clusters of observations in class intervals. It may be necessary to add a class to include the largest observation.

6. Round lower boundaries up and upper boundaries down to establish class limits with the same number of places as the data. Because of adjustments in steps 3 and 5 the final number of classes may differ by one or two from the number established in step 1.
7. Count the number of observations in each class and record the result in a table.

2.3 As we have said earlier, this set of rules represents only one of several satisfactory techniques for establishing class limits when the variable is continuous.

The histogram based on Table 2.11 appears in Figure 2.4(a). The vertical divisions between the bars are drawn above the class boundaries. The values of the class boundaries are shown on the horizontal scale.

Another graphic device for presenting a frequency distribution is the **frequency polygon**. The frequency polygon for the frequency distribution in Table 2.11 appears in Figure 2.4(b), and its relationship to the histogram for the same table appears in Figure 2.4(c). To construct a frequency polygon, we begin by plotting vertically above the midvalue of each class at a height equal to the class frequency. These points are

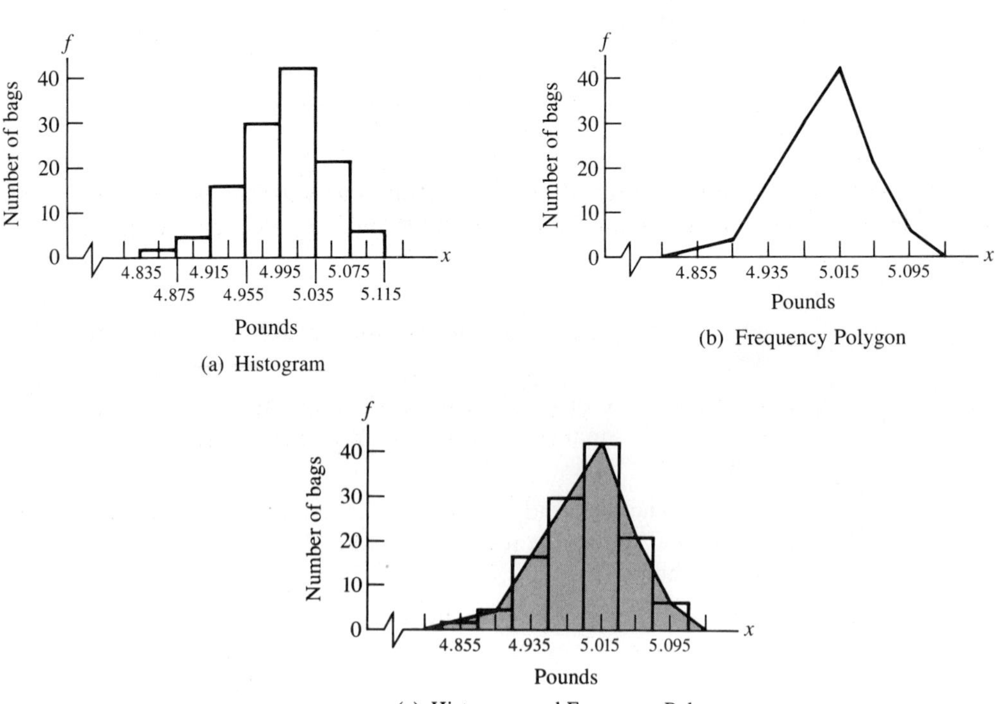

Figure 2.4 *Representations of a frequency distribution*

connected with straight lines. Then, at the ends of the distribution, we draw straight lines to connect the midvalue of the last occupied class with the midvalue of the adjacent empty class and close the figure. Class midvalues are shown on the horizontal scale.

Both the histogram and the polygon in Figure 2.4 emphasize the pattern in the collection of content weights of sugar bags. There is a concentration of weights in the immediate vicinity of 5 pounds and there is tailing off as one moves in either direction from 5 pounds. No weight in the entire collection of 120 observations differs by as much as 0.20 pound from the advertised weight of 5 pounds. When we compare Figure 2.4 with Table 2.9 (the original observations) it is apparent that the pattern in the data receives greater emphasis from the frequency distribution. The difference would be even more pronounced in favor of the latter representation if there were more observations and if they were not arranged in order of magnitude.

Cumulative Frequency Distributions

So far we have organized data into classes by recording the number of observations that have a given value or lie between two values. Now we want to classify data by the *total* number of observations that lie *at* or *below* designated values of the variable. 2.4

Definition In a collection of observations, the **cumulative frequency** F is the *total number* of observations less than or equal to a specified value of the variable.

To illustrate this concept, we will use the same example we used earlier.

Single-Value Classes for Discrete Variables

The information in Table 2.1 on the flaws in 50 sheets of galvanized steel is arranged as a cumulative frequency distribution in Table 2.12. The numbers in the cumulative frequency column (F) are the total numbers of sheets that have no more than the number of flaws shown under x in the same row. For instance, 36 sheets have two or fewer flaws. The values of F are found by adding the frequencies f in Table 2.1 for all classes with x values equal to or less than the given value of x in Table 2.12. Hence 36 is the sum of the 10 sheets with no flaws, the 14 with one flaw apiece, and the

Table 2.12 Cumulative Frequency Distribution

Number of Flaws (x)	Number of Sheets (f)	Cumulative Number of Sheets (F)
0	10	10
1	14	24
2	12	36
3	8	44
4	4	48
5	2	50

12 with two flaws. In other words, a cumulative frequency distribution table is constructed from a frequency distribution table. By the same token, given a cumulative frequency distribution table, we can construct a frequency distribution table by finding the differences in successive values of the cumulative frequency column (F). The difference between the 36 sheets that have two flaws or fewer per sheet and the 24 sheets that have no more than one flaw is 12, the number of sheets that have exactly two flaws per sheet.

A useful device for comparing two or more frequency distributions with widely different numbers of observations is the cumulative relative frequency distribution.

Definition The **cumulative relative frequency** (F') is the proportion of observations that are less than or equal to a specified value of the variable.

The cumulative frequency distribution of Table 2.12 is changed to a cumulative *relative* frequency distribution in Table 2.13. This is accomplished by dividing each of the entries in the F column of Table 2.12 by 50, the total number of observations.

Table 2.13 Cumulative Relative Frequency Distribution

Number of Flaws (x)	Proportion of Sheets (f')	Cumulative Proportion of Sheets (F')
0	0.20	0.20
1	0.28	0.48
2	0.24	0.72
3	0.16	0.88
4	0.08	0.96
5	0.04	1.00

The values of F' tell us the proportion of sheets in the set that have the stated number of flaws (x) or fewer per sheet. For instance, 0.72 is the proportion that have two flaws or fewer per sheet. By subtracting 0.48 from 0.72, we find that 0.24 is the proportion with exactly two flaws per sheet. This is the value of the relative frequency f' associated with two flaws (x) here and in Table 2.4(a). 2.4

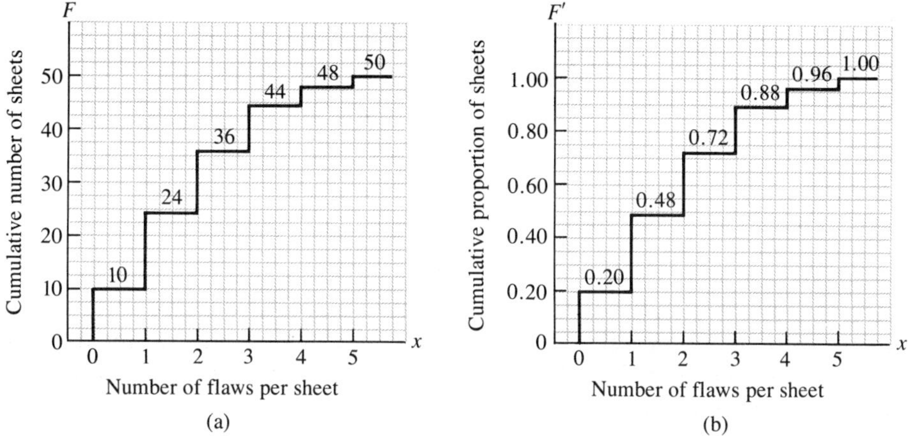

Figure 2.5 *(a) Cumulative frequency step graph; (b) cumulative relative frequency step graph*

The **step graph** in Figure 2.5 is the proper graphic device for cumulative frequency distributions and cumulative relative frequency distributions for observations of discrete variables when values of the variable x have not been grouped into class intervals. The step graph for the cumulative *number* of sheets appears in Figure 2.5(a), and the cumulative *relative* frequency graph appears in (b). From (a) we can see, for instance, that the first 24 sheets (F) have no more than one flaw per sheet (x). Or we can ask how many sheets have three flaws or fewer (x) and read 44 (F) at the top of the vertical step above 3. The heights of the steps are equal to the individual class frequencies (f) in Table 2.1. From (b) we can see that 0.88 (F') have three (x) or fewer flaws each—that is, 88 percent have no more than three flaws apiece. Or we may want to know what proportion have two flaws or fewer (x) and find that 0.72 (F') is the answer. The heights of steps are equal to individual class relative frequencies f' in Table 2.4(a).

Cumulative frequency distributions and cumulative relative frequency distributions are alternative methods for describing collections of data. The first is useful when we want to answer questions such as: *How many* values are less than or equal to a specified value of the variable? The second is useful for answering questions such as: What *proportion* or *percentage* of the observations are less than or equal to a specified value of the variable?

As was the case for the relative frequency distributions and diagrams in Table 2.4 and Figure 2.2, comparisons between two distributions with different numbers

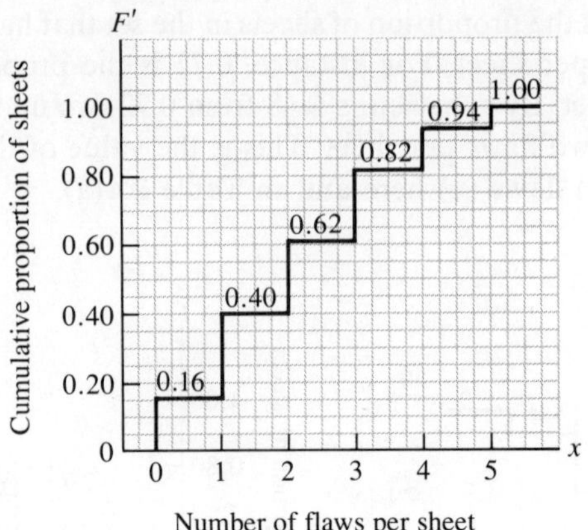

Figure 2.6 *Cumulative relative frequency step graph for new process*

of observations can also be made with cumulative relative frequency step graphs. From the information in Table 2.4(b), we have prepared Figure 2.6 for the new process. Comparing Figure 2.6 with Figure 2.5(b), we see that for the new process, the proportion of flaws less than a specified number remains lower than the similar proportion for the old process all the way up to five flaws. Eighty-eight percent of sheets from the old process, for example, have less than four flaws per sheet while only 82 percent of sheets from the new process have less than four.

When several values of a variable (either discrete or continuous) are grouped into each class, the result is a frequency distribution such as that in Table 2.8. A cumulative frequency distribution based on this information appears in Table 2.14. We see that 24 or fewer ads (x) were observed in each of eight issues (F). By subtracting 8 from 20, we find that between 25 and 30 ads per issue were observed in 12 issues. The cumulative relative frequency distribution table would be formed by dividing each of the entries under F in Table 2.14 by 60.

Table 2.14 *Cumulative Frequency Distribution*

Number of Travel Ads per Issue (x)	Cumulative Number of Issues (F)
13–18	3
19–24	8
25–30	20
31–36	36
37–42	50
43–48	58
49–54	60

A step graph cannot be used to display information for discrete distributions with many-value classes as illustrated in Table 2.14. For example, if we placed the entire vertical rise of 3 for the first class above 13 on the x scale, this would erroneously indicate that exactly 13 ads were observed in three issues. Alternatively, placing the vertical rise of 3 above 18 on the x scale would also create the wrong impression. Rather, we want to indicate that the data are scattered throughout the class, as we did with the histogram in Figure 2.3(b).

Another complication must be resolved before we can graph cumulative frequency distributions for many-value classes. The graph should indicate that the cumulative frequencies shown in a table such as Table 2.14 are not all accounted for until we have passed the upper end of the related class of x values. But the second class in that table, for instance, ends on 24 ads and the third class begins on 25. It isn't apparent where the class ends. We resolve this dilemma by forming the following rule:

Rule | Many-value classes in frequency distributions for discrete variables will be given class boundaries by using the same rules that were used for continuous variables.

Hence the boundaries for the classes in Table 2.14 are 12.5 to 18.5, 18.5 to 24.5, 24.5 to 30.5, and so forth.

The graphic display for many-value classes is called a **cumulative frequency ogive**. The ogive for Table 2.14 is shown in Figure 2.7. Cumulative frequencies F of

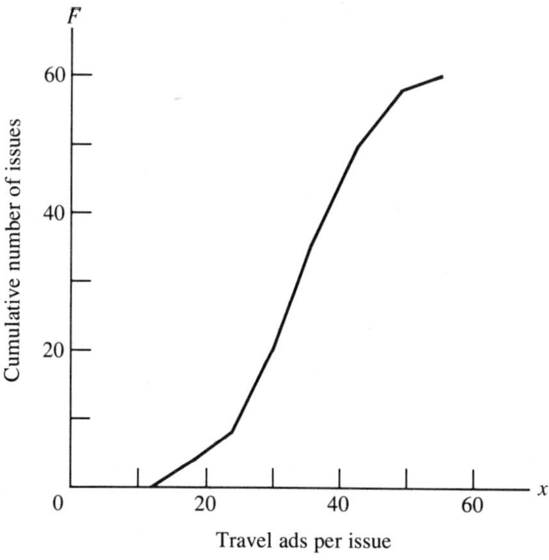

Figure 2.7 *Cumulative frequency ogive*

0, 3, 8, 20, 36, 50, 58, and 60 are plotted vertically above x values of 12.5, 18.5, 24.5, 30.5, 36.5, 42.5, 48.5, and 54.5. The eight points are then connected with straight line segments. As in the case of the histogram, the straight lines imply that the data in a given class are scattered uniformly throughout the class. This implication is at least approximately true.

Preparation of a cumulative *relative* frequency ogive for many-value classes offers no new difficulty. We can simply convert the F (cumulative frequency) scale to an F' (cumulative relative frequency) scale as we did in Figure 2.5.

For observations of a continuous variable, the procedures for preparing cumulative frequency and cumulative relative frequency distribution tables and ogives are identical to those just described for many-value classes based on observations of discrete variables. Since these procedures are the same, we will not illustrate them here with an example for a continuous variable. Instead we will cover this case in Exercise Set B.

Graphic Presentation

The preceding discussion of frequency distributions has shown that tabular presentation of data tends to preserve numerical accuracy while graphic presentation fosters comparison and quick communication of major features. The portion of that discussion devoted to graphic presentation covered construction and use of frequency and relative frequency diagrams, histograms, polygons, step graphs, and ogives. Graphic presentation extends to data other than those contained in frequency distributions, however, and this is an appropriate place for a brief discussion of some of the more important issues and devices in graphic presentation.

When you are faced with the need to illustrate nominal data, the **horizontal bar chart** is often useful, although it is not restricted to this use. Figure 2.8 presents such a chart to illustrate the number of automobile tires a retail outlet has in stock by major type. The lengths of the bars are determined by the vertical grid lines in accordance with the number of tires of each type. Horizontal bar charts are widely used in newspapers, magazines, and other mass media.

Horizontal bar charts and the other graphic devices described up to this point apply to data on which the effect of time is assumed not to matter. The tire count for Figure 2.8 could have been made during a few hours when the store was closed, for instance. In the case of the number of flaws on the 50 sheets of galvanized steel illustrated in Figure 2.1, the count could have been made for a sample of sheets produced in a single production run during which production conditions were essentially stable. One term applied to observations unaffected by the passage of time is **cross-sectional data**.

Time series constitute another broad class of data commanding much interest. The pattern of sales through time is of concern to firms and investors, for example.

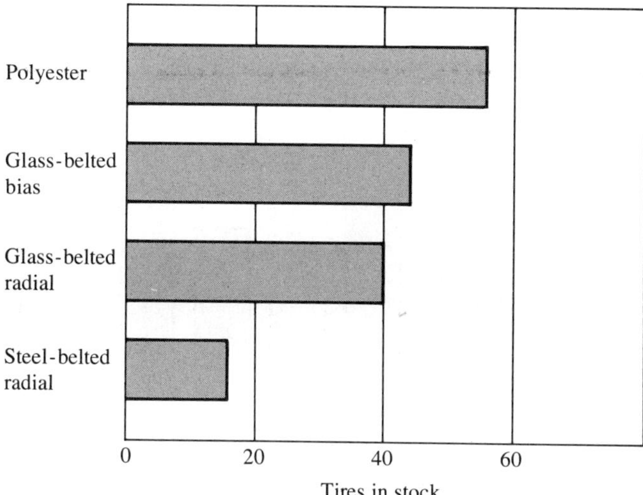

Figure 2.8 *Horizontal bar chart*

In manufacturing, the pattern of change in a dimension such as the diameter of a shaft is important when the goal is to keep the dimension constant. **Vertical bar graphs** and **line graphs** are popular graphic devices for time series. Figure 2.9 uses both a vertical bar graph and a line graph to present the same time series. A vertical bar graph usually is distinguished from a histogram by leaving a small space between the bars in vertical bar graphs but not in histograms. It is also common practice to use a line graph when a large number of time periods are involved. Vertical bar graphs are then reserved for time series with relatively few time periods. A line graph is also more effective for presenting several time series on the same set of axes.

In addition to illustrating simple cross-sectional and time series data, a person working with data often needs to illustrate how the parts of some whole are distributed. The number of freshmen, sophomores, juniors, and seniors in the student bodies of the five high schools in a certain city constitute a case in point. Another example is the composition of home refrigerator sales by type of refrigerator for the past several years in a retail outlet. **Component bar charts** and **component line graphs** (also called belt graphs) are major graphic devices intended for this purpose.

Figure 2.10 shows the component bar chart for the composition of the five high schools in a certain city by number of students in each of the four classes. It is apparent that Central High has the largest total enrollment and is much closer to having an equal number of students in the four classes than is any other school.

Typically, horizontal component bars are used when categories such as schools are being illustrated. Vertical component bars are used when time periods are the identifying characteristic used to distinguish the bars. It is also customary to refer to any type of presentation using horizontal bars as a *chart* and to one featuring vertical bars as a *graph.* The same principle applies in reference to line graphs.

Figure 2.11 is a **relative-component line graph**. Percentages of annual dollar sales are graphed for the four types of refrigerator by year. While it is impossible to

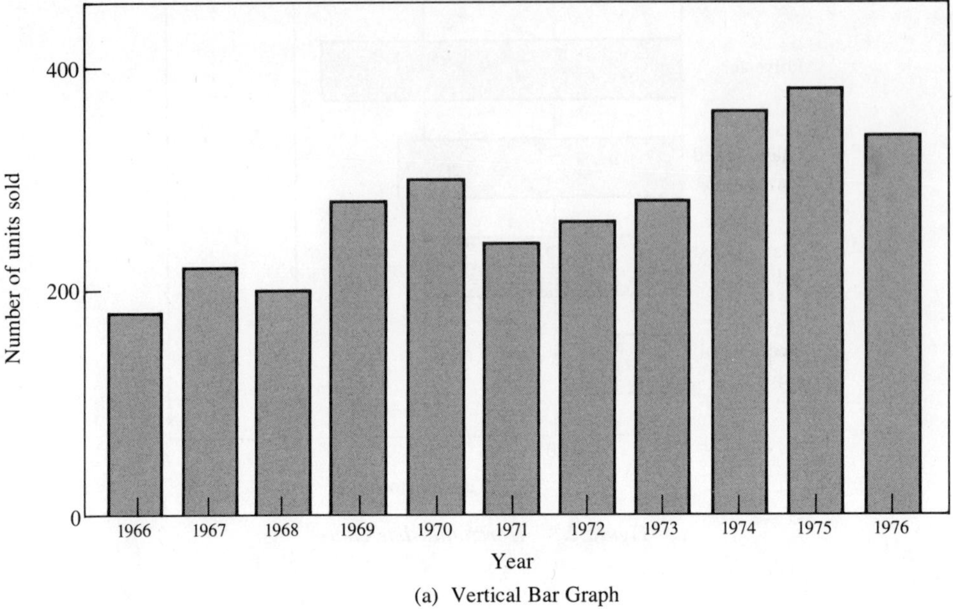

(a) Vertical Bar Graph

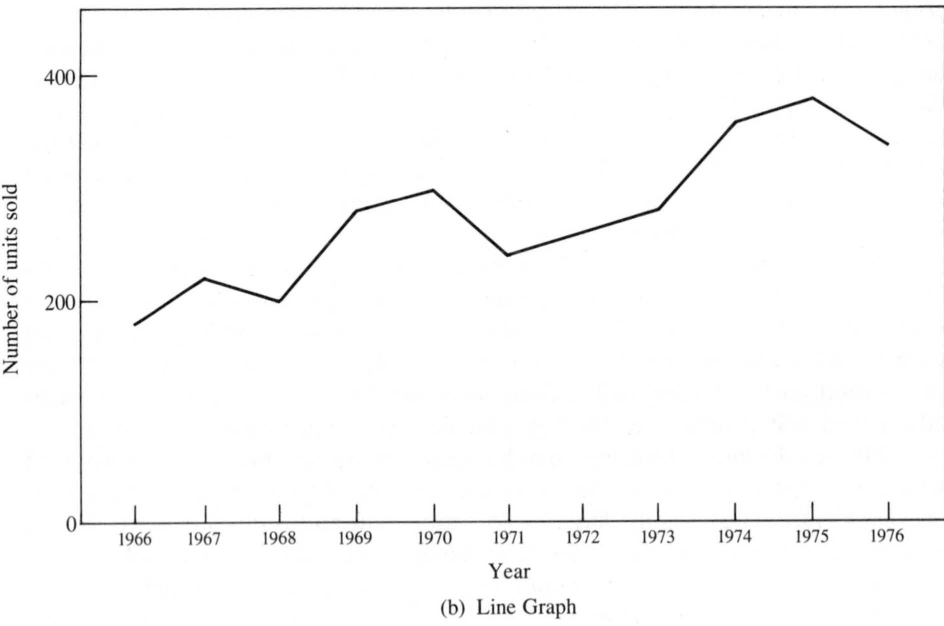

(b) Line Graph

Figure 2.9 *Time series presentations*

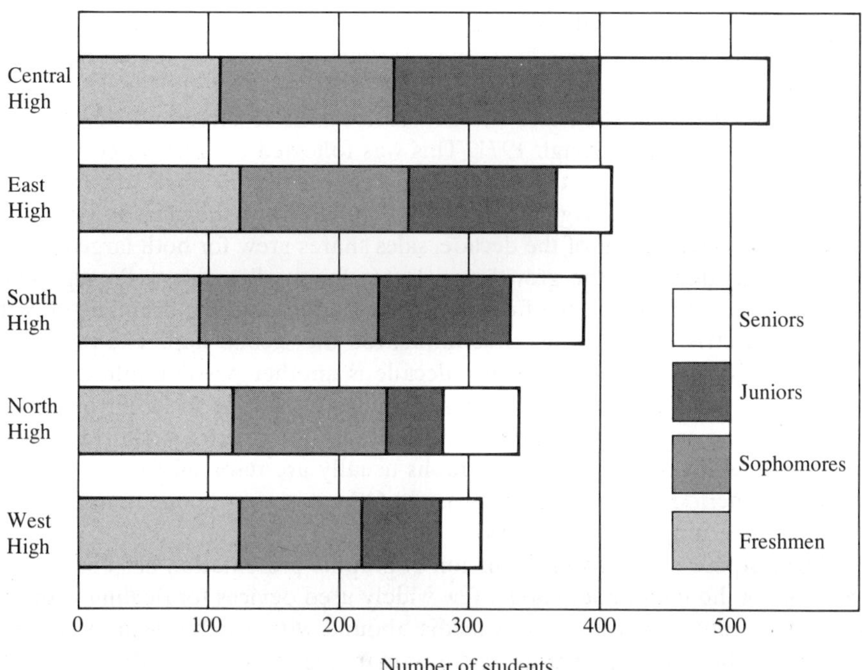

Figure 2.10 *Component bar chart*

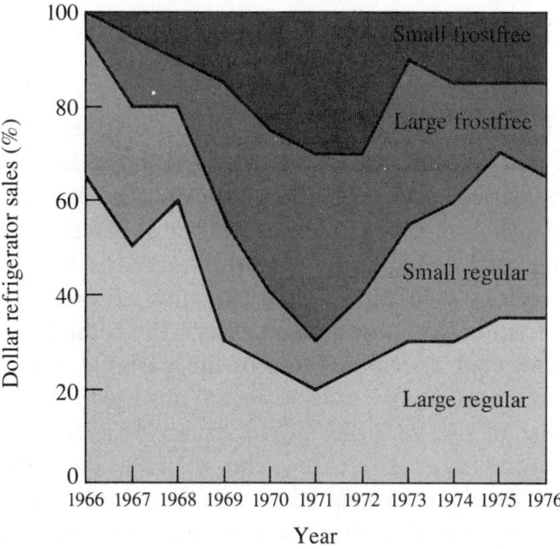

Figure 2.11 *Relative-component line graph*

trace the change in total sales through the 10-year period on a relative graph, the changing composition of sales by type of refrigerator receives far more emphasis than would be shown in a component line graph for the same data. Figure 2.11 shows a marked expansion in the share of sales attributable to large and small frostfree refrigerators from 1966 through 1971. This was followed by a major contraction in sales percentage of large frostfree and a lesser contraction in small frostfree. Meanwhile, in the earlier years, large regular refrigerators shrank the most as a percentage of sales. In the second half of the decade, sales shares grew for both large and small regular refrigerators, but the growth was larger for small regulars. Prestige buying could have been a factor in the first half of the period, and the dramatic growth in the cost of electric energy could have influenced the second half. The reduction in average family size through the entire decade is another possible influence on the patterns, with smaller units increasing in popularity relative to larger ones.

As was the case with simple vertical bar and line graphs, vertical component and vertical relative-component bar graphs usually are reserved for use when only a few time periods are being compared. Otherwise, component and relative-component line graphs are preferred.

The purpose of this short discussion of graphic presentation has been to introduce many of the major issues and a few widely used devices for dealing with them. Technical literature as well as mass media abound with a rich selection of alternatives for graphically emphasizing and summarizing patterns in all sorts of data. Thorough coverage of this topic can, and does, fill entire books. Nonetheless, enough has been said here to make you aware of the subject and to lead you to seek references should you be faced with the need to present data.

Summary

Frequency distributions for collections of observations of continuous variables are composed of many-value classes, all with the same length of class interval if feasible. Class *boundaries* indicate the maximum range for the values in each class. Class *limits* describe the range to the same number of places as the data. Histograms and frequency polygons are alternative graphic devices for illustrating frequency distributions and relative frequency distributions based on continuous variables.

Cumulative frequency distributions present the total number of observations equal to or less than each specified value of the variable. Cumulative relative frequency distributions present the proportion of observations equal to or less than each specified value of the variable. For discrete variables, cumulative frequency distributions and cumulative relative frequency distributions with single-value classes are illustrated graphically with step diagrams. For both discrete and continuous variables, cumulative frequency distributions and cumulative relative distributions with many-value classes are illustrated graphically with ogives.

Frequency distributions and relative frequency distributions, cumulative frequency distributions and cumulative relative frequency distributions furnish alternative ways to present data. Seldom are all four used with the same collection. The selection depends on what is to be emphasized.

Graphic devices are available to illustrate data other than frequency distributions. Cross-sectional data measured on nominal scales can be illustrated with horizontal bar charts. Vertical bar graphs and line graphs are common ways to present time series data. Component bar charts and component line graphs along with their relative counterparts are effective for showing changes in the composition of a total.

See Self-Correcting Exercises 2B.

Exercise Set B

1. Bearing in mind the rules for rounding discussed in Chapter 1, determine which, if any, of the following exact values would *not* be represented by 10.25 when expressed in hundredths:

   (a) 10.247 (d) 10.245 (g) 10.2555
   (b) 10.25005 (e) 10.255 (h) 10.2455
   (c) 10.251 (f) 10.025 (i) 10.258

2. Round each of the following to the nearest tenth, using the round-even rule where necessary:

   (a) 8.251 (c) 8.25001 (e) 8.25
   (b) 8.2501 (d) 8.250001

3. The smallest value listed in Exercise 2 is 8.25, and 8.250001 is the next smallest value listed there. Can you find a value *between* 8.25 and 8.250001? How many such values are there? How would such values be expressed when rounded to tenths?

4. The rounded value 8.3 includes all exact values *between* 8.25 and 8.35, but it does not represent either 8.25 or 8.35. Why?

Exercises 5 through 10 refer to the following frequency distribution of daily high temperatures for the months of June, July, and August of last year in a certain city. The temperatures are recorded in degrees Celsius.

x	f
17–20	1
21–24	8
25–28	21
29–32	31
33–36	17
37–40	12
41–43	2

5. Is the variable discrete or continuous?
6. Has a nominal, ordinal, interval, or ratio measurement scale been applied?
7. According to the table of values obtained in the first part of this chapter using Sturges' formula, what is the recommended number of classes for 92 observations? There are seven classes here. Is this an acceptable number of classes?
8. What is the length of the first class? Are all the classes of equal length?
9. (a) What is the lower class boundary of the first class?
   (b) What role does the upper class boundary of the third class play for the fourth class?
10. In what class are each of the following exact temperature observations recorded: (a) 41°; (b) 40.8°; (c) 40.08°; (d) 16.95°?

Exercises 11 through 20 refer to data obtained at the Fourth Annual Jamestown Juvenile Olympics. The 30 participants in the broad jump leaped the following distances (in meters):

0.80	1.02	1.34	1.76	2.39
0.85	1.05	1.38	1.78	2.41
0.87	1.11	1.46	1.84	2.56
0.92	1.21	1.50	1.93	2.64
1.00	1.27	1.58	1.97	2.66
1.00	1.32	1.75	2.07	2.80

11. (a) According to Sturges' formula, what is an appropriate number of classes for a frequency distribution of these data?
    (b) What is the range of values to be included in the distribution?
    (c) To what decimal accuracy should the interval length be stated? Why?
    (d) What minimum interval length will satisfy the requirements indicated by parts (a), (b), and (c)?
12. Using your results from Exercise 11, construct a frequency distribution for the jump distances, starting with a lower class limit for the first class of 0.80.
13. Give the two class boundaries for each class obtained in Exercise 12.
14. (a) To what decimal accuracy should the class limits have been expressed in Exercise 12? What would happen if the limits had been stated to one less decimal place?
    (b) To what decimal accuracy should the class boundaries have been expressed in Exercise 13?
15. Construct a histogram for the frequency distribution obtained in Exercise 12.
16. (a) What is the midvalue of each class in the frequency distribution of Exercise 12?
    (b) Locate these midvalues on the horizontal axis of your histogram in Exercise 15, and use them to plot a frequency polygon using the same axes as your histogram.
17. Construct a cumulative frequency distribution from the frequency distribution in Exercise 12.
18. Using your distribution from Exercise 17, construct the appropriate graphic device, a cumulative frequency ogive. Pay particular attention to labeling your lines of demarcation on the horizontal axis.

19. Convert the distribution in Exercise 17 to a cumulative relative frequency distribution, and construct a cumulative relative frequency ogive for your results.
20. Compare the ogives obtained in Exercises 18 and 19. How are they alike? How do they differ?
21. A retail ready-to-wear store reported the following sales last year: men's clothing $125,400, women's clothing $93,200, notions $50,700, other $34,600. Why is a horizontal bar chart an appropriate graphical device for illustrating this information?
22. Construct a horizontal bar chart to illustrate the sales data in Exercise 21. Describe the information conveyed by the visual impression.
23. What are cross-section data? What are two types of graphic presentation used to illustrate cross-section data?

Overview

1. For any given class the lower class boundary, upper class boundary, lower class limit, and upper class limit may be arranged in order of increasing size. State this arrangement.
2. Using the discrete data from Exercise 3 of Exercise Set A, construct a cumulative frequency distribution.
3. Construct a cumulative step graph for the distribution of Exercise 2.
4. Convert your distribution from Exercise 2 into a cumulative relative frequency distribution, and construct an appropriate graphic device for your resulting distribution.
5. From the following distribution of weights of rodents in an experimental laboratory, construct a cumulative relative frequency ogive. The weights are given in grams.

x	f
81–85	8
86–90	10
91–95	18
96–100	13
101–105	5
106–110	4
111–115	2

6. Are the data of Exercise 5 discrete or continuous?
7. Why is a cumulative relative frequency ogive appropriate for presenting your results from Exercise 5? Why is a step graph inappropriate?
8. When is a step graph inappropriate for discrete data?

9. When is it appropriate to use each of the following graphic devices? If the device is used for more than one type of distribution, list all types for which it is used.
   (a) Histogram
   (b) Ogive
   (c) Frequency polygon
   (d) Step graph
   (e) Frequency diagram
10. A collection of data are recorded to ten-thousandths of a unit. In a frequency distribution of these data, (a) to what degree of accuracy are class interval limits stated? (b) to what degree of accuracy are interval lengths stated? (c) to what decimal accuracy are class boundaries understood to be?
11. So far we have dealt only with distributions having all intervals of equal length. But sometimes, due to unusual features of the data, an open-ended class at either or both ends of a distribution, or an especially long class at the ends, is all but unavoidable. For instance, a frequency distribution of annual family incomes in the United States might require a relatively long or open interval at one end. Which end, lower or upper income level, would more likely require such treatment? Why?
12. The following table shows the number of inhabitants of a town classified by age for a ten year period.

Inhabitants Classified by Age (in thousands)

Year	Under 20	20 to 60	60 and Over
1967	12.0	23.2	10.7
1968	13.2	25.3	11.9
1969	14.7	28.7	14.4
1970	16.1	32.4	15.6
1971	17.0	34.3	16.8
1972	17.2	36.4	17.3
1973	17.7	38.6	17.5
1974	18.0	38.7	18.4
1975	18.2	39.8	18.4
1976	18.7	41.4	19.6

Construct a component line graph for these data.

13. Construct a relative component line graph for the data in Exercise 12.
14. Interpret the graph in Exercise 13 and compare this with your interpretation of the graph in Exercise 12.

3

Measures of Properties from Frequency Distributions

In the latter half of Chapter 1 we discussed useful properties of collections of data and described measures of central location, such as the mean and the median. We also discussed the variance and standard deviation, two measures of dispersion. The observations in a statistical population were considered individually in finding values for these measures. In Chapter 2 we learned to group a large collection of data into classes and to record the number of observations for each class. We saw that the resulting frequency distribution made the pattern in the statistical population much more apparent than did observations considered individually.

Now we are ready to see how values for measures of properties such as location and dispersion can be found from frequency distributions. We could, of course, go back to the original observations in a large collection and use the methods discussed in Chapter 1. The advent of modern computers has made this approach feasible even with very large collections of data. Nevertheless, there are times when we have access only to the frequency distribution and not to the original observations. There are other instances in which it would take considerably longer to prepare and enter the original observations into the computer than it would to find values for the desired measures from the frequency distribution. For these and many other reasons, knowing how to find these values from frequency distributions is still highly desirable.

Measures of Location

The Mean 3.1

Discrete Data in Single-Value Classes Let's return to the first example about the number of flaws in each of 50 sheets of galvanized steel. The frequency distribution in the two leftmost columns in Table 3.1 is the same as the one in Table 2.2.

Recall that X is the number of flaws per sheet and f is the number of sheets with the stated number of flaws per sheet.

Table 3.1 *The Mean from Single-Value Classes*

Flaws per Sheet (x)	Number of Sheets (f)	fx
0	10	0
1	14	14
2	12	24
3	8	24
4	4	16
5	2	10
	$N = 50$	$\Sigma\,(fx) = 88$

$$\mu = \frac{\sum (fx)}{N} = \frac{88}{50}, \text{ or } 1.76$$

Table 3.1 tells us there are 10 sheets with no flaws, 14 sheets with one flaw each, and so on. Were we to find the mean from the original observations, we would add 10 zeroes to 14 ones and so on to find the sum of all observations, which we would then divide by 50. Alternatively, we can make use of a weighting notion similar to the weighted mean we considered in Chapter 1. We can weight each value of x in Table 3.1 with its frequency of occurrence. This process gives us the third column in Table 3.1, the collection of fx values. Then we can sum these weighted values to obtain 88. It is clear that the result is identical to the sum of the original observations. Dividing the sum of the weighted values by the number of observations (50) gives us a mean of $^{88}/_{50}$ or 1.76 flaws per sheet, the same mean we would have found from the original observations. We can generalize the process just described:

Statement

The basic formula for finding the mean μ of any finite statistical population from its frequency distribution is

$$\blacksquare \quad \mu = \frac{\sum (fx)}{N}, \quad \blacksquare \qquad \textbf{3.1}$$

where x is the value representative of every observation in a given class, f is the number of observations (frequency) in that class, and N is the total number of observations in the population.

Discrete Data in Many-Value Classes In Table 3.2, the two columns to the left comprise the frequency distribution for travel advertisements that was first presented in Table 2.8. This distribution was constructed from the original observations of number of travel advertisements in each of 60 issues of a national magazine listed in Table 2.6. For example, the three observations 14, 15, and 18 account for the frequency count of three in the first class in Tables 2.8 and 3.2. The sum of these three observations is 47, and when this is added to the other observations in Table 2.6 the grand total is 2027. This total divided by 60 yields a mean of 33.783 travel ads per issue for the original data.

Table 3.2 *The Mean from Many-Value Classes for Observations of a Discrete Variable*

Number of Ads per Issue	Number of Issues (f)	Class Midvalues (x)	fx
13–18	3	15.5	46.5
19–24	5	21.5	107.5
25–30	12	27.5	330.0
31–36	16	33.5	536.0
37–42	14	39.5	553.0
43–48	8	45.5	364.0
49–54	2	51.5	103.0
	$N = \Sigma f = 60$		$\Sigma (fx) = 2040.0$

$$\mu = \frac{\sum (fx)}{N} = \frac{2040}{60}, \text{ or } 34$$

Given only the frequency distribution in Table 3.2, we must attempt to find the sum of the observations. In the first class, however, we can't tell from the frequency distribution that the three values in that class are 14, 15, and 18 with a sum of 47. Consequently, we find the *midvalue* of the class (15.5) by averaging the class limits $[(13 + 18)/2]$. Then we assume that the product of the class frequency ($f = 3$) and the class midvalue ($x = 15.5$) is close enough (46.5) to the correct sum of observations in the class for practical purposes (47, which we know in this case). We repeat this process for the other classes and then add all these products to find the approximate grand total $[\Sigma(fx) = 2040]$. The assumption is that the product of the midvalue and frequency very nearly equal the sum of the values in each class. The sum of the frequencies in the second column of Table 3.2 is the number of observations (60). Dividing the approximate total, 2040, by the number of observations, we have a mean of exactly 34 travel ads per issue. This approximates the correct mean of 33.783 closely enough for most purposes.

When we form a frequency distribution with many-value classes, the numbers placed in a given class lose their identity. We replace each of them in a given class with a representative number (the class midvalue) in order to find the mean. This was mentioned on page 38 as a guideline for forming classes. The closer the midvalues are to the correct means of the observations in the classes, the more nearly correct will be the mean calculated from the frequency distribution.

As is the case with single-value classes, we use Equation 3.1 to find the mean of a frequency distribution with many-value classes for observations of a discrete variable. The only modification necessary is the use of the class midvalue to represent every observation in a given class.

Continuous Data in Many-Value Classes In Tables 2.9 and 2.11 we were concerned with observations of values of a continuous variable—the content weights of 120 bags of sugar. The frequency distribution in Table 2.11 is repeated in the first two columns of Table 3.3.

Table 3.3 *The Mean from Many-Value Classes for Observations of a Continuous Variable*

Content Weight (pounds)	Number of Bags (f)	Class Midvalues (x)	fx
4.84–4.87	1	4.855	4.855
4.88–4.91	4	4.895	19.580
4.92–4.95	16	4.935	78.960
4.96–4.99	30	4.975	149.250
5.00–5.03	42	5.015	210.630
5.04–5.07	21	5.055	106.155
5.08–5.11	6	5.095	30.570
	$N =$ 120		$\Sigma(fx) =$ 600.000

$$\mu = \frac{\sum (fx)}{N} = \frac{600{,}000}{120} = 5{,}000$$

As we learned in Chapter 2, when working with a frequency distribution based on observations of a continuous variable we will also be working with many-value classes. In such cases, we use Equation 3.1 and class midvalues in exactly the same manner as we did for observations of a discrete variable in Table 3.2.

The class midvalues can be found by averaging either the class limits or the class boundaries. Then we multiply each midvalue by the frequency in that class and sum the products. In Table 3.3 the grand total of the observations is found to be 600.000. Dividing this by 120 observations, we get a mean of 5.000 pounds per bag from the frequency distribution. Had we summed the original observations in Table 2.9, we would have found a grand total of 599.88 pounds. This would produce a mean of 4.999 pounds per bag, not materially different from the mean obtained from the frequency distribution. In practice, of course, we would not usually find the mean
3.1 from the frequency distribution if we had already found the mean of the original data.

The Median

In Chapter 1, we defined the median to be the middle value in a collection of data when the observations are arranged in order of magnitude. For such a list composed of N observations, we learned that the median is the value of the $[(N + 1)/2]$th observation. Now we consider how to find the median when we are working with frequency distributions.

Single-Value Classes There are 50 observations in the frequency distribution for number of flaws per sheet in Table 3.1. It follows that the median is the 25.5th "observation" $[(50 + 1)/2]$. Beginning with zero flaws, we count the first 10 sheets. Then the next 14 sheets have one flaw apiece. Hence a total of 24 sheets have one flaw or less. The twenty-fifth and twenty-sixth sheets are both in the class with two flaws per sheet. Since the 25.5th "observation" is halfway between the twenty-fifth and twenty-sixth, its value must also be 2 and the median is two flaws per sheet. This is, of course, exactly the same value we would have obtained from the list of original observations placed in order of magnitude. No information is lost when observations of a discrete variable are put in single-value classes.

Another type of situation can arise with single-value classes. For the 40 observations in Table 3.4, $(N + 1)/2$ is 20.5. We see that 20 employees weren't absent and the twenty-first employee was absent 1 day. The median (the value of the 20.5th "observation") lies someplace between zero and one absence. As in the similar case for ungrouped observations in Chapter 1, the median is arbitrarily placed halfway between the two values in question. Hence the median is 0.5 absences for our example.

Table 3.4 *Employee Absences Last Week*

Days Absent (x)	Number of Employees (f)
0	20
1	17
2	3
	40

Many-Value Classes Earlier in this chapter, we saw that the same procedure is used to find the means for frequency distributions with many-value classes whether discrete or continuous variables are involved. The same statement also applies when we are concerned with the median instead of the mean. Nonetheless, we will consider one example for each type of variable.

In Table 3.5, we have a frequency distribution formed from grades on a reading test. The original grades were recorded as discrete integers from 51 to 100. Since there are 47 observations, the median is the grade of the twenty-fourth student when grades are in order of magnitude: $(N + 1)/2 = 24$. The cumulative frequency column tells us that 15 students received grades of 70 or less and that 30 students received grades of 80 or less. The twenty-fourth grade in order of magnitude must therefore be in the class running from grades of 71 to 80. We will refer to the class in which the median falls as the **median class**. Furthermore, we will assume that the class midvalue represents every observation in the class.

Table 3.5 Grade Distribution for a Reading Test

Grades (x)	Number of Students (f)	Cumulative Frequency (F)
51– 60	2	2
61– 70	13	15
71– 80	15	30
81– 90	11	41
91–100	6	47
	47	

It follows that the median grade as determined from this frequency distribution is 75.5, the midvalue of the median class. This is sometimes referred to as the **crude median**. It is obvious that this is not the same as the value we would get from the original observations if we had access to them. The twenty-fourth grade would have to be a whole number. But our approximation is reasonably close. It has a maximum error of half the width of the median class and will usually be closer to the correct value.

For an example based on a continuous variable recall that the frequency distribution in Table 3.3 describes the content weights of 120 bags of sugar. When N is 120, $(N + 1)/2$ is 60.5. Hence the median lies midway between the sixtieth and sixty-first observation in order of magnitude. In Table 3.3, we find that there are 51 observations equal to or less than 4.99 pounds and 93 observations equal to or less than 5.03 pounds. The *median class* runs from 5.00 to 5.03 pounds. Therefore, the *median* is 5.015 pounds (the class midvalue). From the original observations in Table 2.9 we find that the value of the median is 5.00 pounds to two decimal places. The two values agree quite well.

Occasionally with many-value classes the median will not be the midvalue of the median class. In Table 3.6, N is an even number so $(N + 1)/2$ will end in a half unit; that is, the median lies midway between the fifteenth and sixteenth observations. Typical of this situation, the fifteenth observation is at the upper end of one class and the sixteenth observation is at the lower end of the next class. There is no median class. In such cases, we define the *median* to be the value midway between

Table 3.6 Number of Radishes in Each of 30 Bunches

Number of Radishes (x)	Number of Bunches (f)	Cumulative Frequency (F)
8–10	5	5
11–13	10	15
14–16	11	26
17–19	4	30
	30	

the upper limit of the first class and the lower limit of the second. For this example, the median is 13.5 radishes per bunch, as approximated from the frequency distribution. Note that 13.5 is the *class boundary* for these two classes.

Definition In frequency distributions with many-value classes, the **crude median** is either equal to the midvalue of the class in which the $[(N + 1)/2]$th observation falls or is equal to the class boundary if the $[(N + 1)/2]$th observation falls between the upper limit of one class and the lower limit of the next.

This definition covers the two possible cases for distributions with many-value classes. For distributions with single-value classes the procedure for finding the median is essentially the same as that for ungrouped data described in Chapter 1.

The median is a position measure which splits the distribution into two *equal* parts. There is a class of such measures, called **fractiles**, which split a distribution into varying proportions. For example, observations smaller than the first **quartile** constitute the smallest 25 percent of the data in a collection. **Percentiles** (hundredths) and **deciles** (tenths) are other commonly used fractiles.

The Mode

The **mode** is that value which occurs most often in a collection of observations. For example, in Table 3.1 there are 14 steel sheets with one flaw apiece. This is the most frequently recurring value of x. Hence the mode of this distribution is one flaw per sheet.

For distributions with equal intervals and with more than one value in some classes, the same basic concept may be used to find the **modal class**. In other words, the class with the most frequencies in it is the modal class. For the weight distribution in Table 3.3, there are 42 observations in the class 5.00 to 5.03 pounds. Since 42 is the greatest class frequency, the modal class is 5.00 to 5.03 pounds. For frequency distributions with many-value classes, we will define the mode as *the midvalue of the modal class.* Hence for the example just cited, the mode is 5.015 pounds.

The mode is the least commonly used measure of location of the three we have discussed, primarily because the modal value is not precisely defined for some collections of ungrouped data or even for some frequency distributions. For example, for the collection of observations 0,2,3,5,8 every value satisfies the definition of the mode. In a frequency distribution, two or more classes that are adjacent may have the same frequency of observation and this number may be greater than the frequency in any other class. The midvalue of the interval covering these modal classes is the mode. But if the modal classes are not adjacent, the midvalues of all these maximum-frequency classes would satisfy the definition of the mode and we are left with a rather ambiguous situation. Often, however, a better selection of class intervals will result in a smooth pattern peaking at a single mode.

For frequency distributions with many-value classes, we used class midvalues of the median and the modal classes as the median and mode. Equations exist to get somewhat more refined positions for these two measures within their respective classes. Many elementary statistics texts contain these equations and can be consulted on this point by those who wish to pursue this approach.

Summary

Sometimes values of measures of location must be found from frequency distributions. To find the mean from a frequency distribution, we use the class frequencies as weights, multiply them by the representative values of the variable in the classes, sum the products, and divide by the number of observations. Representative values of the variable are the stated values for single-value classes. For many-value classes, class midvalues are the representative values, whether the variable is discrete or continuous. If care has been taken in selecting class intervals, the mean as calculated from a frequency distribution will be nearly the same as that calculated from the original data.

The median, the middlemost value, is the value of the $[(N + 1)/2]$th observation from either end of a frequency distribution. If the median falls within a many-value class, that class is designated the median class and the midvalue of the median class is defined to be the value of the median. If the median falls between classes, the class boundary is defined to be the median. For single-value classes, the median is either a stated value or the mean of two adjacent values.

The mode is the most frequently recurring value in a collection of data. In frequency distributions the modal class is the class with the greatest frequency. The value of the mode is defined as the midvalue of the modal class for many-value classes. For single-value classes the mode is the stated value for the class.

See Self-Correcting Exercises 3A.

Exercise Set A

1. Suppose the mean is determined from grouped data.
   (a) Under what circumstances is the same mean invariably obtained as would have been obtained from the original ungrouped data?
   (b) Under what circumstances is a mean obtained that is likely to differ from the mean that would have been obtained from the ungrouped data? Why?
2. In the formula

$$\mu = \frac{\sum (fx)}{N},$$

what does the x stand for if the mean is being obtained from a frequency distribution with (a) single-value classes? (b) many-value classes?

3. Consider the single-value frequency distribution

x	f
3	9
4	12
5	10
6	5
7	5

   (a) What numerical value is
      (i) f_1? f_2?
      (ii) $\sum f$?
      (iii) N?
      (iv) $f_1 x_1$? $f_2 x_2$?
   (b) Determine $\sum (fx)$.
   (c) Determine μ.
   (d) Determine the median.
   (e) Determine the mode(s).

4. Consider the cumulative frequency distribution

x	F
10	8
11	15
12	23
13	27
14	30

(a) What is N?
(b) Convert this to a frequency distribution.
(c) Determine the mean.
(d) Find the mode(s).
(e) What difficulty with the mode as a measure of central tendency is illustrated by your response to part (d)?

5. The timekeeper for a large industrial firm records the number of times each week that each employee arrives late for work. Her records for a certain employee for last year are summarized below. What is the mean number of times per week that the employee reported late for work? (Leave your answer as a fraction.)

Times late per Week	Number of Weeks
0	32
1	10
2	8
3	2
	52

6. The following cumulative distribution shows the gallons of gas used daily (x) by a groundskeeper. The length of the study was 10 days. What is the mean daily use of gasoline by the groundskeeper?

x	F
0	2
1	6
2	6
3	7
4	9
5	10

7. A local junior college holds evening classes in real estate management. The director of evening studies, wishing to know something about the distribution of the ages of students

in the class, obtains the following data. Ages are recorded to the nearest year. What is the mean age of the students in the class?

Age	Frequency
25–29	4
30–34	6
35–39	6
40–44	2
45–49	2
	20

8. Comparing the measures of central location obtained for a single-value frequency distribution with the same measures for the data in ungrouped form, which of the following, if any, are likely to differ: (a) the means; (b) the medians; (c) the modes?

9. Using the original observations for the broad jump distances preceding Exercise 11 of Set B in Chapter 2, determine: (a) the mean; (b) the median; and (c) the mode.

10. From the frequency distribution obtained from the broad jump data in Exercise 12 of Set B in Chapter 2 and the rules discussed in this section, find: (a) the mean; (b) the median; and (c) the mode. Compare your results with those obtained in Exercise 9.

11. For any class in a frequency distribution the frequency f divided by the population size N is the proportion of all the observations that fall within that class. We could consider this class proportionality as a "weighting" of the class and could derive the formula

$$\mu = \sum \left[\left(\frac{f}{N}\right) \cdot x\right] = \sum (f'x),$$

since

$$\mu = \frac{\sum (fx)}{N} = \frac{f_1 x_1 + f_2 x_2 + \cdots + f_k x_k}{N} = \frac{f_1 x_1}{N} + \frac{f_2 x_2}{N} + \cdots + \frac{f_k x_k}{N}$$

$$= \frac{f_1}{N} \cdot x_1 + \frac{f_2}{N} \cdot x_2 + \cdots + \frac{f_k}{N} \cdot x_k$$

$$= \sum (f'x).$$

Thus if the data are presented in a relative frequency distribution we can compute the mean directly. Use the method just described to find the mean of the following distribution.

x	f'
3–5	0.10
6–8	0.15
9–11	0.40
12–14	0.30
15–17	0.05

12. Convert the following cumulative relative frequency distribution to a relative frequency distribution and find the mean by using the formula given in Exercise 11. Then find the median and modal classes.

x	F'
2–5	0.05
6–9	0.21
10–13	0.56
14–17	0.82
18–21	1.00

13. Using the formula given in Exercise 11, compute the mean for the distribution

x	f'
102	0.08
103	0.06
104	0.22
105	0.30
106	0.30
107	0.03
108	0.01

and determine the median and the mode(s).

Measures of Dispersion

Since the variance and the standard deviation are by far the most frequently used measures of dispersion for frequency distributions, they are the only two we will consider here.

The Variance and the Standard Deviation

3.2 **Discrete Data in Single-Value Classes** Our example again will be the frequency distribution describing the number of flaws per sheet of galvanized steel in Table 3.1. From our discussion in Chapter 1 we recall that the variance σ^2 is the mean of the squared deviations of the observations from their mean μ. With modifications for repeated values, the same concept is the basis for finding the variance from a frequency distribution.

In Table 3.7(a) the first two columns on the left present the frequency distribution encountered in Table 3.1. The mean is shown at the bottom. The third column shows the deviations of the five values of x in the first column from the mean. For example, $x - \mu$ is $5 - 1.76$, or 3.24, for the last class. In the fourth column, the squared deviations of the six values of x from the mean ($\mu = 1.76$) are recorded.

Table 3.7 *The Variance from Single-Value Classes*
(a) The Basic Concept

Flaws per Sheet (x)	Number of Sheets (f)	Deviations ($x - \mu$)	Squared Deviations $(x - \mu)^2$	$f(x - \mu)^2$
0	10	−1.76	3.0976	30.9760
1	14	−0.76	0.5776	8.0864
2	12	0.24	0.0576	0.6912
3	8	1.24	1.5376	12.3008
4	4	2.24	5.0176	20.0704
5	2	3.24	10.4976	20.9952
	$N = 50$			$\Sigma f(x - \mu)^2 = 93.1200$

$$\mu = \frac{\sum (fx)}{N} = \frac{88}{50}, \text{ or } 1.76 \qquad \sigma^2 = \frac{\sum f(x - \mu)^2}{N} = \frac{93.12}{50}, \text{ or } 1.8624$$

$$\sigma = \sqrt{1.8624} = 1.364 \text{ flaws per sheet}$$

(b) The Basic Computation

Flaws per Sheet (x)	Number of Sheets (f)	fx	x^2	fx^2
0	10	0	0	0
1	14	14	1	14
2	12	24	4	48
3	8	24	9	72
4	4	16	16	64
5	2	10	25	50
	$N = 50$	$\Sigma (fx) = 88$		$\Sigma (fx^2) = 248$

$$\mu = \frac{\sum (fx)}{N} = \frac{88}{50}, \text{ or } 1.76 \qquad \sigma^2 = \frac{\sum (fx^2)}{N} - \mu^2 = \frac{248}{50} - \left(\frac{88}{50}\right)^2, \text{ or } 1.8624$$

$$\sigma = \sqrt{1.8624} = 1.364 \text{ flaws per sheet}$$

Contrary to what you may think, we will *not* get the correct value for the sum of the squared deviations if we add the entries in the fourth column. The first value in that column (3.0976) is the squared deviation of 0 from 1.76. But in this class there are 10 sheets with zero flaws apiece and each of these sheets contributes 3.0976 to the sum of the squared deviations. Altogether, the 10 sheets in this class contribute the product of 10 and 3.0976, or 30.976, to the sum of squared deviations. In general, the contribution of each class to this sum is the product of f and $(x - \mu)^2$ and these products are recorded and summed in the right-hand column of Table 3.7(a). When this sum is divided by the number of observations ($N = 50$) the result is 1.8624, the
3.2 variance for the collection of data.

From Appendix H we find that the recorded value of N closest to 1.8624 is 1.86. The table shows 1.364 as the square root of 1.86. Hence the standard deviation from the frequency distribution is 1.364 flaws per sheet.

The approach taken in Table 3.7(a) is always applicable, but it is awkward when the deviations are not convenient values. In Chapter 1, for cases characterized by awkward deviations, another approach was taken. This approach was based on Equation 1.5. Recall that Equation 1.5 is an exact algebraic equivalent of the basic definition of the variance in Equation 1.4. Not only is the variance the mean of the squared deviations: The variance is also the difference between the mean of the squared observations and the square of the mean. Now we merely adapt this alternative concept for use with frequency distributions.

In Table 3.7(b) we show the complete procedure, which includes all the calculations for the mean as well as for the variance. The third column repeats the procedure for finding the mean as shown in Table 3.1. The fourth column records the squares of the different values of x as listed in the first column. The fifth column records the sum of squared observations in each class (fx^2) along with the grand total for all classes: $\sum(fx^2)$. The mean of the squared observations is $^{248}/_{50}$, or 4.96, and the square of the mean (1.76^2) is 3.0976. The variance σ^2 is the difference ($4.96 - 3.0976 = 1.8624$).

Once again, the standard deviation is the square root of the variance, or 1.364 flaws per sheet as determined from Appendix H. In Chapter 1, we had a rule of thumb which said that at least 90 percent of the observations in any collection are within three standard deviations of the mean. By finding the z scores for zero and five flaws per sheet for the distribution in Table 3.7, we can check the statement for this collection. For zero flaws per sheet, we find from Equation 1.7

$$z = \frac{x - \mu}{\sigma}$$

$$= \frac{0 - 1.76}{1.364}, \text{ or } -1.29 \text{ standard deviations.}$$

For five flaws per sheet,

$$z = \frac{5 - 1.76}{1.364}, \text{ or } +2.38 \text{ standard deviations.}$$

Hence no observation in the collection is more than 1.29 standard deviations less than the mean ($\mu = 1.76$) or more than 2.38 standard deviations greater than the mean. We can now state the general relationships for finding the variance of a frequency distribution:

Definition

The **variance** σ^2 of a collection of data as found from a frequency distribution is the mean of the squared deviations from the mean μ. That is,

$$\blacksquare \quad \sigma^2 = \frac{\sum f(x-\mu)^2}{N}. \quad \blacksquare \qquad \textit{3.2(a)}$$

Alternatively, the variance is the difference between the mean of the squared observations and the square of the mean:

$$\blacksquare \quad \sigma^2 = \frac{\sum (fx^2)}{N} - \mu^2 \quad \blacksquare \qquad \textit{3.2(b)}$$

or

$$\sigma^2 = \frac{\sum (fx^2)}{N} - \left(\frac{\sum fx}{N}\right)^2. \qquad \textit{3.2(c)}$$

The standard deviation is, of course, the square root of the variance. The substitution of the appropriate values in Equation 3.2(b) for our example is shown at the bottom of Table 3.7(b).

Many-Value Classes In the first part of this chapter we saw that Equation 3.1 could be used to find the mean from any type of frequency distribution. For single-value classes we substituted for x the values of the variable as listed in the left-hand column of the frequency distribution. For many-value classes, whether the variable was discrete or continuous, we substituted for x the midvalues of the classes.

A direct parallel can be drawn between what we did with Equation 3.1 for many-value classes and what we do with Equations 3.2. If we substitute class midvalues for x in these latter equations, they apply to many-value classes for either discrete or continuous variables.

For the collection of data on advertisements described in Table 3.2, we found the mean μ to be 34 ads per issue. This should result in reasonably convenient numbers for deviations of class midvalues from the mean, which in turn should make Equation 3.2(a) easy to work with. Table 3.8 shows the frequency distribution and the necessary columns to compute the variance of the distribution using Equation 3.2(a). The deviations of class midvalues are found and shown in the fourth column. The squares of deviations appear in the fifth column. The squared deviation for each class is

Table 3.8 *The Variance from Squared Deviations*

Number of Ads per Issue	Class Midvalues (x)	Number of Issues (f)	Deviations ($x - \mu$)	$(x - \mu)^2$	$f(x - \mu)^2$
13–18	15.5	3	−18.5	342.25	1026.75
19–24	21.5	5	−12.5	156.25	781.25
25–30	27.5	12	−6.5	42.25	507.00
31–36	33.5	16	−0.5	0.25	4.00
37–42	39.5	14	5.5	30.25	423.50
43–48	45.5	8	11.5	132.25	1058.00
49–54	51.5	2	17.5	306.25	612.50
		$N = 60$			$\Sigma f(x - \mu)^2 = 4413.00$

$$\mu = 34 \qquad \sigma^2 = \frac{\sum f(x-\mu)^2}{N} = \frac{4413}{60}, \text{ or } 73.55 \qquad \sigma = 8.576$$

multiplied by the number of such squared deviations in the class to get the sum for that class (column 6). Then the grand total of the sixty squared deviations (4413) is found and divided by the number of observations (60) to determine the variance σ^2, which is 73.55. The standard deviation σ is 8.576 ads per issue as determined by averaging adjacent entries for 73.5 and 73.6 in Appendix H. Note that both ends of the distribution are less than 2.5 standard deviations from the mean.

A Shortcut for Many-Value Classes As we just pointed out, Equation 3.1 can be used to find the mean and any of Equations 3.2 can be used to find the variance for any frequency distribution for which class midvalues are available. There is, however, a shortcut which can be used for any frequency distribution in which all class intervals have the same length. The purpose is to reduce the arithmetic required to find the mean and variance.

Part of the information from Table 3.8 is repeated in the first two columns to the left in Table 3.9. To get the third column we arbitrarily select some class midvalue near the center of the distribution as an approximation to the mean and give it the symbol x_a. In our example, the fourth class is near the middle of the distribution so we select 33.5 as x_a. Any other class midvalue will also work. Then we find the number of *class intervals* that each class midvalue deviates from x_a. This is the value of d for each class. In the fourth column the products of the frequencies f from column 2 and the values of d from column 3 appear.

To find the mean by the shortcut method, we substitute in

$$\blacksquare \qquad \mu = x_a + i\frac{\sum (fd)}{N}, \qquad \blacksquare \qquad 3.3$$

where i is the length of each class interval (6 for our example). The substitutions are shown at the bottom of Table 3.9. As expected, the mean μ turns out to be 34 ads per issue as it was in Table 3.2.

Table 3.9 Shortcut for Finding the Mean and Variance

Midvalues	f	d	fd	d^2	fd^2
15.5	3	−3	−9	9	27
21.5	5	−2	−10	4	20
27.5	12	−1	−12	1	12
33.5	16	0	0	0	0
39.5	14	1	14	1	14
45.5	8	2	16	4	32
51.5	2	3	6	9	18
	$N = 60$		$\Sigma(fd) = +5$		$\Sigma(fd^2) = 123$

$$\mu = x_a + i\frac{\sum (fd)}{N}$$

$$= 33.5 + 6\left(\frac{5}{60}\right), \text{ or } 34$$

$$\sigma^2 = i^2\left\{\frac{\sum (fd^2)}{N} - \left[\frac{\sum (fd)}{N}\right]^2\right\}$$

$$= 6^2\left[\frac{123}{60} - \left(\frac{5}{60}\right)^2\right]$$

$$= 36\left(\frac{7355}{3600}\right), \text{ or } 73.55$$

To find the variance by the shortcut method described here, we substitute in

$$\sigma^2 = i^2\left\{\frac{\sum (fd^2)}{N} - \left[\frac{\sum (fd)}{N}\right]^2\right\}. \qquad \textbf{3.4}$$

In Table 3.9 columns 5 and 6 show how to find the sum of the squared values of *d*. The substitutions for our example appear at the bottom right of Table 3.9. The variance by this approach (73.55) is the same as that found by the approach in Table 3.8. The shortcut methods for finding the mean and the variance for a frequency distribution *must* produce the same values for these measures as do any other methods—so long as all class intervals are of equal length. These shortcut methods also apply to single-value classes, although the saving in effort is usually negligible.

Skewness

A frequency distribution which is symmetrical with respect to its median has no **skewness**. That is, it has no tendency to tail off in one direction from the median any differently than it does in the other direction. In Figure 3.1(a) the part of the distribution to the left of the median is a mirror image of the part to the right. This is what is meant by symmetry in this context. Note also that the values of the mean, median, and mode are all equal for the distribution in Figure 3.1(a). In general, the

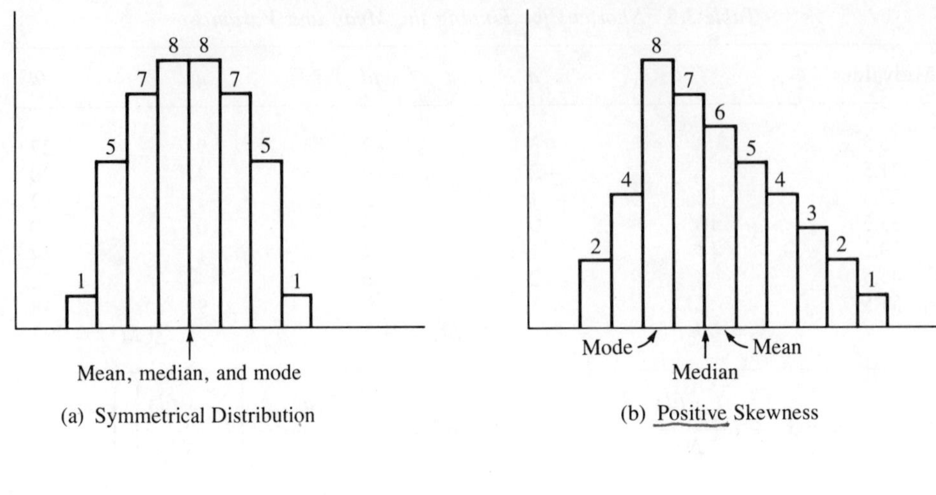

(a) Symmetrical Distribution (b) Positive Skewness

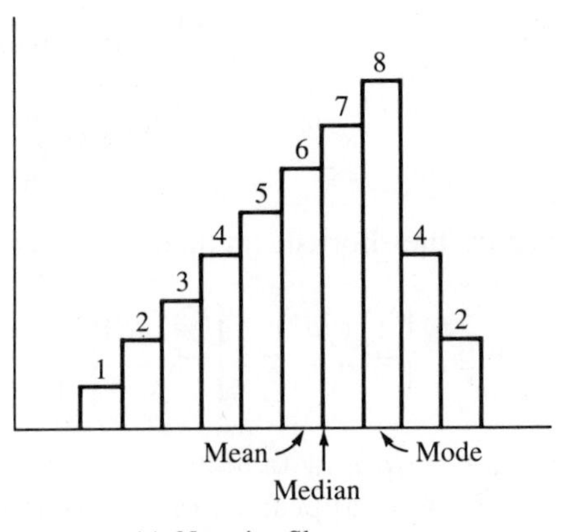

(c) Negative Skewness

Figure 3.1 *Symmetrical and skewed distributions*

values of these measures of central location coincide for a symmetrical distribution that has a single mode.

In Figure 3.1(b), the frequency distribution tails off further to the right of the median than it does to the left. A distribution with its longer tail extending to the right is said to be *skewed to the right.* Alternatively, it is said to have *positive skewness* because the long tail extends in the positive direction of the variable from the median. In skewed distributions the values of the mean, the median, and the mode are usually not the same. Instead, the mean is most affected by the extreme observations and is pulled farthest in the direction of the longer tail. The mode is least affected. The median is subject to intermediate effect and typically lies about two-thirds of the distance from the mode to the mean, although this relationship is only approximately correct.

Just the opposite of what we said for the distribution in Figure 3.1(b) is the case for the distribution in Figure 3.1(c). Here the long tail extends toward the left, in the negative direction from the median. This distribution is therefore said to be *skewed to the left,* or to have *negative skewness,* or to be *negatively skewed.* For such distributions the mean usually is less than the median, which is in turn less than the mode. Again, the median lies approximately two-thirds of the distance from the mode to the mean.

We have already become familiar with measures of central location and dispersion. As you may suspect, measures of skewness have also been devised. In the majority of cases, however, current practice does not report values of any measure of skewness to describe frequency distributions. Instead, the graph of the frequency distribution or verbal reference to the type of skewness present, or both, are reported. Because of this practice, we will not discuss any of the more common measures of skewness.

Summary

As measures of dispersion, values for the variance and standard deviation can be determined from frequency distributions. One approach is to find the variance by finding the weighted mean of the squared deviations from the mean of the data (μ). In this approach, the frequencies in the classes are the weights. Another approach is to find the weighted mean of the squared observations and subtract the square of the mean (μ^2) from that quantity. A third approach, applicable only when class intervals are of equal length, uses deviations in class interval units and usually reduces calculation a great deal.

A new property of collections of observations was introduced—skewness, the lack of symmetry with respect to the median. Distributions with their longer tails extending to the right or in the positive direction from the median are skewed to the right or have positive skewness. Distributions with their longer tails extending to the left are skewed to the left or have negative skewness.

See Self-Correcting Exercises 3B.

Exercise Set B

1. Determine the numerical value of

(a) $\dfrac{\sqrt{16}}{4}$ (b) $\sqrt{\dfrac{16}{4}}$

2. (a) Name two measures of dispersion.
   (b) How are these two measures related?
   (c) Which of these measures is stated in the same units as the original data? How is the other one stated?

3. Suppose two populations A and B, of equal size, have the same mean but A has a greater standard deviation. How would this fact be reflected in histograms or frequency polygons for the two distributions?

4. Determine the mean, the variance, and the standard deviation for each of the following distributions. Compare your results.

(a)

x	f
3	4
4	4
5	4
6	4
7	4
	20

(b)

x	f
3	3
4	6
5	2
6	6
7	3
	20

(c)

x	f
3	2
4	4
5	8
6	4
7	2
	20

5. From your results in Exercise 4, determine which of the distributions is most "spread out." Which has the least dispersion? Justify your conclusions.

6. If the variance exceeds 1, the variance is larger than the standard deviation. (Why?) In each of the following cases, state which is larger, the variance or the standard deviation: (a) $\sigma^2 = 1$; (b) $\sigma^2 < 1$.

7. For the daily gallon usage given in Exercise 6 of Exercise Set A, determine the standard deviation using (a) the basic concept method of Table 3.7(a); and (b) the basic computation equation of Table 3.7(b).

8. Compare your results in Exercise 7(a) and 7(b). Are your answers the same? Should you expect them to be? Why or why not?

9. Using Equation 3.2(a), find σ^2 and σ for the distribution

x	f
10–13	6
14–17	5
18–21	5
22–25	4

10. For the distribution in Exercise 9, find σ^2 and σ using Equation 3.2(b).

11. Compare your answers in Exercises 9 and 10. Are they the same? Should they be the same? Which method required messier computations? Why?

12. Use Equations 3.3 and 3.4 to find the mean and variance of the distribution in Exercise 9.

13. From Equation 3.4 give an equation for determining the standard deviation using the shortcut method.

14. Applying your results from Exercise 13, determine the standard deviation of the distribution of broad jump distances obtained in Exercise 12 of Set B in Chapter 2.

15. Given the frequency distribution

x	f
0–2	2
3–5	5
6–8	1
9–13	2

determine μ and σ using the methods of either Table 3.7(a) or 3.7(b). Suppose you were to use Equation 3.3 and your result from Exercise 13. Are your values for μ the same? Should they agree? Are your values for σ the same? Should they agree? What has gone wrong?

16. The shortcut method may be used only under what conditions? If these conditions are not met, what must be done?

17. (a) Consider the distributions in Exercise 4. What sort of skewness, if any, does each possess?
    (b) Consider the distribution given in Exercise 5 of Set A. What sort of skewness, if any, is present? Justify your conclusion.

18. Look at the distribution of ages in Exercise 7 of Set A. Can you determine what sort of skewness it possesses just by looking at it? Now determine the median and mode(s) and compare them with the mean found in that exercise. Does this comparison bear out your visual conclusion as to the type of skewness present?

19. (a) When the mean exceeds the median of a frequency distribution, what sort of skewness usually is present?
    (b) When the mean equals the median, what sort of skewness usually is present?

20. The June daily sales for a particular year, recorded to the nearest dollar, for a corner newspaper stand can be summarized in tabular form as follows:

Daily Sales (x)	Number of Days (f)
0–2	10
3–5	5
6–8	10
9–11	5
	30

(a) If the mean of daily sales is \$5, what is the variance?
(b) What is the standard deviation? Why is the standard deviation preferable to the variance as a measure of dispersion?

Overview

1. (a) In the population

   3,2,1,7,5,3,8,2,4,4,7,3

   find: (i) the mean; (ii) the median; (iii) the mode.

   (b) If the population given in part (a) had 13 members instead of 12 and that thirteenth member was 8, state how each of the following would differ from the same measures obtained in part (a): (i) the mean; (ii) the median; (iii) the mode.

2. Give two formulas for finding the mean of any frequency distribution, and give one which can be used only if all the intervals in the distribution are of the same length.

3. For the distribution with single-value classes,

x	f
8.0	3
8.5	7
9.0	11
9.5	9
10.0	6
10.5	4

   find, by any appropriate method, the

   (a) Mean
   (b) Median
   (c) Mode
   (d) Variance
   (e) Standard deviation
   (f) Type of skewness present

4. For the distribution

x	f
2.1–3.4	7
3.5–4.8	5
4.9–6.2	3
6.3–7.6	8
7.7–9.0	2

   find, by any appropriate method, the

   (a) Mean
   (b) Median
   (c) Mode
   (d) Variance
   (e) Standard deviation
   (f) Type of skewness present

5. In a symmetrical distribution:

   (a) How are the mean, median, and mode(s) related?

(b) What is the skewness?

(c) Is the standard deviation large or small?

6. Suppose that in a company with 100 employees, all of whom earn less than $30,000 annually, a new super-directorship is created with an annual salary of $45,000. How will the mean be affected? The median? The mode? The standard deviation? Which measure of central tendency is likely to be most affected? Which will be least affected?

7. A nominal scale is assigned to the property "fur color" in a genetic study of rodents. The color black is assigned the number 1, white the number 2, and agouti (mottled) the number 3. The following frequencies are obtained for the fur colors in the first generation under study:

x	f
1	15
2	12
3	13

For this collection, find each of the following numerical values and determine what color it represents: (a) the mean; (b) the median; (c) the mode.

8. In reference to Exercise 7, why are your answers to parts (a) and (b), perfectly good numerical values though they be, of no sensible value in this study? This difficulty arises from the use of a nominal scale, which is unavoidable here. When a nominal scale must be used, the only meaningful measure of central tendency is the mode.

9. In a survey of collegiate consumer attitudes toward brand names, the same beer was served under four different names, Boozo, Muskelunge, Thor, and Twinkle, to each member of a team of 100 eager volunteer tasters at a large university. Boozo was the choice of 44, Muskelunge received 16 votes, Thor was chosen by 37, and Twinkle was selected as tops by 3. What is the (a) mean choice? (b) Median choice? (c) Modal choice?

10. Here is a relative frequency diagram showing the number of sales per week for a mutual fund salesman:

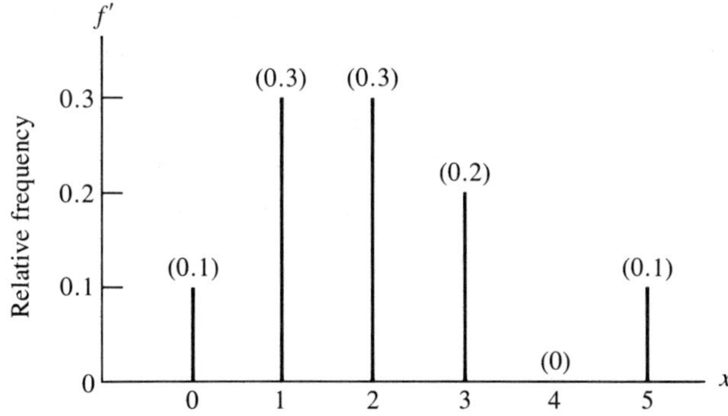

What is the mean number of sales per week?

Glossary of Equations

3.1 $$\mu = \frac{\sum(fx)}{N}$$

The arithmetic mean of a frequency distribution is the sum of the products of the class midvalues and frequencies divided by the number of observations.

3.2(a) $$\sigma^2 = \frac{\sum f(x-\mu)^2}{N}$$

The variance of a frequency distribution is found by summing the products of the squared class midvalue deviations from the mean and the frequencies associated with them and then dividing this sum by the total number of observations. This equation is useful when the mean is obvious and the deviations are small whole numbers.

3.2(b) $$\sigma^2 = \frac{\sum(fx^2)}{N} - \mu^2$$

The variance of a frequency distribution is the difference between two quantities. The sum of the products of class frequencies and squared midvalues divided by the total number of observations is the first quantity. From this quantity is subtracted the square of the arithmetic mean of the distribution.

3.3 $$\mu = x_a + i\frac{\sum(fd)}{N}$$

The mean of a frequency distribution with equal class intervals is the assumed mean plus the product of the class interval length and the mean class interval deviation from the assumed mean.

3.4 $$\sigma^2 = i^2\left\{\frac{\sum(fd^2)}{N} - \left[\frac{\sum(fd)}{N}\right]^2\right\}$$

The variance of a frequency distribution with equal class intervals is the product of the square of the class interval length and the difference between the mean squared class interval deviation and the square of the mean deviation.

Part Two

Background for Statistical Inference

4

Probability

Part One was concerned with *descriptive statistics.* Given a collection of observations—for example, on years of formal education of persons enrolled in an adult education program—how could we effectively describe the central tendency and dispersion in years of education? We found we could do this with summary measures such as the mean and median (for central tendency) and the standard deviation (for dispersion). Also, we learned to use a table of the frequency distribution as well as graphic presentations of the frequency distribution to describe the location and spread of the variable.

Part Three will be devoted to *statistical inference.* The objective in statistical inference is to reason from characteristics that we observe from a partial collection of observations to the corresponding characteristics for the entire collection of concern to us. We might have the years of education of 1500 out of 40,000 persons in a national adult education program. One important problem in inference would then be: What does the mean years of education of the 1500 persons tell us about the mean years of education of the entire 40,000?

If you are tempted to answer "nothing for sure" and "probably something" to this question, you are correct as far as that goes. An inference—reasoning from particulars to the general—can never be certain. It is subject to uncertainty, and the task in statistical inference is to quantify this uncertainty. How do we make a statement about the average years of education of the 40,000 adult students and how much reliance can we place on such a statement, or inference? We need background in probability to understand both the underlying concepts and the mechanics of statistical inference—so we turn now to that subject.

Sets and Events

The language and logic of **sets** is a necessary preliminary to probability. **4.1**
Emphasis on sets is common in high school mathematics, but we will review what we need in order to proceed. We do this by way of an example.

A coffee dispensing machine has four buttons, or selections. It will serve *coffee black, coffee with cream, coffee with sugar,* or *coffee with cream and sugar.*

Definition | A **set** is a well-defined collection of entities or elements within some frame of reference.

Suppose over a period of time that 20 servings were dispensed to 16 customers (4 customers purchased two servings each). We could be concerned with the 16 customers. That is one frame of reference. On the other hand, we might fix our attention on the servings. That is another frame of reference. In the one case we are concerned with a set of 16 persons, and in the other with a set of 20 servings. We can imagine two experiments that could be constructed, each within one of these frames of reference.

	Customers	Servings
Experiment:	observe a customer at the vending machine	observe a serving dispensed by the machine
Relevant set:	16 customers	20 servings

Definition | The set of all possible outcomes of an experiment is called a **sample space.**

Notice that the concept of a sample space can correspond to units of observation as that term was used in Chapter 1. In the experiments identified above the sample spaces are the 16 customers and the 20 servings. Each could be identified by a time ordering, that is, the 1st to the 16th customer and the 1st to the 20th serving. However, sample space can also refer to outcomes of observations. If a coin is tossed once, the sample space is {head, tail}. If an experiment is defined as selecting one of 16 customers and observing the customer's sex, the sample space is {male, female}.

Definition | An **event** is a subset of a sample space.

Events are defined according to what interests us in a given frame of reference. For example, we might be interested in the age as well as the sex of the vending machine customers. Suppose we are interested in the ingredients added to the coffee in a serving. It is common to use capital letters to identify various events. We could let $C =$ the event that a serving contains cream and let $S =$ the event that a serving contains sugar. In set language, the event C includes the *elements* (servings) in the *sample space* (the 20 servings) that exhibit the *characteristic* C (contain cream). What about the servings that do not contain cream?

Definition The **complement** of an event is the set of all elements in a sample space that do not satisfy the event. If the event is designated as A, its complement is termed A'.

In our example, C' would represent the event that a serving did *not* contain cream and S' would stand for the event that a serving did *not* contain sugar.

An important concept is the *intersection* of two events. An intersection is involved when we speak of a serving containing cream and sugar.

Definition The **intersection** of two events, A and B, is the set of all elements in a sample space that satisfy both the event A and the event B. The intersection of A and B is denoted by $A \cap B$.

If we speak of a serving containing cream *or* sugar we are talking about a *union* of two events. The use of the word *or* here is inclusive—that is, it really means *and/or.* The union of the event *cream* with the event *sugar* means *cream or sugar or both.* Thus the union of two events contains the intersection of the two events.

Definition The **union** of two events, A and B, is the set of all elements in a sample space that satisfy either the event A or the event B or both. The union of A and B is denoted by $A \cup B$.

So far we have developed a precise way of referring to events. In our coffee machine example, the various events could be represented by a Venn diagram as in Figure 4.1. Here the enclosed rectangle represents the sample space. The numbers 1 through 20 represent the successive servings dispensed. They are the *points* in the sample space. The left-hand circle represents the event C (cream) and the right-hand circle represents the event S (sugar). The overlap between the circles is thus the intersection cream *and* sugar ($C \cap S$). The shaded portion of the left-hand circle represents the event *cream and no sugar* ($C \cap S'$) and the shaded part of the right-hand circle stands for the event *sugar and no cream* ($S \cap C'$). The black coffee button produces the event $C' \cap S'$, which is portrayed by the space not within either circle but within the rectangle. We can see that servings 4, 14, and 17 were black coffee.

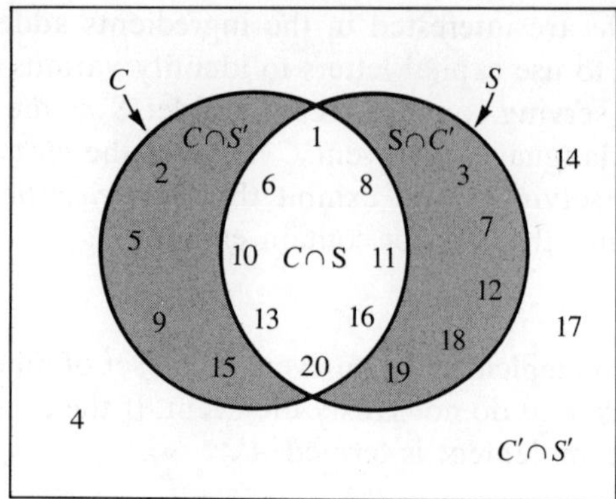

Figure 4.1 *Venn diagram for ingredients added to coffee*

A sample space was earlier defined as the set of all possible outcomes of an experiment, and an event was defined as a subset of a sample space. It is also necessary to regard *all* the elements in a sample space as an event—that is, the sample space is a subset of itself. Further, the subset that contains no elements in the sample space—called the **null set**—is also recognized as an event. We will denote an event that is not possible by the letter ○ with a slash through it, that is, $\varnothing$.

Definition Two events, A and B, are **mutually exclusive** if their intersection is not possible, that is, if $A \cap B = \varnothing$.

In the sample space of servings, the event *cream* and the event *sugar* are not mutually exclusive because we can have cream and sugar together. The events defined by the intersections shown in Figure 4.1 are mutually exclusive, however. They are the servings produced by the four buttons. They are, in symbols, $C \cap S'$, $C \cap S$, $S \cap C'$, and $C' \cap S'$. We cannot in one serving (outcome of the experiment) get *both* black coffee ($C' \cap S'$) and coffee with cream ($C \cap S'$). The intersection of these events
4.1 is impossible. In symbols, $(C \cap S') \cap (C' \cap S') = \varnothing$.

Views of Probability

4.2 We encounter probability statements in much of our everyday experience. Here are some examples:

1. The probability of heads in a coin toss is ______.

2. The probability that a voter prefers candidate A is ______.
3. The probability that it will rain tomorrow is ______.

Each of these statements involves an *experiment* and an *event*—a subset of the possible outcomes of the experiment. If the number 1 were given to complete any of the statements, we would regard the event as certain to happen; if zero were given, we would regard the event as certain *not* to happen. In giving a probability, we are assigning a number to an event. The mathematical definition of probability is a series of rules that must be followed in making such assignments:

Rules

1. The probability assigned to any event cannot be negative or greater than 1.0.
2. The total probability assigned to the sample space must equal 1.0; that is, the experiment will have *some* outcome.
3. Probabilities must be assigned so that the probability of the union of any two events is the sum of their separate probabilities minus the probability assigned to their intersection.
4. The null set, $\varnothing$, is assigned zero probability.

Note that these rules say nothing about the basis for establishing probabilities or how we are to interpret and use them. There are three viewpoints that are commonly used in establishing and interpreting probabilities—equally likely sample points, relative frequency, and degree of belief.

Equally Likely Sample Points

An early use of probability was in developing strategies in games of chance. Typically, the outcomes of experiments in such games—for example, the six sides of a die which may face up, the 38 positions on which a roulette wheel may come to rest, the two sides of a coin which may face up, the 52 cards which may be drawn from a standard full deck—are regarded as equally likely. This may be thought of as an assumption or as a definition of an honest chance mechanism. In either case, we establish probabilities by counting the number of sample points (elements) satisfying any event and relating this number to the number of points (elements) in the sample space.

Rule

When the elements in a sample space are regarded as equally likely, we obtain the probability of an event A by taking the ratio of the number of elements satisfying the event to the total number of elements in the sample space.

Using $P(A)$ for probability of A, $N(A)$ for number of elements satisfying A, and $N(\text{U})$ for number of elements in the sample space, we have

$$\blacksquare \quad P(A) = \frac{N(A)}{N(\text{U})}. \quad \blacksquare \qquad \textbf{4.1}$$

Let's return to our 20 coffee servings and suppose we could design an experiment so that it was equally likely we would observe any one of the servings. Suppose, as in Figure 4.1, there were three servings of black coffee, four servings of coffee with cream only, five servings of coffee with sugar only, and eight servings with sugar and cream. We have

$$N(C' \cap S') = 3 \qquad N(C' \cap S) = 5$$
$$N(C \cap S') = 4 \qquad N(S \cap C) = 8$$
$$N(\text{U}) = 20.$$

We can see also from Figure 4.1 that $N(C) = 12$ and $N(S) = 13$. Then, for the probability of the event *cream,* we have

$$P(C) = \frac{N(C)}{N(\text{U})} = \frac{12}{20} = 0.60.$$

For the event *sugar,* we have

$$P(S) = \frac{N(S)}{N(\text{U})} = \frac{13}{20} = 0.65,$$

and for the event *sugar and cream,* we have

$$P(S \cap C) = \frac{N(S \cap C)}{N(\text{U})} = \frac{8}{20} = 0.40.$$

For the event *sugar or cream,* we count all the points within the two circles.

$$P(S \cup C) = \frac{N(S \cup C)}{N(\text{U})} = \frac{17}{20} = 0.85.$$

It is clear that assigning probabilities by Equation 4.1 will satisfy the first two of the mathematical properties stated earlier. Let us check the third property. It requires that

$$P(S \cup C) = P(S) + P(C) - P(S \cap C).$$

This relation is satisfied by the probabilities established above. What happens is that the eight servings with sugar *and* cream are counted in both $N(S)$ and $N(C)$. Thus in adding $P(S)$ and $P(C)$ those eight servings are counted twice. The equation is maintained when these same eight servings are subtracted once:

$$P(S \cup C) = P(S) + P(C) - P(S \cap C)$$

$$\frac{5+4+8}{20} = \frac{5+8}{20} + \frac{4+8}{20} - \frac{8}{20}$$

$$\frac{17}{20} = \frac{13}{20} + \frac{12}{20} - \frac{8}{20}$$

$$0.85 = 0.65 + 0.60 - 0.40.$$

Probabilities assigned by counting sample points satisfy the third mathematical requirement. It is a requirement of consistency in the probabilities of related events that is always maintained when probabilities are established directly by counting elements in the underlying sample space of the experiment. It is often called the **general addition law** of probability.

Relative Frequency

The notion of an experiment involves operations in a set of circumstances that could be repeated over and over again. You can toss a coin or roll a die again and again in essentially the same manner. You can think of the next day's weather following all days when the barometer drops more than one point. You can think of repeating over and over the experiment of selecting a voter from a registration list in connection with public opinion polling.

The probability of an event is viewed by many as the *long-run relative frequency* of the event as the underlying experiment is repeated over and over. To use this definition to establish probabilities, we need information from a large number of repetitions. Strictly speaking, a finite number of repetitions only provides an approximation to a true probability because "long-run" implies an indefinite number of repetitions of an experiment. Such approximations are useful, however, and how to make them and measure their accuracy is an important subject in statistical inference–Part Three of this text.

Using equally likely sample points gave us a way of establishing probabilities in certain circumstances, but it did not provide an interpretation of probability. Relative frequency provides a way of interpreting such probabilities. If we assign $1/6$ to the event that a die toss will result in two dots facing up, we can interpret that as meaning that the relative frequency of a 2 as the die is cast over and over will approach one-sixth.

Degree of Belief

Another way of assigning probability to an event is according to the betting odds you would give in favor of the event occurring–that is, on your *degree of belief* in the proposition that the event will occur. If you are willing to bet even money on "heads" on the toss of a particular coin, you regard the chances of heads equal to the chances of tails. For you, $P(H) = 0.50$. Another person, who perhaps knows the coin has two heads, would assess $P(H)$ as 1.0. A third person, who knows the coin has both a head and a tail but believes the tosser can influence the outcome, might assess the probability of heads as 0.3. These possibilities point out that probability, as a degree of belief, is subjective. People with different beliefs assign different probabilities to an event, often for quite subjective reasons. The events to which probabilities are assigned by degree of belief can be unique–that is, beliefs are not restricted to repetitive events. This way of assigning probability has a long history and has been adopted by people particularly interested in individual decision making.

Addition Laws for Probability

Suppose a security analyst is considering the price 6 months from now of a stock he holds in some of his customer accounts. He assigns the following probabilities to the events shown below:

Event	Probability
A: Price lower than today	0.65
B: Price same as today	0.05
C: Price higher than today	0.30
	1.00

Note that the assignment is a subjective one reflecting a pessimistic outlook for the stock price. The assignment does meet the first three mathematical requirements, however:

1. None of the events listed has been assigned a probability that is negative or greater than 1.0.

2. The total probability assigned to the possible outcomes (sample space) is 1.0.
3. The probability of the union of any two events is the sum of their separate probabilities minus the probability assigned to their intersection.

In this case, the events are mutually exclusive. The events $A \cap B$, $A \cap C$, and $B \cap C$ are null sets and are assigned the probability zero. Therefore, in connection with the third requirement there will be no double counting when we add, for example, $P(A)$ and $P(B)$ in obtaining $P(A \cup B)$. Thus we can state two addition rules for probability:

Statement

General Addition Law: The probability of the union of two events is the sum of the probabilities of the two events minus the probability of their intersection:

$$P(A \cup B) = P(A) + P(B) - P(A \cap B). \tag{4.2}$$

Statement

Special Addition Law: The probability of the union of two *mutually exclusive* events is the sum of the probabilities of the two events:

$$P(A \cup B) = P(A) + P(B). \tag{4.3}$$

These laws hold for any assignments of probabilities, whether made on the basis of equally likely sample points, relative frequency, or degree of belief. The addition laws can be used in two ways:

1. To check the consistency of assignments of probabilities to events
2. To calculate the probabilities of events to which probabilities have not been directly assigned from probabilities of events which have been assigned

Suppose a friend tells you that the probability she will be home this evening is 0.50, the probability she will be watching television is 0.30, the probability she will be at home watching television is 0.20, and the probability she will be either at home or watching television is 0.40. All these events have been assigned probabilities between 0 and 1, but this is not a consistent set of probability assignments because

$$P(H \cup T) \neq P(H) + P(T) - P(H \cap T)$$
$$0.40 \neq 0.50 + 0.30 - 0.20.$$

On the other hand, if your friend had assigned only the first three probabilities, you could have calculated the probability that should be assigned to $P(H \cup T)$ from

$$P(H \cup T) = P(H) + P(T) - P(H \cap T)$$

$$= 0.50 + 0.30 - 0.20$$

4.2 $$= 0.60.$$

Summary

Probability is an assignment of numbers to events that are defined as subsets of the outcomes of an experiment. Probabilities can be assigned to events by considering the underlying experimental outcomes as equally likely. Or they can be estimated by reference to actual relative frequencies of occurrence of the events. They can also be assigned as personal measures of belief that the events will in fact occur. Regardless of how you assign probabilities to events or what interpretation you give them, the numbers you assign must meet four requirements: (1) the probability of any event must be between 0 and 1; (2) the probability assigned to the sample space must be equal to 1; (3) the probability of the union of any two events must be equal to the sum of their individual probabilities less the probability of their intersection; (4) the null set must be assigned zero probability. The third requirement constitutes a general addition law for probabilities. A special form of the addition law applies to events that are mutually exclusive. Two events are mutually exclusive if their intersection is the null set.

See Self-Correcting Exercises 4A.

Exercise Set A

1. Twelve undergraduate students in a university got together to discuss the year's activities for the local Simulation Society. The students are identified by $\{i = 1,2,\ldots,12\}$. Upon inquiring about the students, the meeting's organizer found that the set of business majors was $\{1,2,3,4,5,6\}$, the set of mathematics majors was $\{7,8,9\}$, the set of liberal arts majors was $\{7,8,9,10,11,12\}$, the set of upper-division students was $\{1,4,7,9,10\}$, and the set of students who were members of the society was $\{1,3,7,8\}$. Using the symbols B for business, M for mathematics, L for liberal arts, U for upper division, and S for members of the society, list the following subsets:

$$S \cup L; \quad B \cap U; \quad U \cup S; \quad U \cap S; \quad B \cap M; \quad L \cap M'; \quad U' \cap S'.$$

2. For any set with n elements, there are 2^n possible subsets. For example, if a set has three members, then there are 2^3 or 8 possible subsets.
   (a) For the set $\{a,b\}$, list all possible subsets.
   (b) List all possible subsets of the set $\{2,4,6,8\}$.
   (c) How many elements does the null set have? List all possible subsets of the null set.
   (d) How many subsets of a set with six elements are possible?
   (e) What set is a subset of every set?

3. For any event A in a sample space U, $\text{U} = A \cup A'$. Using this fact and rule 3 for the assignment of probability, determine a formula for $P(A')$ in terms of $P(A)$.

4. For two sets A and B, what must be true in order that $A \cap B = A \cup B$? Even if this condition is not met, which of $A \cap B$ and $A \cup B$ is necessarily a subset of the other?

Exercises 5 through 9 refer to the following experiment. Select a card randomly from a regular deck of 52, and record the suit and denomination. The sample space is denoted by U. Let A be the event that the card is an ace, H that the card is a heart, S that it is a spade, and F that it is a face card (jack, queen, king).

5. (a) How many elements are there in the sample space?
   (b) Determine:
      (i) $N(A)$
      (ii) $N(H \cap S)$
      (iii) $N(A \cap H)$
      (iv) $N(A')$
      (v) $N(F)$
      (vi) $N(F \cap H)$
      (vii) $N(F \cup A)$
      (viii) $N(H' \cup S')$
      (ix) $N(H' \cap S')$
      (x) $N(A \cap A')$

6. Describe, without using negative phrases, the following sets: (a) $H' \cap S'$; (b) $H' \cup S'$; (c) $H' \cap A$.

7. Which of the following pairs of sets are mutually exclusive?
   (a) A,S
   (b) $A,\varnothing$
   (c) S,S'
   (d) F,H
   (e) S,H
   (f) F,U

8. Give correct union or intersection notation, as required, for the following events:
   (a) The card is an ace or a face card.
   (b) The card is neither a spade nor a face card.
   (c) The card is the ace of spades.
   (d) The card is an ace and a face card.
   (e) The card is a spade which is an ace or a face card.

9. Assuming that each card in the deck has the same probability of being drawn, determine:
   (a) $P(A)$
   (b) $P(A')$
   (c) $P(A \cap A')$
   (d) $P(H' \cap S')$
   (e) $P(A \cup F')$
   (f) $P(F)$
   (g) The probability of drawing the ace of hearts
   (h) The probability of drawing the two of diamonds
   (i) The probability of drawing a red card
   (j) The probability of obtaining a denomination less than 7

10. What view of probability might the speaker be taking in each of the following assignments of probability?
    (a) "Since more boys are born than girls, there's a better than even chance that our baby will be a boy."
    (b) "The probability of precipitation tonight is 10 percent and tomorrow 30 percent."
    (c) "With my luck, I've got about a 99 percent chance of flunking that test!"
    (d) "Filly Buster is a sure bet to win the Triple Crown this year."
    (e) "Let's see: if I buy one ticket and they sell a hundred, my probability of winning is 0.01."
    (f) "You're more likely to roll a 7 with two dice than to roll any other single number."
11. Three students were asked to assign subjective probabilities to the events A and B, the union of which is the sample space. Which of the following probability assignments are not valid? Why?
    (a) $P(A) = 1.0$, $P(B) = 0.5$, $P(A \cap B) = 0.5$.
    (b) $P(A) = 0.3$, $P(B) = 0.5$, $P(A \cap B) = 0.2$.
    (c) $P(A) = 0.2$, $P(B) = 0.7$, $P(A \cap B) = 0.3$.

Joint, Marginal, and Conditional Probability

4.3 The elements in a sample space correspond to the *units of observation* in a statistical study. *Events* correspond to characteristics of these units that are of interest to us. In the example of flaws on 50 galvanized steel sheets (Table 2.1), the 50 sheets can be considered as elements in the sample space of the experiment *select 1 sheet from the 50 sheets.* If we can do that in a way that makes each element (sheet) in the sample space equally likely, then we would assign to the event that the sheet selected has exactly one flaw the probability

$$P(\text{one flaw}) = \frac{N(\text{sheets with one flaw})}{N(\text{sheets})} = \frac{14}{50} = 0.28.$$

Often we are interested in two characteristics of the observational units. This was the case in the coffee machine example. We are interested in whether or not a serving contains cream and whether or not a serving contains sugar. We can tabulate the elements in that sample space (20 servings) in the following way:

	Sugar (S)	No Sugar (S')	Total
Cream C	8	4	12
No cream C'	5	3	8
Total	13	7	20

Now if we use equally likely sample points for establishing probabilities, we can state probabilities for various events, some of which we worked with earlier. First, there are the event intersections produced by the buttons on the machine:

$$P(C \cap S) = \frac{N(C \cap S)}{N(\text{U})} = \frac{8}{20} = 0.40$$

$$P(C \cap S') = \frac{N(C \cap S')}{N(\text{U})} = \frac{4}{20} = 0.20$$

$$P(C' \cap S) = \frac{N(C' \cap S)}{N(\text{U})} = \frac{5}{20} = 0.25$$

$$P(C' \cap S') = \frac{N(C' \cap S')}{N(\text{U})} = \frac{3}{20} = 0.15.$$

These **joint** probabilities, so-called because the events refer to the joint occurrence of two characteristics defined by the intersections, can be assembled in a table as follows:

	S	S'
C	0.40	0.20
C'	0.25	0.15

The events defined by the cells (intersections) in the joint probability table are mutually exclusive. If the joint probabilities are added by rows (see the special addition law for mutually exclusive events), we obtain the probabilities $P(C)$ and $P(C')$. That is,

$$P(C) = P(C \cap S) + P(C \cap S') = 0.40 + 0.20 = 0.60$$

$$P(C') = P(C' \cap S) + P(C' \cap S') = 0.25 + 0.15 = 0.40.$$

These are the same answers we would obtain by counting sample points from the original table. Similarly, if we add the joint probabilities by columns we obtain

$$P(S) = P(C \cap S) + P(C' \cap S) = 0.40 + 0.25 = 0.65$$

$$P(S') = P(C \cap S') + P(C' \cap S') = 0.20 + 0.15 = 0.35.$$

Probabilities of the type we have just calculated are called **marginal** probabilities because they appear at the margins of the following kind of table:

	S	S'	Total
C	0.40	0.20	0.60
C'	0.25	0.15	0.40
Total	0.65	0.35	1.00

The joint probabilities together with the two sets of marginal probabilities can be seen to represent nothing more than the division of the entries in our original two-way table by the total number of elements in the sample space.

Often we may want to restrict the sample space of an experiment in accordance with a condition of interest to us. We may want to find the probability that a serving contains cream *given* that it contains sugar. Once this restriction is stated, the sample space of interest becomes the servings containing sugar rather than all servings. This probability can be obtained directly from our original tabulation of sample points or from Figure 4.1 as

$$P(C \text{ given } S) = \frac{N(C \cap S)}{N(S)} = \frac{8}{13} = 0.615$$

because the restricting condition allows us to replace $N(\mathrm{U}) = 20$ with $N(S) = 13$. Similarly,

$$P(C' \text{ given } S) = \frac{N(C' \cap S)}{N(S)} = \frac{5}{13} = 0.385.$$

If we restrict our interest to servings with cream, we can ask what is the probability that a serving contains sugar given that it contains cream. From the original table we can see this is

$$P(S \text{ given } C) = \frac{N(S \cap C)}{N(C)} = \frac{8}{12} = 0.667,$$

and in a similar way

$$P(S' \text{ given } C) = \frac{N(S' \cap C)}{N(C)} = \frac{4}{12} = 0.333.$$

The standard way to indicate a conditional event is with a vertical bar. Thus $B|A$ means the event that B occurs *given* that A occurs. Another way of saying this is the event that B occurs *when* A occurs. Thus the event following the bar is the restricting condition, and $B|A$ is read as *B given A*. Conditional probabilities may be obtained from joint and marginal probabilities by the following equation.

Definition

The probability of event B **conditional** on event A is the probability of the intersection $A \cap B$ divided by the marginal probability of A:

$$P(B|A) = \frac{P(A \cap B)}{P(A)}. \qquad \textbf{4.4}$$

In terms of counting equally likely sample points, Equation 4.4 can be rationalized by noting that

$$P(A \cap B) = \frac{N(A \cap B)}{N(\mathrm{U})} \quad \text{and} \quad P(A) = \frac{N(A)}{N(\mathrm{U})}.$$

Then Equation 4.4 reduces to

$$P(B|A) = \frac{N(A \cap B)}{N(\mathrm{U})} \div \frac{N(A)}{N(\mathrm{U})} = \frac{N(A \cap B)}{N(A)},$$

which is the way we would directly establish $P(B|A)$ by counting equally likely sample points.

A restatement of Equation 4.4 allows us to formulate the **general multiplication rule** for probabilities:

Statement

General Multiplication Law: The probability of the joint occurrence of two events is the product of the marginal probability of the first event and the conditional probability of the second event given the first:

$$P(A \cap B) = P(A) \cdot P(B|A). \qquad \textbf{4.5}$$

The general multiplication law will allow us to calculate joint probabilities when we have the necessary marginal and conditional probabilities for inputs into the relation of Equation 4.5.

Independence

Consider two events, A and B, with non-zero probabilities. The events are said to be independent if the conditional probability of A given B is equal to the marginal probability of A.

Definition Two events, A and B, are **independent** if $P(B|A) = P(B)$. It will also then be true that $P(A|B) = P(A)$.

When two events are independent, it means that the probability we have established that one of them will occur does not change with the knowledge that the other has occurred. For example, consider a game in which a die is rolled and a coin is tossed. Let B be the event that the die shows 1 and A the event that the coin shows heads. Does knowing that heads occurred on a play of the game affect our belief in the proposition that the 1 also occurred? There is no reason to suppose so, and we would assign

$$P(B) = 1/6, \qquad P(B|A) = 1/6.$$

Also, our belief that heads occurred is not affected by knowledge that a 1 occurred on a play of the game. We would assign

$$P(A) = 1/2, \qquad P(A|B) = 1/2.$$

When events are independent, a special form of the multiplication rule applies. Recall that the general multiplication rule is

$$P(A \cap B) = P(A) \cdot P(B|A).*$$

If the events are independent, $P(B|A) = P(B)$. Moreover:

Statement Special Multiplication Law: The probability of the joint occurrence of two independent events is the product of the marginal probabilities of the two events:

4.3 ■ $P(A \cap B) = P(A) \cdot P(B)$. ■ *4.6*

Calculating Probabilities

4.4 The probability laws presented earlier are used to calculate probabilities that are not known from probabilities that are known. They are restated here in symbols:

*Note that $A \cap B$ is the same event as $B \cap A$, and it is true that $P(B \cap A) = P(B) \cdot P(A|B) = P(A \cap B)$.

General addition law:

$$P(A \cup B) = P(A) + P(B) - P(A \cap B)$$

Special addition law—mutually exclusive events:

$$P(A \cup B) = P(A) + P(B)$$

General multiplication law:

$$P(A \cap B) = P(A) \cdot P(B \mid A)$$

Special multiplication law—independent events:

$$P(A \cap B) = P(A) \cdot P(B)$$

The importance of probability calculations to the study of statistics is twofold. First, the processes and interpretation of results in statistical inference call for an understanding of probability calculations and the language of probability. Second, you should begin to see that in calculating probabilities of events we are beginning to study the behavior of samples drawn from populations. Consider the occurrence of five heads on five successive tosses of a fair coin. We start with the probability of a head on a single toss, which we assign as 0.50. This represents the long-run relative frequency of heads in repeated trials of the experiment *toss a fair coin.* It is the proportion of heads in an infinite population of heads and tails. The experiment *toss the coin five times* is equivalent to taking a random sample* of five observations from an infinite population of 50 percent heads and 50 percent tails. The answer to the probability of tossing five heads in a row tells us the long-run relative frequency of this event in repeated experiments of drawing five observations from the population. You should see the similarity between this probability and the probability of drawing five Democrats in a sample of five from a large group of voters of whom 50 percent are Democrats.

Intersections of Independent Events

Learning to recognize common situations is helpful in probability calculations. The coin-tossing problem mentioned above is one of an intersection of independent events. Consider two tosses and the probability that both yield heads. The sample space is the possible outcomes of the two tosses considered together, because the experiment is *toss the coin twice.* Let $P(A)$ be the probability of heads on the first

*The strict meaning of a *random* sample appears in Chapter 13.

toss and $P(B)$ be the probability of heads on the second toss. The events are independent—the probability of heads on the second toss does not depend on whether heads occurs on the first toss. Thus $P(A) = 0.50$ and $P(B) = 0.50$ for an honest coin, and

$$P(A \cap B) = P(A) \cdot P(B) = 0.50(0.50) = 0.25.$$

We might have used H_1 for heads on the first toss and H_2 for heads on the second. Then we would have written

$$P(H_1 \cap H_2) = P(H_1) \cdot P(H_2) = 0.50(0.50) = 0.25.$$

For the probability of heads on each of three successive tosses, the special multiplication law would give

$$P[(H_1 \cap H_2) \cap H_3] = P(H_1 \cap H_2) \cdot P(H_3)$$
$$P(H_1 \cap H_2 \cap H_3) = 0.25(0.50) = 0.125.$$

A Venn diagram can be used to show that $H_1 \cap H_2 \cap H_3$ is the same as $(H_1 \cap H_2) \cap H_3$. In Figure 4.2, the shading shows $H_1 \cap H_2$ as the overlap between the H_1 and H_2 circles. The overlap of this shaded area and the H_3 circle is the overlap among all three circles.

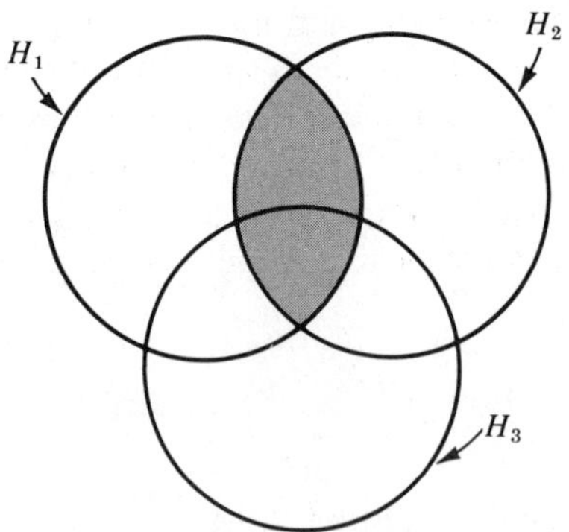

Figure 4.2 *Venn diagram for triple intersection*

Extending the same principle, we can see that the probability of five heads in as many tosses is

$$P(H_1 \cap H_2 \cap H_3 \cap H_4 \cap H_5) = P(H_1) \cdot P(H_2) \cdot P(H_3) \cdot P(H_4) \cdot P(H_5)$$
$$= 0.50(0.50)(0.50)(0.50)(0.50) = 0.03125.$$

Intersection of Conditional Events

To select two members to be responsible for setting up their sound system, the names of five members of a rock group are placed in a hat and the drummer will draw first one name and then a second name without replacing the first. There are three singers in the group. What is the probability that two singers will be selected? Here the probability that a singer will be selected on the second draw clearly depends on whether a singer is selected on the first draw. If a singer is not selected on the first draw, there are three singers out of the four names left, but only two singers out of four names are left if a singer is selected on the first draw. We could write

$$P(S_2 | S_1) = 2/4 \qquad P(S_2 | S_1') = 3/4.$$

In order to draw two singers, a singer must be selected on the first draw *and* a singer must be selected on the second draw. Applying the general multiplication rule, we can write

$$P(S_1 \cap S_2) = P(S_1) \cdot P(S_2 | S_1)$$

$$P(S_1 \cap S_2) = \frac{3}{5}\left(\frac{2}{4}\right) = \frac{6}{20} = 0.30.$$

Suppose a draw of three is made. What is the probability that all three are singers?

$$P[(S_1 \cap S_2) \cap S_3] = P(S_1 \cap S_2) \cdot P(S_3 | S_1 \cap S_2)$$

$$P(S_1 \cap S_2 \cap S_3) = \frac{3}{5}\left(\frac{2}{4}\right)\left(\frac{1}{3}\right) = \frac{6}{60} = 0.10.$$

The principle of this extension is the same as the one shown earlier. You should be able to put $P(S_3 | S_1 \cap S_2)$ in words and verify the probability of $1/3$ for that event.

For another example, consider a state university in which 20 percent of the students are seniors (S). Forty percent of the seniors are from out of state (O). If a student is selected at random, what is the probability that he or she will be a senior from out of state?

$$P(S \cap O) = P(S) \cdot P(O | S)$$

$$= 0.20(0.40) = 0.08.$$

Unions of Mutually Exclusive Events

For the experiment of drawing a sheet at random from the 50 galvanized steel sheets in Table 2.1, the probabilities of various numbers of flaws are

x = number of flaws	0	1	2	3	4	5
$P(x) = f/N$	0.20	0.28	0.24	0.16	0.08	0.04

If we want to refer to the event that the sheet selected has exactly one flaw, we can do so by writing $x = 1$. Similarly, $x = 2$ is the event that the sheet selected has exactly two flaws. To specify the event *at least* two flaws, we write $x \geq 2$. *More than* two flaws is $x > 2$, *no more than* two flaws is $x \leq 2$, and *less than* two flaws is $x < 2$.

The events shown in the table above are mutually exclusive. If the sheet selected has exactly one flaw ($x = 1$), it cannot have also exactly two flaws ($x = 2$) and so forth. The probability of less than two flaws, $P(X < 2)$, is the probability of the union of the mutually exclusive events $x = 0$ and $x = 1$. Applying the special addition rule for mutually exclusive events, we can write

$$\begin{aligned} P(X < 2) &= P[(X = 0) \cup (X = 1)] = P(X = 0) + P(X = 1) \\ &= 0.20 + 0.28 = 0.48. \end{aligned}$$

An application like this is so straightforward that writing out $P[(X = 0) \cup (X = 1)]$ is hardly necessary. To get used to the inequality signs ($>, <, \geq, \leq$), try verifying for yourself that

$$\begin{aligned} P(X > 3) &= 0.12, \qquad P(X \geq 4) = 0.12, \\ P(X < 3) &= 0.72, \qquad P(X \leq 3) = 0.88. \end{aligned}$$

On checking the last of these you may have been tempted to use the complement of $P(X \leq 3)$, namely $P(X > 3)$, which is the same as $P(X \geq 4)$:

$$\begin{aligned} P(X \leq 3) &= 1 - P(X > 3) \\ &= 1 - [P(X = 4) + P(X = 5)] \\ &= 1 - (0.08 + 0.04) = 0.88. \end{aligned}$$

This is permitted because the probability assigned to the total sample space is 1.0.

Unions of Non-Mutually Exclusive Events

Suppose a 70 percent foul shooter is awarded two foul shots in a basketball game. What is the probability that he will score on *at least one* of the shots? Let event A be a score on the first shot and event B be a score on the second shot. Since either of these events satisfies the event *a score on at least one* of the shots, we want the union $A \cup B$. But A and B are not mutually exclusive–that is, they can both occur in the experiment *take two foul shots.* The general addition law where the probability of intersection $A \cap B$ is subtracted out applies:

$$P(A \cup B) = P(A) + P(B) - P(A \cap B).$$

If A and B are independent events, we can get $P(A \cap B)$ from $P(A) \cdot P(B)$. If we assume that is the case here, we have

$$P(A) = 0.70, \qquad P(B) = 0.70$$

$$P(A \cup B) = P(A) + P(B) - P(A) \cdot P(B)$$

$$= 0.70 + 0.70 - (0.70)(0.70)$$

$$= 1.40 - .49 = 0.91.$$

The problem of at least one of a series of repetitive trials producing a specified event is a common one, so it is worthwhile showing another approach to it. In the case above, notice that there is only one way we will *not* get at least one score from the two-shot opportunity—and that is if *both* shots fail. If A and B (successes) are independent, so are A' and B' (failures). We may write

$$P(A \cup B) = 1 - P(A' \cap B')$$

The probability of failing to score on any shot is $1 - 0.70 = 0.30$. Thus $P(A') = 0.30$ and $P(B') = 0.30$ and

$$P(A \cup B) = 1 - 0.30(0.30) = 1 - 0.09 = 0.91.$$

Let's consider another problem. An archer takes four shots at a target. The probability that she scores a bulls-eye on any shot is 0.10. What is the probability that she scores at least one bulls-eye? Let H stand for hitting the bulls-eye. Then

$$P(H_1 \cup H_2 \cup H_3 \cup H_4) = 1 - 0.90(0.90)(0.90)(0.90)$$

$$= 1 - 0.6561 = 0.3439.$$ 4.4

Summary

When two events are being considered, we may want to refer at different times to a joint probability, a marginal probability, or a conditional probability. A joint probability refers to the intersection of two events, marginal probability refers to the union of the two events, and conditional probability refers to the occurrence of one event *when* the other occurs. The two events are said to be independent when their conditional probabilities are equal to their marginal probabilities. The events are dependent when the probabilities of one event change (or differ) depending on whether the other event occurs.

The relations between joint, marginal, and conditional probabilities are incorporated in four probability laws—a general and a special (for mutually exclusive events) addition law and a general and special (for independent events) multiplication law. The laws can be used as formulas for calculating probabilities for events that were not directly assigned probability from probabilities that have been assigned. In most instances the laws are used for calculating probabilities of more complex events from the probabilities known for simpler events. Gaining facility with probability calculations requires a fair amount of practice.

See Self-Correcting Exercises 4B.

Exercise Set B

1. From the probability table on page 102, describe the following, first in set notation and then in words:
   (a) The event which has a joint probability of 0.40
   (b) The event which has a marginal probability of 0.40
2. For any two events A and B, we may state that

$$B = (A \cap B) \cup (A' \cap B).$$

   Verify this with a Venn diagram.
3. Suppose A and B are two arbitrary events.
   (a) Under what condition(s) does $P(B \mid A) = P(B)$?
   (b) If $P(B \mid A) = P(B)$, is it necessarily true that $P(A \mid B) = P(A)$?
4. (a) Are mutually exclusive events dependent or independent?
   (b) If A and B are mutually exclusive, what is (i) $P(A \mid B)$? (ii) What is $P(B \mid A)$?
5. Complete the following statement correctly: "Equation 4.5 may be rewritten as $P(A \cap B) = P(B) \cdot$ ______."
6. If A and B are arbitrary events with $P(A) = 0.5$, $P(B) = 0.2$, and $P(A \cup B) = 0.55$, determine:
   (a) $P(A \cap B)$
   (b) $P(A \mid B)$
   (c) $P(A' \mid B)$
   (d) $P[(A \cup B)']$
7. If A and B are independent events and $P(A) = 0.4$ and $P(B) = 0.25$, determine:
   (a) $P(A \cap B)$
   (b) $P(A \mid B)$
   (c) $P(A \cup B)$
   (d) $P(A' \mid B)$

8. If A and B are mutually exclusive events and $P(A) = 0.4$ and $P(B) = 0.25$, determine:
   (a) $P(A \cap B)$
   (b) $P(A \mid B)$
   (c) $P(A \cup B)$
   (d) $P(A' \mid B)$

9. Complete the following two-way table correctly:

	x	x'	
A	15	7	___
A'	___	5	___
	___	___	36

10. Complete the following table by providing the missing joint and marginal probabilities:

	x	x'	
A	___	0.4	___
A'	0.3	___	___
	___	0.5	___

Exercises 11 through 17 refer to the following table, where A, X, Y, and Z all represent events in a sample space and the numerical values in the table are the numbers of sample points in the respective joint events:

	X	Y	Z
A	6	5	3
A'	4	7	0

11. How many joint events are shown in the table? List them.

12. How many marginal events are shown in the table? List them.

13. Determine the following numerical values:
   (a) $N(\text{U})$
   (b) $N(A)$
   (c) $N(Z)$
   (d) $N(A \cap Y)$
   (e) $N(A' \cap Z)$

14. What is the numerical value of
   (a) $P(A)$?
   (b) $P(Z)$?
   (c) $P(A \cap Z)$?
   (d) $P(A' \cap Z)$?

15. From the given table construct a similar table listing probabilities rather than integers in the appropriate positions.

16. (a) How many marginal probabilities did you obtain in the column margin of your table in Exercise 15? What is their sum?

(b) How many marginal probabilities did you obtain in the row margin? What is their sum?

17. Using the results obtained in Exercise 14 and an appropriate formula for calculating conditional probability, determine:

(a) $P(A \mid Z)$
(b) $P(A' \mid Z)$
(c) $P(Z \mid A)$
(d) $P(Z \mid A')$

18. At the first meeting of a retirement community social group, an acting president is to be selected by lot: each person's name is printed on a slip of paper and all the slips are placed in a box which is then shaken. A name is subsequently drawn, and that person thereby becomes the temporary officeholder. Forty percent of those present at the meeting are under 65 years of age, and the rest are 65 or older. Seventy percent of those present are women, and 25 percent are men under the age of 65. Assuming that each slip has the same probability of being drawn, and letting W represent the event that a woman's name is selected and S the event that the person chosen is under 65, enter the appropriate joint and marginal probabilities in the table:

	W	W'	
S	___	___	___
S'	___	___	___
	___	___	1.00

19. With reference to Exercise 18, describe in words the following events:

(a) S'
(b) $W \cap S'$
(c) $W' \cap S$
(d) $W \cap W'$

20. Again with reference to Exercise 18 and the results obtained there, determine:

(a) $P(S')$
(b) $P(W \cap S')$
(c) $P(W \mid S')$
(d) $P(S' \mid W)$
(e) $P(W' \mid S)$

21. An experiment consists of rolling a fair die six times and recording the numbers that turn up.

(a) Is the result of each roll dependent on or independent of a result on any previous roll(s)?
(b) What is the probability of rolling six 1s?
(c) What is the probability of *not* rolling six 1s?
(d) What is the probability of rolling six 3s?
(e) What is the probability of rolling six 1s *or* six 3s?
(f) What is the probability of rolling six 1s *and* six 3s?

Overview

1. A gathering of 20 business managers included 12 members of the local Chamber of Commerce. All 12 were engaged in selling at either the wholesale or retail level. Three

sold at both wholesale and retail, and five at retail only. For the set of members of the Chamber of Commerce, construct a two-way classification table and find

$$N(R), \quad N(R'), \quad N(W), \quad \text{and } N(W'),$$

where R stands for selling at retail and W stands for selling at wholesale.

2. Among nonmembers of the Chamber of Commerce at the gathering in Exercise 2,

$$N(R) = 6, \quad N(W)=4, \quad N(W')=4, \quad N(R')=2, \quad \text{and } N(R \cap W) = 2.$$

Find $N(R \cup W)$ and $N(R' \cup W')$.

3. Draw a Venn diagram for the situation of Exercises 1 and 2 using C (for chamber members), R, and W for your three circles. Label the four mutually exclusive overlaps, or intersections, in the diagram with the appropriate set designations and find the number of business managers in each intersection.

4. Using only rules of simple arithmetic, derive Equation 4.5 from Equation 4.4.

5. Apply Equation 4.5 to obtain a formula for determining $P(X \cap Y)$.

6. Determine a formula for the marginal probability $P(A)$ in terms of $P(A \cap B)$ and $P(B \mid A)$ for Equation 4.4.

7. Under what condition(s) may we say that for two sets A and B, $P(A \cup B) = P(A) + P(B)$? Otherwise, what equation must be used to determine $P(A \cup B)$?

8. Draw a Venn diagram for A, B, and C in which $P(A \cap B) = 0$. Then express $P(C)$ as a sum of probabilities of mutually exclusive intersections. Given, further, that $P(A' \cap B' \cap C) = 0$, $P(A' \cap B' \cap C') = 0.2$, $P(A \cap C) = 0.1$, $P(C') = 0.6$, $P(A) = 0.4$, and $P(B) = 0.4$, find $P(B \cap C)$.

9. If $P(A) = 0.4$, $P(B) = 0.2$, and $P(A \cup B) = 0.5$, determine:
   (a) $P(A \cap B)$
   (b) $P(A \mid B)$
   (c) $P(B \mid A)$
   (d) $P(A' \mid B)$

10. If A and B represent two events, determine whether A and B are independent when
   (a) $P(A) = 0.45$, $P(B) = 0.4$, and $P(A \cap B) = 0.18$.
   (b) $P(A) = 0.4$, $P(B) = 0.4$, and $P(A \cup B) = 0.4$.
   (c) $P(A) = 0.4$, $P(B) = 0.2$, and $P(A \cap B) = 0.8$.
   (d) $P(A) = 0.2$, $P(B) = 0.7$, and $P(A \cap B) = 0$.

11. In the definition of independent events, it is stated that if $P(B \mid A) = P(B)$, then $P(A \mid B) = P(A)$. Starting with the assumption that $P(B \mid A) = P(B)$ and applying Equations 4.4 and 4.5, prove that $P(A \mid B)$ must equal $P(A)$ as stated in the definition.

12. An experiment consists of rolling two fair dice simultaneously and recording the numbers that turn up. Find the probability that
   (a) Two 1s turn up.
   (b) Two 3s turn up.
   (c) A pair (the same number on each) turns up.
   (d) A pair does *not* turn up.

13. An experiment consists of selecting from an urn one ball from the following collection: five black balls numbered 1 through 5, five red balls numbered 1 through 5, and three white balls numbered 1 through 3. Let B represent the event that a black ball is selected, R that a red one is, W that a white one is, and E that an even-numbered one is.

(a) How many points are there in the sample space?

(b) Determine:

(i) $P(B)$	(vi) $P(B \cup R')$
(ii) $P(W \cap E)$	(vii) $P(E \cap R')$
(iii) $P(B \cup E)$	(viii) $P(E \mid R)$
(iv) $P(B \cup W)$	(ix) $P(R \mid E)$
(v) $P(B \cap R')$	(x) $P(E' \mid E)$

14. Suppose the experiment in Exercise 13 were revised so that a ball is selected and then, without replacement of the first ball in the urn, a second one is selected. Let B_i represent the selection of a black ball on the ith draw, R_i a red, W_i a white, and E_i an even-numbered one.

(a) Is the result of the second selection dependent on or independent of the first selection? Why?

(b) State in words what is meant by the following events:

(i) $R_1 \cap R_2$	(iv) $R_1' \cup E_1'$
(ii) $R_1 \cup R_2$	(v) $R_1' \cap E_1'$
(iii) $(R_1 \cap E_1) \cap (R_2 \cap E_2)$	(vi) $R_1 \cup B_1$

(c) An alternative representation for $R_1 \cup B_1$ is W_1'. Determine an alternative representation for (i) $W_2 \cup B_2$; (ii) $E_1 \cup E_2$; (iii) B_2'.

(d) Determine:

(i) $P(R_1 \cap R_2)$	(iv) $P[(R_1 \cap E_1) \cap (B_2 \cap E_2)]$
(ii) $P(R_1 \cap R_2')$	(v) $P(B_1' \cap B_2')$
(iii) $P(R_1 \cup R_2')$	(vi) $P[E_1' \cap B_1) \cap (E_2 \cap B_2)]$

(e) Give symbolic representation for the event that a red, even-numbered ball is selected on the first draw and a white, odd-numbered one on the second.

15. In an opinion poll in a four-year college, 1000 students were asked to classify their political philosophy as conservative (C) or liberal (L). In addition, each student polled was asked to identify her or his class as freshman, sophomore, junior, or senior (Fr, So, Jr, Sr), and the pollster recorded the student's sex (M, F). The results are presented in the following table. The number in each cell represents the number of students who are classified according to the marginal notations in the table. For example, the cell in the upper left-hand corner indicates that 25 of the 1000 students are classified as male, freshmen, and conservative.

	Male		Female		
Class	Conservative	Liberal	Conservative	Liberal	Total
Fr	25	200	25	150	400
So	25	25	50	100	200
Jr	75	25	50	50	200
Sr	125	25	25	25	200
Total	250	275	150	325	1000

(a) Are the events corresponding to the cells in the table mutually exclusive?

(b) From the table given, determine a probability for each cell in the table below:

	$M \cap C$	$M \cap L$	$F \cap C$	$F \cap L$
Fr				
So				
Jr				
Sr				

(c) Let $A = \text{Fr} \cap M \cap C$ and $B = \text{Fr} \cap F \cap C$. In words, describe the following events: (i) A; (ii) B; (iii) $A \cap B$; (iv) $A \cup B$.

(d) Find the probabilities for the table below. Explain the rationale behind your work.

	C	L
Fr		
So		
Jr		
Sr		

(e) In words, describe the events Fr, C, $C \cap \text{Fr}$, $C|\text{Fr}$.

(f) Let $X = \text{Fr} \cap C$ and $Y = \text{Fr} \cap L$. Find $P(X \cup Y)$ and explain how you found this result. What is the meaning of the event $X \cup Y$?

(g) If $P(C) = 0.4$ and $P(\text{Fr}) = 0.4$, what is the value of $P(C \cup \text{Fr})$? What law did you use to find this value? In words, explain the meaning of the event $C \cup \text{Fr}$.

Glossary of Equations

4.1 $P(A) = \dfrac{N(A)}{N(\text{U})}$

When the elements in a sample space are regarded as equally likely, the probability of an event A is obtained by finding the ratio of the number of elements satisfying the event to the number of elements in the sample space.

4.2 $P(A \cup B) = P(A) + P(B) - P(A \cap B)$

The probability of the union of two events is the sum of the probabilities of the two events minus the probability of their intersection.

4.3 $P(A \cup B) = P(A) + P(B)$	The probability of the union of two *mutually exclusive* events is the sum of the probabilities of the two events.
4.4 $P(B \mid A) = \dfrac{P(A \cap B)}{P(A)}$	The probability of event B conditional on event A is the probability of the intersection $A \cap B$ divided by the marginal probability of A.
4.5 $P(A \cap B) = P(A) \cdot P(B \mid A)$	The probability of the joint occurrence of two events is the product of the marginal probability of the first event and the conditional probability of the second event given the first.
4.6 $P(A \cap B) = P(A) \cdot P(B)$	The probability of the joint occurrence of two *independent events* is the product of the marginal probabilities of the two events.

5

Sampling and Sampling Distributions

In Part One we learned how to describe some characteristics of statistical populations. Remember that a statistical population is the *complete* collection of data of interest in a study. In most situations one has only a sample of the statistical population and must use the sample data to learn about the population. Here are a few examples:

1. Sample surveys in marketing research to learn how buyers use a product
2. Tests of a new drug by selected physicians to estimate its effectiveness
3. Agricultural experiments to estimate the effectiveness of new materials or methods
4. Public opinion polls to learn how people feel about current issues
5. Testing selected items from a production line to learn about the quality of all items turned out
6. Trying out a new instructional method on selected students to estimate its effectiveness with all students of a particular kind

In the chapter on probability the idea of an *experiment* was emphasized. Each of the sample studies just mentioned should be thought of as an experiment in the sense that we can imagine the study being conducted over and over again—just as we can think of repeating the experiment of selecting a serving from among 20 servings produced by a coffee machine and observing the ingredients in the serving. The idea of conducting a sample study over and over includes selecting the units of observation and observing a result. In our examples a result might be the average length of time *selected subscribers* keep *Newsweek* magazine, the percentage of *persons polled* who think the president is doing a good job, and the average breaking strength of tennis *racket frames selected* for testing.

The *idea* of repetitions of a sample study is critical in developing an understanding of statistical inference. To examine the ideas in detail, the first half of this chapter goes into a critical question: What would happen if such an experiment were repeated over and over? Beginning in Part Three we use these ideas to develop and explain the methods for telling what we have learned about a population from the one experiment, or sample study, that has actually been carried out.

5.1 Sample Space and Random Variable

To begin our study of sampling experiments we will set up a model population. Then we will talk about repeated experiments involving that population. In doing this we will have complete knowledge about our population. This makes it possible to understand the "what would happen if" questions that represent the theory behind statistical inference.

A Finite Population

A store handles a line of table radios. In the line are both AM and FM models with both regular and digital clocks. The store makes a profit of \$6 on each of the AM models, \$12 on the FM regular clock model, and \$18 on the FM digital clock model. Yesterday the store sold 12 table radios as follows:

4 AM regulars @ \$6:	profit = \$ 24
3 AM digitals @ \$6:	profit = \$ 18
4 FM regulars @ \$12:	profit = \$ 48
1 FM digital @ \$18:	profit = \$ 18
	\$108

The store's total profit was \$108 on 12 radios, or an average profit of \$9 per radio. In Chapter 2 we would have described this population as follows:

x	f
\$ 6	7
\$12	4
\$18	1

$$\mu = \frac{\sum (fx)}{N} = \frac{7(6) + 4(12) + 1(18)}{12} = \frac{108}{12} = 9 \text{ dollars}$$

$$\sigma^2 = \frac{\sum [f(x - \mu)^2]}{N} = \frac{7(6 - 9)^2 + 4(12 - 9)^2 + 1(18 - 9)^2}{12} = \frac{180}{12} = 15$$

$$\sigma = \sqrt{15} = 3.87 \text{ dollars}$$

Thus our population of profits on radios sold has a mean of $9 and a standard deviation of $3.87.

Sample Space

Imagine that the store employs a very lazy bookkeeper. The sales manager comes in this morning and counts 12 sales slips for table radios. She asks the bookkeeper, "How much did we make on these?" The bookkeeper shuffles the sales slips and picks one out. Seeing that it was for an AM digital model (and knowing the profit for the different models), he replies, "Six dollars each." The bookkeeper's procedure represents an experiment. We want to examine that experiment now in probability terms.

The 12 sales slips are the elements in the sample space. Assigned to each of these elements are two numbers—a selection probability and a value of a variable of concern in the experiment. Neither of these numbers in this case appears on the slip, but the association is made. The selection probabilities are assigned to the sales slips on an equally likely sample points basis and reflect a belief in the fairness of the bookkeeper's shuffling process. We will suggest a better selection procedure in the next section. The values of the variable (profit) were assigned to the elements (slips) by a rule based on the bookkeeper's knowledge—if the slip shows an AM regular, profit is $6; if the slip shows an FM regular, profit is $12; and so on.

Definition | A **random variable** is a rule for assigning numerical values to the events of a sample space.

Table 5.1 portrays the association of selection probabilities and profit values with the elements in the sample space of our experiment. Shown at the right is a listing of values of the variable (profit) along with their total probabilities. Thus the probability of the event *profit = $6* is $^7/_{12}$, the probability of the event *profit = $12* is $^4/_{12}$, and the probability of the event *profit = $18* is $^1/_{12}$.

The list of values assigned by the random variable rule along with their probabilities is called the **probability distribution of the random variable**. The variable

Table 5.1 Sample Space and Random Variable

Element	Model	Selection Probability	Value of Random Variable	x	$P(x)$
Slip #2	AM reg	$1/12$	6		
Slip #4	AM reg	$1/12$	6		
Slip #5	AM reg	$1/12$	6		
Slip #9	AM reg	$1/12$	6	6	$7/12 = 0.5833$
Slip #6	AM dig	$1/12$	6		
Slip #8	AM dig	$1/12$	6		
Slip #11	AM dig	$1/12$	6		
Slip #1	FM reg	$1/12$	12		
Slip #3	FM reg	$1/12$	12		
Slip #7	FM reg	$1/12$	12	12	$4/12 = 0.3333$
Slip #10	FM reg	$1/12$	12		
Slip #12	FM dig	$1/12$	18	18	$1/12 = 0.0833$
					1.0 0.9999

(profit in our case) is often called the random variable although that term is also used for the rule by which the values are originally assigned (as above).

The probability distribution of the random variable in our experiment—select a sales slip and report the profit for the model shown on the slip to the sales manager—looks just like a relative frequency table for our population. Except for the use of the symbol $P(x)$ for probability of x instead of f/N for relative frequency, it is outwardly the same. However, the relative frequency table refers to results from *enumerating* the $N = 12$ observations in the population. The bookkeeper could have constructed it if he had taken the time. The probability distribution represents our view of "what would happen if" the experiment conducted by the bookkeeper were repeated over and over. The probability of $7/12$ means that $7/12$, or 58.3 percent of the time in a long series of such repetitions,* the bookkeeper would report "six dollars each" to the manager, 33.3 percent of the time $12 would be reported, and 8.33 percent of the time $18 would be reported. In working with the probability distribution we will use the fractions with denominators of 12 because it is easier than working with the decimal form of the probabilities.

Mean and Variance of a Random Variable

If the distribution of our random variable looks like our population, its mean and variance will be the same as the population. All that has changed is that the

*Technically a series without limit. Remember we are speaking in terms of theoretical ideas.

Table 5.2 Mean and Variance of a Random Variable

x	$P(x)$	$x \cdot P(x)$	$(x - \mu)^2$	$(x - \mu)^2 \cdot P(x)$
6	$7/12$	$42/12$	9	$63/12$
12	$4/12$	$48/12$	9	$36/12$
18	$1/12$	$18/12$	81	$81/12$
		$108/12$		$180/12$

$$\mu = 108/12 = 9 \qquad \sigma^2 = 180/12 = 15$$

weights for different values of the variable (profit) are now probabilities rather than frequencies—long-run relative frequencies for repetitions of an experiment instead of numbers of observations in the population.

In Table 5.2 the mean and variance of our random variable have been calculated. The formulas are

$$\blacksquare \quad \mu = \sum [x \cdot P(x)] \quad \blacksquare \tag{5.1}$$

and

$$\blacksquare \quad \sigma^2 = \sum [(x - \mu)^2 \cdot P(x)] \quad \blacksquare \tag{5.2}$$

Not surprisingly, $\mu = 9$ and $\sigma^2 = 15$, just as before. The mean of a random variable is also called its **expected value**. Viewing probabilities as long-run relative frequencies, we can interpret the mean of a random variable as the average value for the variable as the number of repetitions is increased without limit. If we take a degree-of-belief view of probability, we would just say that our belief is so distributed among the events—\$6, \$12, and \$18 profits—as to produce an expectation, or mean, of \$9. In this case a \$9 profit can never occur, but nevertheless \$9 is the expected value in the sense of average.

Parameters and Statistics

Earlier we talked about gaining knowledge about populations from the partial information present in a sample. Our lazy bookkeeper might like to think he gained some information about the average profit per radio by noting the profit on *one* of the radios sold. Actually he conveyed very little of any use to the sales manager, but he did make an effort, however minimal. We will not consider selection of one unit of observation from all units of interest to be an acceptable sampling method. But it has served to illustrate random variables.

Definitions A **parameter** is a measure of a characteristic of the data composing a statistical population. A **statistic** is a measure of a characteristic of the data composing a sample.

In our radio profit example, we have already calculated two parameters–the mean profit μ and the standard deviation of profits σ. The mean is a measure of central tendency and the standard deviation is a measure of dispersion in the profits data. The business of inference is to say something useful about a parameter whose value is unknown by using a statistic from a sample. We exclude samples of one observation and begin our discussion of statistics (in the sense of the definition above) by talking about a sample of two observations. What we will see is that a statistic–because it is the result of an experiment that we can visualize repeating over and over–can vary from sample to sample. Understanding and measuring this variability
5.1 are central to statistical methods.

5.2 Probability Distribution of the Sample Mean

Again to develop ideas as best we can, we will pursue our abbreviated model situation of the lazy bookkeeper. We know the true mean profit per radio sold yesterday, but the bookkeeper and the sales manager do not. We are going to design a sampling method for the bookkeeper, and then with the advantage of our complete knowledge about the population we will see what would happen if it were repeated over and over. It will take much longer to do this than it would take the bookkeeper to go through all the slips and find out what we already know. But sampling *is* a practical alternative–if not the only one–in realistic situations.

Random Sampling

The bookkeeper's method of selecting a sales slip was not really a good one. We do not know enough about the bookkeeper to assume that his shuffling was fair, and it was on that assumption that we constructed the probability distribution of the random variable for his experiment of selecting one slip. Even if we rule out some deliberate deception, there is no positive reason to assume that the sample points in the experiment were equally probable. What we recommend as an improvement is a randomizing device that will provide some assurance that the equally likely assumption is prudent. If the bookkeeper was willing to take a sample of two slips, here is our recommendation: Number the slips (face down) from 1 through 12. Then toss a coin and roll a die. Note the result. Then toss the coin and roll the die again. Note the result. Use the following table to select the slips to look at:

Result	Slip No.	Result	Slip No.
$H \cap 1$	1	$T \cap 1$	7
$H \cap 2$	2	$T \cap 2$	8
$H \cap 3$	3	$T \cap 3$	9
$H \cap 4$	4	$T \cap 4$	10
$H \cap 5$	5	$T \cap 5$	11
$H \cap 6$	6	$T \cap 6$	12

We can have some confidence in the fairness of this process and be more assured that an assumption of equally probable sample points is reasonable. After all, we might have considerable experience with the particular coin and die on which to support the conclusion that the probability of heads is $\frac{1}{2}$ and the probability of each die face is $\frac{1}{6}$. Suppose that we do. Then the probability of selecting each slip through each toss of the coin and roll of the die is $(\frac{1}{2})(\frac{1}{6}) = \frac{1}{12}$. Note that we could select the same slip twice.*

Sample Space for $n = 2$

Our experiment involves two draws in the manner described. The sample space for the experiment consists of 144 points as shown in Figure 5.1. These points represent the possible sequences of sales slips that may be selected by our sampling procedure. The points along the "northwest-to-southeast" diagonal are the points that signify selecting the same slip twice. Also notice that in the count of sequences we view selecting slip 1 on the first draw and slip 2 on the second draw as a different outcome than selecting slip 2 on the first draw and slip 1 on the second. This leads us to the product rule:

Rule If event A has $N(A)$ outcomes and event B has $N(B)$ outcomes, then the number of ordered ways in which A and B can occur is $N(A) \cdot N(B)$.

The product rule could have been applied to count the 144 ways depicted in Figure 5.1. Event A would be the first draw and event B the second; $N(A) = 12$ and $N(B)$

*The sampling here is called sampling *with replacement.*

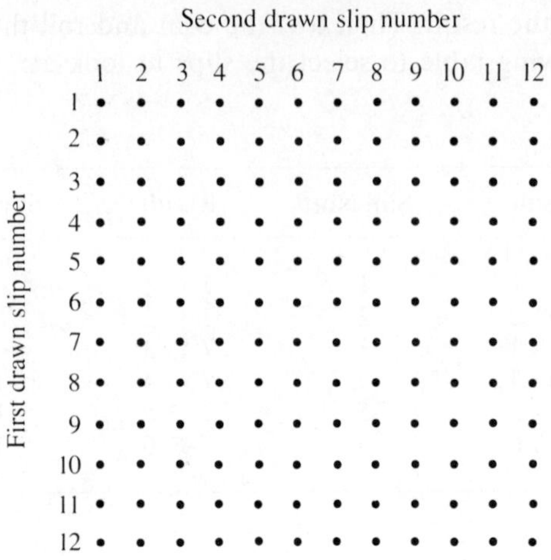

Second drawn slip number

First drawn slip number	1	2	3	4	5	6	7	8	9	10	11	12
1	•	•	•	•	•	•	•	•	•	•	•	•
2	•	•	•	•	•	•	•	•	•	•	•	•
3	•	•	•	•	•	•	•	•	•	•	•	•
4	•	•	•	•	•	•	•	•	•	•	•	•
5	•	•	•	•	•	•	•	•	•	•	•	•
6	•	•	•	•	•	•	•	•	•	•	•	•
7	•	•	•	•	•	•	•	•	•	•	•	•
8	•	•	•	•	•	•	•	•	•	•	•	•
9	•	•	•	•	•	•	•	•	•	•	•	•
10	•	•	•	•	•	•	•	•	•	•	•	•
11	•	•	•	•	•	•	•	•	•	•	•	•
12	•	•	•	•	•	•	•	•	•	•	•	•

Figure 5.1 *Sample space for selection of 2 sales slips from 12 sales slips*

$= 12$; and the number of ordered ways in which the two draws can occur (together) is $12 \cdot 12 = 144$.

The Sample Mean

If our bookkeeper is going to draw two slips, he might have in mind averaging the two associated profit values and reporting that result to the sales manager. The manager could then multiply that figure by the number of slips she counted (12) to answer her question of how much profit was made from sales of table radios yesterday. If the figure given actually coincided with the average profit from all 12 sales, she would have the correct answer to her question because

$$N\mu = \sum x.$$

However, if the average for the two profit figures drawn differs from μ, her figure for total profit will be correspondingly in error. We will concentrate on the sample averages that could be produced in the sample of two profits.

Definition

The **sample mean** of a sample of n observations from a statistical population is

$$\blacksquare \quad \overline{x} = \frac{\sum_{i=1}^{n} x_i}{n}. \quad \blacksquare \qquad 5.3$$

Keep in mind that *sample* (singular) refers to one collection of n observations and *samples* (plural) refers to many collections–each composed of n observations. This use is different from the everyday use of the word in phrases such as "he gave me a free sample of a new roll-on deodorant," which would mean one and not several roll-on units.

The values of the sample mean which result from the different first-drawn and second-drawn profit values that are possible are shown in Table 5.3.

Table 5.3 *Values of the Sample Mean $\bar{x}$ Resulting from a Draw of Two Observations*

If First x Drawn Is	And Second x Drawn Is 6	12	18
6	$\bar{x} = 6$	$\bar{x} = 9$	$\bar{x} = 12$
12	$\bar{x} = 9$	$\bar{x} = 12$	$\bar{x} = 15$
18	$\bar{x} = 12$	$\bar{x} = 15$	$\bar{x} = 18$

The probabilities of the events depicted by Table 5.3 can be calculated by the multiplication law for independent events:

$$P(x_1 \cap x_2) = P(x_1) \cdot P(x_2).$$

Here x_1 is the profit noted from the first draw and x_2 is the profit noted from the second. The events are independent if the outcomes of the randomizing mechanism—tossing a coin and throwing a die—are independent from one occasion to another.

One convenient way to organize the calculation of the joint probabilities needed is to construct another table in the manner of Table 5.3. In Table 5.4 we put the probability distribution of X for a single draw at both the top and left margins of the table. This is the probability distribution developed earlier and applies to *any* draw by virtue of independence. Then we apply the multiplication law for independent events by multiplying the probabilities in each row and column to obtain the probability for the event in the cell at the intersection of that row and column. For example, the probability for the event $x_1 = 6$ *and* $x_2 = 12$ is $7/12 \cdot 4/12 = 28/144$.

From Tables 5.3 and 5.4 we can see that some of the sequences of x_1 and x_2 lead to the same value for the sample mean. For example, three sequences lead to $\bar{x} = 12$. They are $x_1 = 18$ and $x_2 = 6$, $x_1 = 12$ and $x_2 = 12$, and $x_1 = 6$ and $x_2 = 18$. Since these sequences are mutually exclusive events, we can add their probabilities to obtain the probability of a sample mean of 12 (dollars). These additions are shown

Table 5.4 Multiplication of Marginal Probabilities to Obtain Probabilities of Intersections

x_1	$P(x_1)$	6 $^7/_{12}$	12 $^4/_{12}$	18 $^1/_{12}$	x_2 $P(x_2)$
6	$^7/_{12}$	$^{49}/_{144}$	$^{28}/_{144}$	$^7/_{144}$	
12	$^4/_{12}$	$^{28}/_{144}$	$^{16}/_{144}$	$^4/_{144}$	
18	$^1/_{12}$	$^7/_{144}$	$^4/_{144}$	$^1/_{144}$	

in Table 5.5. The final result is what we have been after—the probability distribution of the sample mean.

Table 5.5 Probability Distribution of the Sample Mean for $n = 2$

Sample Mean $\bar{x}$	Probability $P(\bar{x})$		
6	$^{49}/_{144}$		
9	$^{56}/_{144}$	=	$^{28}/_{144} + ^{28}/_{144}$
12	$^{30}/_{144}$	=	$^7/_{144} + ^{16}/_{144} + ^7/_{144}$
15	$^8/_{144}$	=	$^4/_{144} + ^4/_{144}$
18	$^1/_{144}$		

With our advantage of complete knowledge about the population, we are able to see the *errors* that might be made in using the sample mean as a substitute for the population mean. We see that the probability of a sample mean of \$6 is $^{49}/_{144} = 0.3403$. If the sample survey were to give this result and the sales manager used $\bar{x} = \$6$ in place of the unknown (to her) population mean, she would *underestimate* the population mean by \$3. Her estimate of yesterday's total profit on table radio sales would be off accordingly. She would use

$$N \cdot \bar{x} = \text{estimated total profit}$$

$$12 \cdot \$6 = \$72,$$

while we know the actual total profit was $12(\$9) = \$108$.

Definition 5.2

The **sampling distribution** of a statistic is the probability distribution of the statistic for samples of n observations from the population.

Characteristics of the Sampling Distribution of the Mean 5.3

To study the sampling distribution of the mean further, let's construct the distribution for a random sample of three draws from our model population. For three draws a table like Table 5.3 would have a third dimension. It would be a cube with $3 \cdot 3 \cdot 3 = 27$ compartments. The underlying sample space of sales-slip sequences would have $12 \cdot 12 \cdot 12 = 1728$ elements. There are several aids for doing the necessary probability calculations, and some of them are introduced in the second half of this chapter. The principles are no different from the ones we have already shown, however. Since we want to concentrate on the sampling distribution, we just show the final result in Table 5.6.

Table 5.6 *Sampling Distribution of $\bar{X}$ for $n = 3$ Showing Calculation of the Mean and Variance of $\bar{X}$*

$\bar{x}$	$P(\bar{x})$	$\bar{x} \cdot P(\bar{x})$	$(\bar{x} - \mu_{\bar{X}})^2$	$(\bar{x} - \mu_{\bar{X}})^2 \cdot P(\bar{x})$
6	$343/1728$	$2058/1728$	9	$3087/1728$
8	$588/1728$	$4704/1728$	1	$588/1728$
10	$483/1728$	$4830/1728$	1	$483/1728$
12	$232/1728$	$2784/1728$	9	$2088/1728$
14	$69/1728$	$966/1728$	25	$1725/1728$
16	$12/1728$	$192/1728$	49	$588/1728$
18	$1/1728$	$18/1728$	81	$81/1728$
		$15{,}552/1728$		$8640/1728$

$$\mu_{\bar{X}} = 15{,}552/1728 = 9.0 \qquad \sigma^2_{\bar{X}} = 8640/1728 = 5.0$$

The mean and variance of $\bar{X}$ are calculated in the same way as for any other random variable. We weight values of the random variable by their probabilities and sum the products to obtain the mean. We weight squared deviations of each value from the mean by the probability for the value and sum the products to obtain the variance. This follows Equations 5.1 and 5.2 except that the random variable is $\bar{X}$ instead of X. To distinguish the mean and variance of the sample mean from the mean and variance of X, we put the subscript $\bar{X}$ after the μ and σ^2 when we are referring to a sampling distribution.

If you calculate the mean and variance of $\bar{X}$ from the sampling distribution for $n = 2$ tabulated in Table 5.5, you will find that $\mu_{\bar{X}} = 9.0$ and $\sigma^2_{\bar{X}} = 7.5$. In fact, if we constructed the sampling distribution of the mean for $n = 4$ and for $n = 5$ and calculated their means and variances, we would find the additional results given in Table 5.7.

Table 5.7 *Mean, Variance, and Standard Deviation of $\bar{X}$ for Random Samples of Different Size from Population with $\mu = 9$ and $\sigma^2 = 15$*

Sample Size n	Mean $\mu_{\bar{X}}$	Variance $\sigma^2_{\bar{X}}$	Standard Deviation $\sigma_{\bar{X}}$
2	9.0	7.50	2.74
3	9.0	5.00	2.24
4	9.0	3.75	1.94
5	9.0	3.00	1.73

Regardless of the sample size, the expected value of the sample mean, $\mu_{\bar{X}}$, is equal to the population mean. The variance of $\bar{X}$ and its square root—the standard deviation of $\bar{X}$—diminish with increasing sample size. The relationships are as follows:

Statement The mean (or expected value) of the sampling distribution of the mean is equal to the mean of the population sampled:

$$\blacksquare \quad \mu_{\bar{X}} = \mu. \quad \blacksquare \tag{5.4}$$

Statement The standard deviation of the sampling distribution of the mean is equal to the standard deviation of the population sampled divided by the square root of the sample size:

$$\blacksquare \quad \sigma_{\bar{X}} = \frac{\sigma}{\sqrt{n}}. \quad \blacksquare \tag{5.5}$$

The standard deviation of the sample mean is also called the standard error of the sample mean or, more briefly, the **standard error of the mean**. The reason for the term standard *error* is that the dispersion of the sampling distribution of the mean reflects the errors that might be encountered in using the sample mean in place of the population mean—that is, in using $\bar{x}$ as an *estimate* of μ. This follows from the fact that $\mu_{\bar{X}} = \mu$, so that in looking at deviations between $\bar{x}$ and $\mu_{\bar{X}}$ we are looking at differences between $\bar{x}$ and μ. These differences, $\bar{x} - \mu$, are called sampling errors. Their mean (or expected value) is zero. The standard deviation of these *errors* comes naturally to be called the standard error of the mean. Notice that as the sample size increases, the standard error of the mean diminishes. The diminishing sampling error associated with increasing sample size is sometimes called the *law of large numbers.*

A significant fact about Equations 5.4 and 5.5 is that they do away with having to calculate the mean and standard deviation of a sampling distribution in the way we did in Table 5.6. In fact, if all we want (or need) is these two values ($\mu_{\bar{X}}$ and $\sigma_{\bar{X}}$), we don't have to go through the calculations to get the probability distribution of $\bar{X}$. With our complete knowledge about the 12 table radio sales, we could compute

$$\mu_{\bar{X}} = \mu = 9.0 \text{ dollars}$$

$$\sigma_{\bar{X}} = \frac{\sigma}{\sqrt{n}} = \frac{\sqrt{15}}{\sqrt{3}} = \sqrt{5} = 2.24 \text{ dollars.}$$

A use we have already made of mean and standard deviation of population distributions can be carried over to the mean and standard deviation of sampling distributions. This is the generalization that at least 90 percent of the values in a distribution will lie within three standard deviations of the mean of the distribution. Applied to a sampling distribution, the same generalization is that the probability of a sample mean within three standard errors of the population mean is at least 0.90. For the sampling distribution for $n = 3$ in the sales-slip problem, this range is

$$\begin{array}{rcl} \mu - 3\sigma_{\bar{X}} & \text{to} & \mu + 3\sigma_{\bar{X}} \\ 9.0 - 3(2.24) & \text{to} & 9.0 + 3(2.24) \\ 9.0 - 6.72 & \text{to} & 9.0 + 6.72 \\ \$2.28 & \text{to} & \$15.72 \end{array}$$

Thus, without deriving the detailed sampling distribution, we know that the probability is at least 0.90 that the mean of a sample of three slips will be within \$6.72 of the population mean. If you refer back to Table 5.6, the actual probability for means between \$2.28 and \$15.72 can be found to be 1715/1728, or 0.9925.

5.3

Summary

This half chapter addressed a major question: What would happen if a sampling experiment to estimate a population mean were repeated over and over again. The discussion was carried out in terms of an example population where we had complete information about the population but the conductor of the survey and user of the survey information did not. We found that if the sampling process was properly designed we could use the laws of probability to develop the probability distribution of the sample mean in a proposed survey. The idea of a probability distribution of a (sample) statistic underlies statistical inference. The sample measure, or statistic, is regarded as a random variable, and its probability distribution is called a sampling distribution.

In our example, the sampling distribution of the mean allowed us to make statements about probabilities of errors that the sales manager could make in using the sample mean as an estimate of the underlying population mean. In general, the sampling distribution of a statistic displays the behavior, in probability terms, of a sample measure. In particular, we found that for a random sample of size n the expected value of the sample mean is equal to the mean of the population sampled. Moreover, we found that the standard deviation of the sample mean is equal to the standard deviation of the population divided by the square root of the sample size. The standard deviation of the sample mean is also called the standard error of the mean.

Using a general rule about distributions we can say that the error in the mean of a random sample will be less than three standard errors of the mean with probability at least 0.90. This is a weak statement because the probability is an inequality. In most cases, as it was in our example, the probability will be greater than that. Three important questions remain. The first is how we can make stronger probability statements about sampling error of a mean when the population standard deviation is known. The second question is whether we can make similar statements wholly on the basis of sample information. That is, if *we* didn't have complete knowledge about the population, could the bookkeeper say anything about the accuracy of an estimate of the population mean based on a single sample mean? A third question is what the bookkeeper can know about the accuracy of the mean of a sample he *proposes* to take. An answer to the last question would allow the bookkeeper to balance his laziness against the chance that the sales manager would get him fired for providing inaccurate information once she discovered it. We begin to examine such questions in succeeding chapters.

See Self-Correcting Exercises 5A.

Exercise Set A

1. If we know that for some population $\mu = 8$ and for some sampling distribution of the mean $\mu_{\bar{X}} = 8$, can we determine the sample size? If so, what is it? If not, why not?

2. If we know that for some population $\sigma = 12$ and the standard error of the mean is 6, can we determine the sample size? If so, what is it? If not, why not?
3. If it is known that the standard error of the mean is 2 for a sample size of 8, what is the standard deviation of the original population?
4. For any sample size, the original distribution and the sampling distribution of the mean have the same mean and so, in a sense, are "centered" at the same location. How do the two distributions differ, and how does this difference vary as the sample size increases?
5. Give the names of four parameters we studied in Chapters 1 to 3. Does the symbol $\bar{x}$ represent a parameter or a statistic? Why?
6. If a traveler has a choice of nine airlines for a convenient flight from O'Hare to Kennedy, and a choice of four convenient connections to Heathrow, in how many ways can the trip from O'Hare to Heathrow via Kennedy be scheduled?
7. For a population with $\mu = 12$ and $\sigma = 10$, determine $\mu_{\bar{X}}$ and $\sigma_{\bar{X}}$ for sample sizes of 2, 4, 25, and 100. Enter the correct values in the appropriate positions in the table:

n	$\mu_{\bar{X}}$	$\sigma_{\bar{X}}$
2		
4		
25		
100		

8. In testing concentrations of a pollutant in the vicinity of a certain industrial plant, analysis of the air is made hourly over a 12-hour period at a specific site near the plant. The hourly concentrations (in parts per million) of the pollutant are recorded as follows:

Time	6	7	8	9	10	11	12	13	14	15	16	17
Concentrations	0.16	0.15	0.16	0.18	0.18	0.19	0.20	0.22	0.22	0.23	0.21	0.20

(a) If the human tolerance level of concentration is considered to be 0.20 parts per million, assign a random variable to each simple event (single element) of the sample space by the following rule: If the reading does not exceed the tolerance level, zero is assigned; if the reading does exceed the danger level, the amount by which it exceeds that level is assigned.

(b) Determine the probability distribution of the random variable and then find the expected value and the variance.

(c) If one reading is selected at random, what is the probability that

(i) The random variable is zero?

(ii) The reading exceeds 0.20?

(iii) The random variable exceeds 0.20?

(iv) The random variable exceeds zero?

(v) The tolerance level is exceeded?

9. To determine the amount of progress in reading made in the past year by 10 fifth graders, their results on an achievement test in May are compared with their results on a similar

test the preceding May. The following scores for students A to J were recorded as "achieved grade level" for each test:

	A	B	C	D	E	F	G	H	I	J
Current test	6.0	5.8	6.2	4.9	5.5	8.8	5.8	3.1	6.2	6.3
Earlier test	4.8	4.6	5.7	3.8	4.6	6.4	5.0	2.4	5.0	4.9

(a) Assign values of a random variable by associating with each student the difference between this and the previous year's test scores.

(b) What does the random variable represent? Why might the teacher be more interested in this random variable than in the original test scores?

(c) Determine the probability distribution of the random variable.

(d) Why might the teacher find a random variable value of less than 1.0 disappointing? What is the probability of a random variable less than 1.0?

(e) What is the expected value of the random variable? Should the teacher be heartened or disappointed by this result? Justify your conclusion.

10. With reference to Exercise 9, the teacher, concerned that student F, obviously very bright, and student H, who seems to have severe learning disabilities, might unduly be affecting the "average" improvement, decides to look at the scores of only the other eight students. Find the probability distribution of the random variable for these eight and the expected value of the random variable. How should the teacher feel about these results? Why?

11. The Highway View Tourist Home has five double rooms, and the owners, both retired statisticians, have found from 2 years of experience that the number of rooms rented on any given night follows approximately the distribution given below:

x (Number of Rooms Rented)	$P(x)$
0	0.25
1	0.15
2	0.20
3	0.18
4	0.12
5	0.10

(a) How do you think the values for $P(x)$ have been obtained?

(b) What is the expected value of the random variable?

(c) Over the period of a year (365 days), how many room rentals can the owners expect?

12. Referring to Exercise 11 and assuming that the demand for rooms at the Highway View is independent of the season and day of the week, what is the probability that on two randomly selected nights

(a) The Highway View has the No Vacancy sign displayed both nights?

(b) The *average* number of rooms rented on these two nights is 3?

(c) No rooms are occupied on either night?

13. The law firms in a large city have the following probability distribution of number of partners. For a random sample of size $n = 2$, find the probability distribution of the average number of partners per firm.

Number of partners	2	3	4	5
Probability	0.4	0.3	0.2	0.1

14. Find the mean and variance of the population in Exercise 13. Calculate the mean and variance of the probability distribution of the sample mean from the distribution you obtained.

15. Consider a deck of 16 cards consisting of the jack, queen, king, and ace of the four suits. Construct a representation of the sample space for the experiment *draw a card.*

16. Suppose in connection with the card draw in Exercise 15 that jack, queen, king, and ace count 1, 2, 3, and 4 points, respectively, and that an extra point is scored if a spade or heart is drawn. Assign point scores and probabilities to the sample points and find the probability distribution of the point score for a single draw.

17. Calculate the mean and variance of the probability distribution in Exercise 17.

Aids in Probability Calculations

Some techniques useful in calculating probabilities are probability trees and formulas for permutations and combinations. These do not involve any new principles beyond the four probability laws presented in Chapter 4. The probability tree is just a convenient way to visualize the structure of many problems. It helps in organizing a problem and its associated calculations. Permutation and combination formulas permit us to count readily the number of ways an event of interest can occur. When the sample space of an experiment goes beyond a number of outcomes that can be easily listed, these formulas are a welcome convenience.

Probability Trees 5.4

Consider a three-game playoff between two teams. Let A represent the event *team A wins a game* and B the event *team B wins a game.* We will attach numerical subscripts 1, 2, and 3 to indicate the first, second, and third games in the series. The logical possibilities for the series can be diagrammed in the sequential tree of Figure 5.2.

Each path along the tree represents a possible outcome for the series. For example, the path $A_1 \longrightarrow B_2 \longrightarrow A_3$ means A wins the first game, B wins the second,

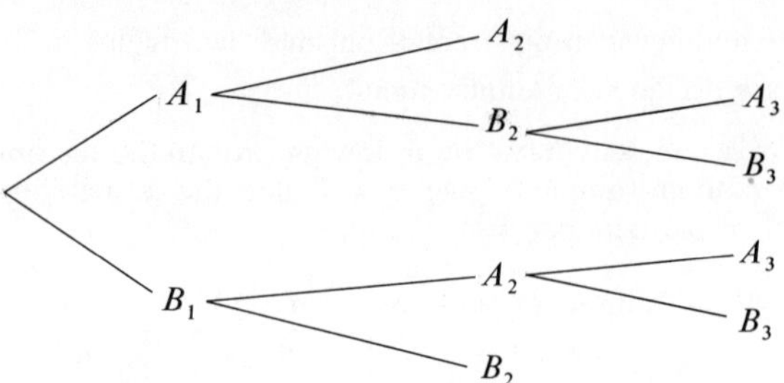

Figure 5.2 *Tree diagram for a playoff series*

and A wins the third game (and the playoff!). This is the event $A_1 \cap B_2 \cap A_3$. If we have probability values for the events along the branches, we can apply the appropriate form of the multiplication law to work out the probability of the sequence corresponding to any path along the tree.

Suppose the probabilities for the first game are 0.6 that A wins and 0.4 that B wins, and let us assume that the psychology of the series is such that winning a game increases the probability of winning the next game by 0.1. The probabilities are placed on the branches as in Figure 5.3. For example, 0.7 is $P(A_2 | A_1)$. Each probability on a branch is the probability for the event the branch leads to—*conditional* on the events on the path up to that point. The events do not have to refer to a time sequence, although they do in this example.

The tree helps us to keep track of the conditional probabilities and identify specified events. Suppose our problem is to find the probability that A wins the playoff. We need only identify the paths that terminate with A (A must win the last game to win the series). Then multiply the probabilities appearing on the path leading up to each such terminal. Each of those paths is a mutually exclusive way for A to win the series, so we can sum the path products to obtain the desired probability. These calculations, shown in Figure 5.3, lead to the result: $P(A \text{ wins series}) = 0.648$.

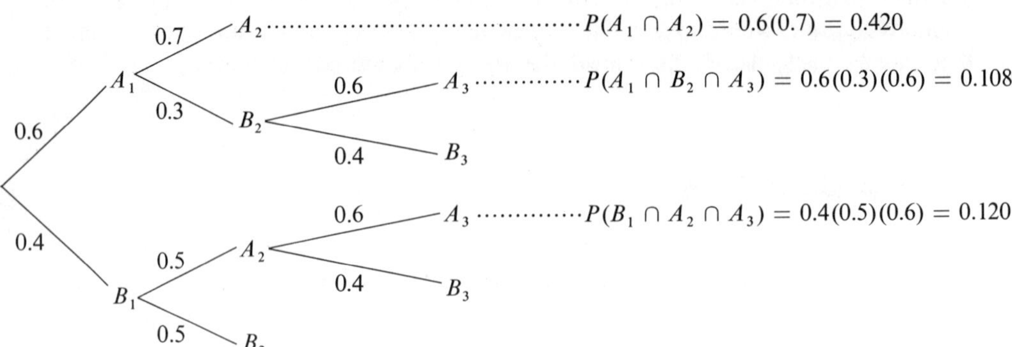

Figure 5.3 *Use of a tree diagram to calculate probability of an event*

A type of problem that the probability tree helps to clarify might be termed the *reverse conditional probability problem.* This occurs when our interest is focused on a restricted sample space corresponding to particular terminal events. Suppose you have been out of town and on returning home a friend tells you that B won the series. You ask, "How many games?" Your friend replies that all he knows is that B won. If you accepted the initial probability assignments, what should be your belief (probability) that the series lasted only two games? In accordance with the general definition of conditional probability, this probability is

$$P(\text{series is 2 games} \mid B \text{ wins series}) = \frac{P(\text{series is 2 games} \cap B \text{ wins series})}{P(B \text{ wins series})}.$$

The event *B wins the series* is our restricted sample space, and it includes all the paths on the tree diagram that terminate in B. Since these are mutually exclusive ways for B to have won, we add the products for those paths. The numerator refers to those events in the denominator that also satisfy the requirement that the series lasts two games. There is only one, namely $B_1 \cap B_2$. Thus the required probability is

$$\frac{0.200}{(0.072 + 0.080 + 0.200)} = 0.5682.$$

Figure 5.4 shows the component probabilities.

The calculations to find a sampling distribution can be shown in probability tree form. If the draws are independent, the probabilities for a given event on successive draws do not change. Multiplication of probabilities along paths is now an application of the special multiplication law. Figure 5.5 develops the same joint probabilities as Table 5.4. To keep track of the sample means, we sum the x values along each path and divide by n to obtain the sample mean for that path. Then it is necessary, as before, to sum the probabilities for different event sequences (paths) leading to the same value of the sample mean.

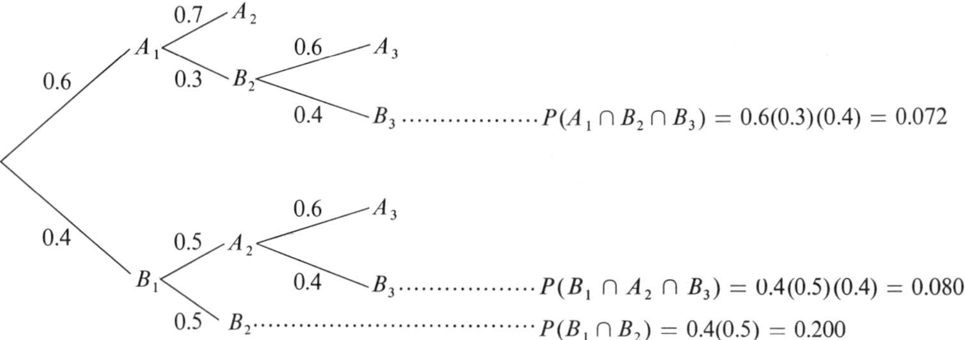

Figure 5.4 *Mutually exclusive paths defined by a terminal event*

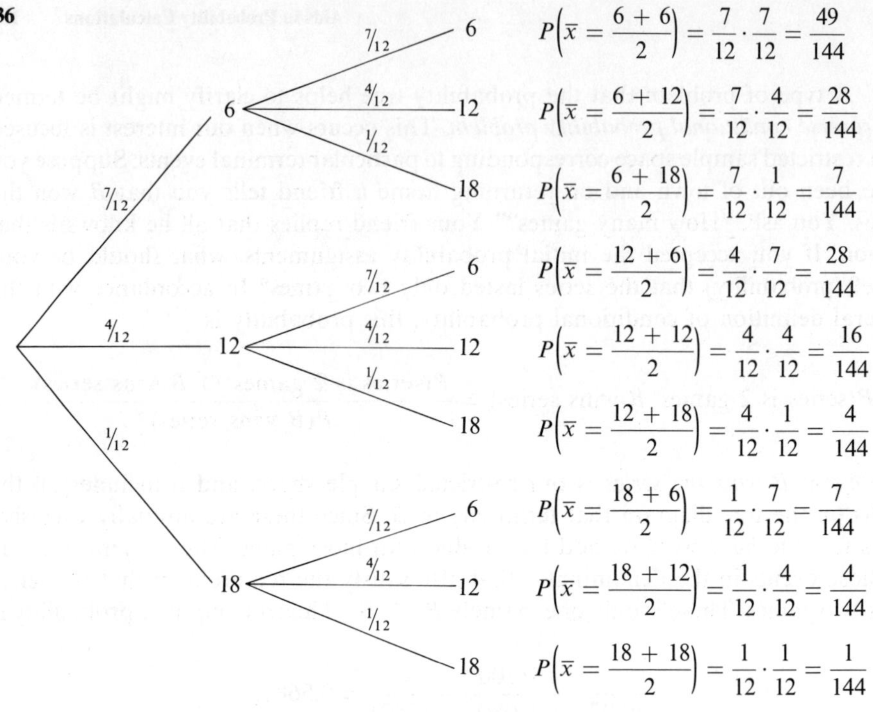

Figure 5.5 *Probability tree for sampling distribution of a mean*

Table 5.8 *Extension of Tabular Method to Obtain Joint Probabilities for Three Draws*

$(x_1 + x_2)$	$P(x_1 + x_2)$	6 7/12	12 4/12	18 1/12	x_3 $P(x_3)$
12	49/144	343/1728 $\bar{x} = 6$	196/1728 $\bar{x} = 8$	49/1728 $\bar{x} = 10$	
18	56/144	392/1728 $\bar{x} = 8$	244/1728 $\bar{x} = 10$	56/1728 $\bar{x} = 12$	
24	30/144	210/1728 $\bar{x} = 10$	120/1728 $\bar{x} = 12$	30/1728 $\bar{x} = 14$	
30	8/144	56/1728 $\bar{x} = 12$	32/1728 $\bar{x} = 14$	8/1728 $\bar{x} = 16$	
36	1/144	7/1728 $\bar{x} = 14$	4/1728 $\bar{x} = 16$	1/1728 $\bar{x} = 18$	

The tree diagram can be extended to three draws by another division of the branches. Our example would lead to 27 final branches, which would require a fairly large piece of paper. A more compact way to extend the number of draws is shown in Table 5.8. There the possible sums from the first two draws along with their probabilities are placed at one margin, and the probability distribution of x for the third draw is placed at the other margin. Multiplying the pairs of marginal probabilities now gives the probabilities for the event sequences $(x_1 + x_2) \cap x_3$ represented by the cells in the table. We must add the probabilities for cells that lead to the same $x_1 + x_2 + x_3$ sum and hence the same mean. Then all that remains is to list the probability table for $\bar{x}$. If you do these steps from Table 5.8, you will obtain the sampling distribution given in Table 5.6. 5.4

Permutations and Combinations 5.5

It is useful to be able to count the number of elements in various sample spaces. If any of N elements can be selected on any one draw, the number of elements in the sample space for n draws is N^n. This was the case in the sales-slip example, where $N = 12$ and there were $12^2 = 144$ different sequences for a sample of size 2. The sample of three sales slips generated a sample space of $12^3 = 1728$ elements.

Suppose we have 50 student names on a class enrollment list. The individual names are cut out and placed in an urn. We can consider sampling from the urn *with replacement.* This means that after each draw, the name drawn is replaced before the next draw. Thus any element (name) can be selected on any draw and the number of elements in the sample space is $N^n = 50^n$.

We might also sample the urn *without replacement.* This means that the element selected on any draw is not replaced before the next draw. Elements selected on a given draw cannot then be selected on succeeding draws. A little thought will verify that when $N = 50$ and sampling is without replacement, there will be 50(49) elements in the sample space for $n = 2$; 50(49)(48) elements for $n = 3$; 50(49)(48)(47) elements for $n = 4$; and so on. The reasoning here is that any one of N elements can be selected on the first draw; then only $N - 1$ are left to select from on the second draw, $N - 2$ remain for the third draw, and so forth. When sampling is without replacement, the number of elements in the sample space for a sample size of n is

$$N(N - 1)(N - 2) \cdots (N - n + 1).$$

For four draws without replacement from 50 elements, $N = 50$ and $n = 4$ and there are four terms in the product, that is, 50(49)(48)(47). The last term is $50 - 4 + 1 = 47$, or $(N - n + 1)$ as above.

Let us continue with another example of sampling without replacement. Suppose you have five musical recordings you wish to arrange on one tape. In how many ways can you arrange them? Note that we have ruled out repetitions of the same recording—each must be heard only once when the tape is played and they all must be heard. Each way that you could arrange the recordings is called a **permutation** of the five elements. It is just one of the ways of ordering five elements. The number of such ways is called the *number of permutations* of the five elements. You know from our earlier examples that the number of permutations of the five recordings is

$$(5)(4)(3)(2)(1) = 120.$$

One way to visualize this is to think of five positions in the tape program—first, second, third, fourth, and fifth. We are talking about the number of ways of *assigning* the five recordings *A, B, C, D,* and *E* to these five positions. One arrangement is portrayed in Figure 5.6. There we see the five elements (A,B,C,D,E) and the positions (1,2,3,4,5) to which the recordings can be assigned. Only one such assignment is shown, but the number of different *assignments* that are possible is 120. You can assign any one of the five recordings to the first position, and having done that you can assign any of the four remaining recordings to the second position and so forth.

Suppose you have room on your tape for only three recordings. Once you have selected any three recordings, there will be a number of ways of ordering the selected recordings on the tape. You have two problems in determining the number of different programs you can present:

1. How many different collections of three recordings can be made from the five?
2. How many ways are there of *assigning* the three recordings in any collection to the positions (first, second, third) on the program tape?

The first question involves what is called the number of combinations of three elements selected from five elements. In this problem it is not hard to list all the combinations, or collections. There are 10 of them:

$$ABC \quad ABD \quad ABE \quad ACD \quad ACE$$
$$ADE \quad BCD \quad BCE \quad BDE \quad CDE$$

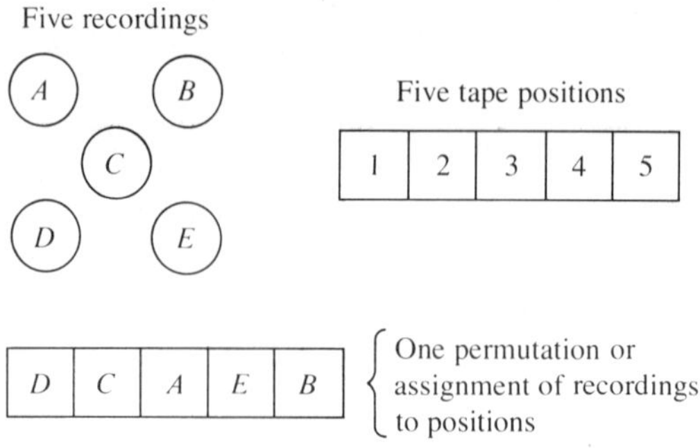

Figure 5.6 *An assignment of recordings to tape positions*

From what you know about permutations, there are 3(2)(1) = 6 ways of assigning any combination of three recordings to the three positions available on the tape. Thus there are 10 combinations × 6 permutations per combination = 60 different programs you can have. Notice in this count that a different way of assigning a collection is counted as a different program. The six permutations of the combination *ABC* are

$$ABC \quad ACB \quad BAC \quad BCA \quad CAB \quad CBA$$

We see that there are 60 ways of *selecting and then assigning* three from five elements. This total is called the number of permutations of five elements three at a time. Each permutation is a different assignment to three positions. The number of permutations can be arrived at by considering that any one of the five elements (recordings) can be assigned to the first position. Given that assignment, any of four remaining elements can be assigned to the second position, and after that, any one of three elements can be assigned to the third position. Thus, a direct way to find the number of permutations of five elements three at a time is

$$5(4)(3) = 60.$$

Formulas for (the number of) permutations and (the number of) combinations of n elements selected from N elements involve the mathematical quantity **factorial.** Factorial of any positive integer N is defined as

$$N! = N(N-1)(N-2)\cdots(1),$$

that is, the product of N and all lesser positive integers. The symbol ! stands for factorial, and $N!$ is read as "N factorial." Both 1! and 0! are defined to be 1.0. The formulas for permutations and combinations are

$$\blacksquare \quad {}_NP_n = \frac{N!}{(N-n)!} \quad \blacksquare \qquad 5.6$$

and

$$\blacksquare \quad {}_NC_n = \frac{N!}{(N-n)!n!}. \quad \blacksquare \qquad 5.7$$

In Equation 5.6, ${}_NP_n$ is read as the number of permutations of n elements selected from N elements, and ${}_NC_n$ in Equation 5.7 is read as the number of combinations of n elements selected from N elements. Some say, to be more brief, the number of permutations of N things n at a time and the number of combinations of N things n at a time. In our example we were concerned with

$$_5C_3 = \frac{5!}{(5-3)!3!} = \frac{5!}{2!3!} = \frac{5(4)(3)(2)(1)}{2(1) \times 3(2)(1)} = 10,$$

$$_3P_3 = \frac{3!}{(3-3)!} = \frac{3!}{0!} = \frac{3(2)(1)}{1} = 6,$$

5.5

$$_5P_3 = \frac{5!}{(5-3)!} = \frac{5!}{2!} = \frac{5(4)(3)(2)(1)}{2(1)} = 60.$$

Summary

Probability trees and formulas for counting combinations and permutations are useful tools in solving probability problems. Probability trees enable us to visualize the mutually exclusive sequences, or paths, that are possible in a series of events. After placing known probabilities on the branches of the tree, the probabilities required in a problem can be calculated with the aid of the logic of the tree diagram. In working with permutations and combinations it is important to distinguish what is counted in either case. When we count combinations, we count the number of different *collections* of n elements that can be assembled from N elements. When we count permutations of n elements drawn from N elements, we count the number of *assignments* that can be made to n positions from among N elements.

See Self-Correcting Exercises 5B.

Exercise Set B

1. Referring to Exercise 12 of Set A, use a tree diagram to list all possible outcomes of the two-night rental sequence. How many terminal branches does your diagram have?
2. Use the tree diagram of Exercise 1 to determine the probability that, for the two-night sequence, the average number of room (a) rentals is 4 and (b) vacancies is 4.
3. Assuming that the probability for the birth of a boy is 0.51 and that of a girl is 0.49, use a tree diagram to determine the probability that, in a three-child family with no multiple births,

(a) Exactly two of the children are girls.

(b) At least two of the children are girls.

4. You have two quarters and time to play the "Double-Your-Money" slot machine only four times before your bus leaves. The machine takes one quarter on each play, and when it pays it releases two quarters. Unknown to you, the probability that the machine pays off on any one play is $\frac{1}{5}$. Construct a tree diagram by placing W (for win) and L (for lose) in appropriate positions and entering the correct probability along each portion of each path.

5. Refer to your diagram from Exercise 4:

   (a) Circle those path terminals which represent your losing money (that is, ending up with fewer quarters than your original two).

   (b) What is the probability of ending up with no quarters?

   (c) What is the probability of ending up with more than two quarters?

   (d) What is the probability of breaking even from this sequence of plays?

6. A strategy sometimes suggested in gambling is to bet half your cumulated winnings on the next bet if you are ahead and bet double the amount of your cumulated losses on the next bet if you are behind. Assume a \$10 initial bet on a fair coin toss and follow this strategy through two more tosses (three tosses in all). Is it a paying strategy?

7. A man has four bills in his wallet in denominations of \$1, \$5, \$10, and \$20. He selects one bill randomly and then selects a second bill without returning the first. Find the probability distribution of the total amount of money drawn.

8. Weekly demand for replacement part 062 at a supply facility has been found to be equally likely for the number of demands 0, 1, and 2. Tabulate the sample space for total demands over a 2-week period and (assuming independence) find the probability distribution of total demands for the 2-week period.

9. Extend your results in Exercise 8 to find the probability distribution of total demand for a 3-week period (again assuming independence).

10. One hundred stocks in a portfolio were classed as either cyclical or growth stocks. Of 60 cyclical stocks, 45 had increased in price in the past month, while 15 of the 40 growth stocks had increased in price. If a stock is selected at random from the portfolio, what is the probability that it will be

    (a) A cyclical stock that has not increased in price?

    (b) A stock that has increased in price?

    (c) Given that the stock selected has not increased in price, what is the probability that it is a cyclical stock?

11. A furniture salesperson found that her performance (good versus bad) was independent of the weather (sunny or cloudy). The probability that she has a good day is 0.40, and the probability that the weather is sunny is 0.60. Find:

    (a) The probability that the weather is sunny and the salesperson has a bad day.

    (b) The probability that the weather is sunny when the salesperson has a bad day.

12. A TV manufacturer trims his sets in either black or grey, and he can use either a walnut or a maple stain. A consumer survey showed that 20 percent of consumers desired a black trim with a walnut stain and that half of all consumers desired a black trim or a walnut stain. If it has already been decided to trim 30 percent of the sets in black, what proportion of the sets should be given a walnut stain?

13. The probability that a particular football team will run its "green" offense is 0.60 when a play begins inside its own 30-yard line and 0.25 otherwise. The probability that the

fullback carries the ball given the "green" offense is 0.50 and 0.30 otherwise. If 20 percent of the team's offensive plays begin inside its own 30-yard line, what is the marginal probability that the fullback carries the ball?

14. In a game of dice the roller wins if the sum of the spots on two dice thrown total 7 or 11 on the first roll. The nonroller wins if the sum of the spots is 2, 3, or 12 on the first roll. Find the probability that
    (a) The roller wins on the first roll.
    (b) The nonroller wins on the first roll.
    (c) The game is undecided on the first roll.

15. A production inspector is checking quality control for two processes, *A* and *B*, where process *B* follows process *A*. She knows that 11 percent of the production units going through the two processes are rejected after the completion of process *B*. The rejection rate for process *A* is 10 percent, but 80 percent of those units are corrected in process *B*. With the aid of a tree diagram, determine the rejection rate for process *B* for those production units that are acceptable after process *A*.

16. Refer to Table 5.8 in determining your answers to parts (a) and (b).
    (a) Find the probability that $x_1 + x_2 + x_3$ equals (i) 18; (ii) 36; (iii) 48.
    (b) Find the probability that $\bar{x}$ for three draws is (i) 6; (ii) 9; (iii) 12; (iv) 54.

17. Evaluate each of the following:
    (a) 3!
    (b) 7!/0!
    (c) 10!
    (d) ${}_5C_2$
    (e) ${}_7C_1$
    (f) ${}_{10}P_4$
    (g) ${}_{10}P_{10}$
    (h) ${}_{10}P_1$
    (i) ${}_NC_3$
    (j) ${}_NP_5$

18. Each of 15 students in a speech class must address the class once within the next 2 weeks.
    (a) In how many ways may the speakers be scheduled?
    (b) Eight students are to be chosen by lot to give their speeches next week. How many different eight-member selections are possible? In how many different ways may their speeches be scheduled?

19. The college dining hall manager has 86 recipes for hamburger dishes. In how many ways may four hamburger dishes be scheduled for Monday, Wednesday, Thursday, and Saturday of next week if no recipe is to be repeated?

20. A research and development manager must create a three-person team that possesses the skills of a geologist, an engineer, and a mathematician. Available are two geologists, eight engineers, and three mathematicians. How many different combinations can be named for the team?

21. Three successive draws are to be made from an urn containing chips numbered 1 through 10. How many different sample sequences are there if (a) sampling is with replacement? (b) If sampling is without replacement?

22. Express the result in Exercise 21(b) as the product of the number of different sample combinations and the number of permutations per sample combination.

23. To establish the potential success of a product, it has been decided to sell it in three cities on a test basis. The marketing research department has selected seven cities as possible test areas. How many different groups of test cities can be finally selected?

24. If the test marketing in Exercise 23 can be carried out in only one city at a time, how many different schedules can be set up for the three cities finally selected?

25. What is the meaning of the product of your answers in Exercises 23 and 24?

Overview

1. What does the symbol $\mu_{\bar{X}}$ denote? How does it differ from μ?

2. What does the symbol $\sigma_{\bar{X}}$ denote? How does it differ from σ?

3. For the sample space with elements *A, B, C,* and *D,* list all possible ordered samples of size 2
   (a) If sampling is done with replacement.
   (b) If sampling is done without replacement.

4. Let the assignment of a random variable be made to the sample space of Exercise 3 on the following basis:

$$A \leftrightarrow 2$$
$$B \leftrightarrow 7$$
$$C \leftrightarrow 8$$
$$D \leftrightarrow 3$$

   (a) Find the expected value of the random variable.
   (b) For each of the samples you obtained in part (a) of Exercise 3, find the sample mean. Then form the frequency distribution of these sample means and from it determine $\mu_{\bar{X}}$ and $\sigma_{\bar{X}}$.
   (c) Compare your values for $\mu_{\bar{X}}$ and $\sigma_{\bar{X}}$ from part (b) with those that would be obtained applying Equations 5.4 and 5.5, respectively. Do these values agree? Should they?
   (d) For each of the samples you obtained in part (b) of Exercise 3, find the sample mean. Then form a frequency distribution of these means and from it determine $\mu_{\bar{X}}$ and $\sigma_{\bar{X}}$.
   (e) Compare your values for $\mu_{\bar{X}}$ and $\sigma_{\bar{X}}$ with those you would obtain applying Equations 5.4 and 5.5, respectively. Your results for $\sigma_{\bar{X}}$ should differ because Equation 5.5 is applicable only when sampling is done *with replacement.* Draws without replacement are not independent. The appropriate formula for sampling without replacement is

$$\sigma_{\bar{X}} = \sqrt{\frac{\sigma^2}{n}\left(\frac{N-n}{N-1}\right)},$$

   where n is the sample size and N is the number of elements in the population.

5. Use the formula given in part (e) of Exercise 4 to find $\sigma_{\bar{X}}$ for samples of size 5 and a population of size 25 having a standard deviation of 12, if sampling is done without replacement.

6. What happens to the "correction factor" $(N - n)/(N - 1)$ in the formula for sampling without replacement given in Exercise 4(e) if N is very large and n is comparatively small? For example, if
   (a) N is 1000 and n is 5?
   (b) N is 1,000,000 and n is 100?

   This behavior of the correction factor sometimes leads statisticians to ignore it for large populations unless the sample is a significant part of the population.

7. Construct a tree diagram to indicate what can happen when two fair dice are rolled once. Use this diagram to obtain a probability distribution for the *sum* of the numbers that turn up on the dice. If you were betting on a sum to turn up when two fair dice are rolled, what would you select as the most likely to turn up?

8. Janet Eager is told by a prospective employer, "You are on trial for a year. One mistake is all right, but two mistakes and you're out!" Janet figures that there are four critical decisions to make during the year, and the probability that she will make any one of them correctly is $4/5$, but she doesn't want to run more than a one-in-five chance of not lasting out the year. Assume independent events and draw a tree diagram for the process. Should Janet take the position?

9. A salesman figures that the probability is 0.6 that he will sell a prospect on the first call and that on each subsequent call on a prospect who has not been sold, the probability of selling drops by 0.1. If the salesman is willing to make up to three calls on a prospect, what is the probability of selling to him? If the salesman persists beyond this, how much better can he do?

10. A union committee of six people is to be chosen from the supervisory employees in a factory. Although there are 10 male and 3 female supervisors, the union leadership wants an equal split of male and female committeepersons. How many different committees can be formed?

11. A box contains three red marbles and three blue marbles. Use tree diagrams to show the probability that two successive draws from the box will yield different colored marbles if
    (a) The drawing is with replacement.
    (b) The drawing is without replacement.

12. There are three positions open and 10 candidates–6 men and 4 women. Assume equal selection probabilities and use combinations to find
    (a) The probability that three men will be selected.
    (b) The probability that one man and two women will be selected.

13. Which of ${}_NC_n$ and ${}_NP_n$ cannot exceed the other? Why? For what value of n must ${}_NC_n$ equal ${}_NP_n$? Why?

14. An alternative formula for determining the number of permutations of N things taken n at a time is

$${}_NP_n = N(N - 1)(N - 2) \cdots (N - n + 1).$$

    Demonstrate, substituting 8 for N and 3 for n, how this formula could be obtained from Equation 5.6 for ${}_8P_3$ using appropriate cancellation of factors.

15. A department store window dresser is planning an all-blue display of fabric bolts, trims, and buttons. The store has, in various tones of blue, 18 bolts of different fabrics, 20 different kinds of trims, and 12 styles of buttons. The dresser plans to display 6 bolts, 12 styles of trim, and 6 different types of buttons.

(a) In how many ways can the dresser choose the following:
   (i) Fabrics to be displayed?
   (ii) Trims to be shown?
   (iii) Types of buttons to be used?
   (iv) All the components of the display together?
(b) How many times must the product rule be used to obtain the answer to question (iv) above?

16. If the window dresser in Exercise 15 decided to double the number of bolts to be displayed, how would this affect the answers to part (a) of that exercise?

Glossary of Equations

5.1	$\mu = \sum [x \cdot P(x)]$	The mean (or expected value) of a random variable is the summation of products of values of the random variable and their probabilities.
5.2	$\sigma^2 = \sum [(x - \mu)^2 \cdot P(x)]$	The variance of a random variable is the summation of squared deviations of values of the random variable from their mean, weighted by their corresponding probabilities.
5.3	$\bar{x} = \dfrac{\sum_{i=1}^{n} x_i}{n}$	The sample mean of a sample of n observations from a statistical population.
5.4	$\mu_{\bar{X}} = \mu$	The mean (or expected value) of the sampling distribution of the mean is equal to the mean of the population sampled.
5.5	$\sigma_{\bar{X}} = \dfrac{\sigma}{\sqrt{n}}$	The standard deviation of the sampling distribution of the mean is equal to the standard deviation of the population sampled divided by the square root of the sample size.
5.6	${}_NP_n = \dfrac{N!}{(N-n)!}$	The number of permutations of n elements selected from N elements.
5.7	${}_NC_n = \dfrac{N!}{(N-n)!n!}$	The number of combinations of n elements selected from N elements.

6

The Binomial Distribution

In the first half of Chapter 5 we developed the basic concept of a sampling distribution. The example we used was a finite population of 12 dollar profit figures for 12 table radios that were sold yesterday by a small store. We developed the theory for sampling with replacement, which led to basic formulas that are also applied when sampling without replacement from large populations. Modifications to accommodate without replacement sampling are detailed in Chapter 13.

This chapter introduces a sampling distribution–the **binomial distribution**–which applies to an important special case of sampling with replacement. The special case is sampling from a population of *nominal scale* data, in which all that concerns us about the units of observation is the presence or absence of a particular characteristic. As we learned in Chapter 1, it is common to assign the value 0 to absence of the characteristic and 1 to presence of the characteristic. Thus the population–the *complete* collection of data of interest–is a collection of 0s and 1s. We call this a **binary**, or two-number, population.

In the radio sales-slip problem the population had three numbers, \$6, \$12, and \$18. That was done mainly for convenience–especially in making it fairly easy to derive the probability distribution of the sample mean. The intention was to illustrate the fundamental ideas that apply to sampling from populations with *any* number of different possible values of x. Indeed, these ideas apply as well in the special case of two possible values of x, a binary population. Because the binary population is such a common special case, it deserves the special treatment that is usually accorded it in textbooks.

6.1 Binary Populations

Whenever units of observation are classified into two categories, a binary population is under investigation. In a public opinion poll respondents may be asked to *agree* or *disagree* with a stated viewpoint; in testing a new drug, physicians will classify patients into those who benefited and those who did not benefit from the treatment. In a local labor force survey, an investigator may be interested in estimating the percentage of persons in the labor force who are unemployed.

The U.S. Census Bureau and the U.S. Bureau of Labor Statistics classify the labor force into six major occupational groups. Objectives of a community survey could include determining the proportions (or percentages) of the labor force in each major occupational group. While this is a sixfold breakdown, it can also be viewed one at a time as six binary populations. The labor force classed as professional, technical, and kindred workers (one class) or not is one binary population; the labor force classed as clerical, sales, and kindred workers or not is another binary population; and so on through the six major classes. At any one time we can restrict our considerations to the event of belonging or not belonging to a particular class. Thus, study of sampling from binary populations is basic to statistical methods for dealing with classification data in general.

The prevalence of an attribute among units of observation relevant to a study is measured by the proportion of units having the attribute. If we assign the value $x = 0$ to units not possessing the attribute and $x = 1$ to units possessing the attribute, then the proportion having the attribute is the mean value of x in the population. The overall proportion is also the probability of the event $x = 1$ in repeated random draws from the population. Suppose we are considering 2000 automobiles in a community, 500 of which are equipped with catalytic converters to reduce pollutants. This is the entire population of interest. In Table 6.1 we show the frequency table for the population and calculate the population mean using Equation 3.1 from Chapter 3. We also express the population in probability terms and calculate the expected value of X from Equation 5.1 in Chapter 5.

Table 6.1 *Mean of a Binary Population Expressed in Frequency and Probability Terms*

Attribute	x	f	fx	$P(x)$	$x \cdot P(x)$
Not equipped	0	1500	0	0.75	0
Equipped	1	500	500	0.25	0.25
Total		2000	500		0.25

$$\mu = \frac{\Sigma(fx)}{N} = \frac{500}{2000} = 0.25 \qquad \mu = \Sigma[x \cdot P(x)] = 0.25$$

Here we see that the population mean is the probability of observing an equipped vehicle. The mean is given the symbol π (Greek lowercase pi) in statistical work with attributes.

Definition The **mean of a binary population** is equal to π, the probability of the attribute assigned the value 1.

■ $$\mu = \pi = P(X = 1)$$ ■ ***6.1***

Because the binary population has such a simple structure, the population standard deviation can be expressed and calculated by a special formula. In Table 6.2, the variance and standard deviation are calculated for our example population in probability terms. The right-hand side of the table shows the calculation in terms of π, up to the point of weighting squared deviations by probabilities.

Table 6.2 *Variance and Standard Deviation of a Binary Population*

x	$P(x)$	$(x-\mu)^2$	$(x-\mu)^2 \cdot P(x)$	x	$P(x)$	$(x-\mu)^2$	$(x-\mu)^2 \cdot P(x)$
0	0.75	0.0625	0.046875	0	$1-\pi$	π^2	$\pi^2(1-\pi)$
1	0.25	0.5625	0.140625	1	π	$(1-\pi)^2$	$(1-\pi)^2\pi$

$$\begin{aligned} \sigma^2 &= \sum[(x-\mu)^2 \cdot P(x)] \\ &= 0.1875 \\ \sigma &= \sqrt{0.1875} = 0.433 \end{aligned}$$

The variance is the sum of the weighted squared deviations that appear in the right-hand column of Table 6.2. A little factoring of the result in terms of π yields

$$\begin{aligned} \sigma^2 &= \pi^2(1-\pi) + (1-\pi)^2\pi \\ &= [\pi(1-\pi)][\pi + (1-\pi)] \\ &= \pi(1-\pi). \end{aligned}$$

Knowing this, we could have calculated the variance originally from

$$\sigma^2 = \pi(1-\pi) = 0.25(0.75) = 0.1875.$$

Definition The **variance of a binary population** is the product of π and $1-\pi$, where π is the probability that $X = 1$:

■ $$\sigma^2 = \pi(1-\pi).$$ ■ ***6.2***

Figure 6.1 shows three different binary populations. The first is the one we have been working with—a binary population with $\pi = 0.25$. The second is a binary population with $\pi = 0.50$, as might represent preference for one of two candidates

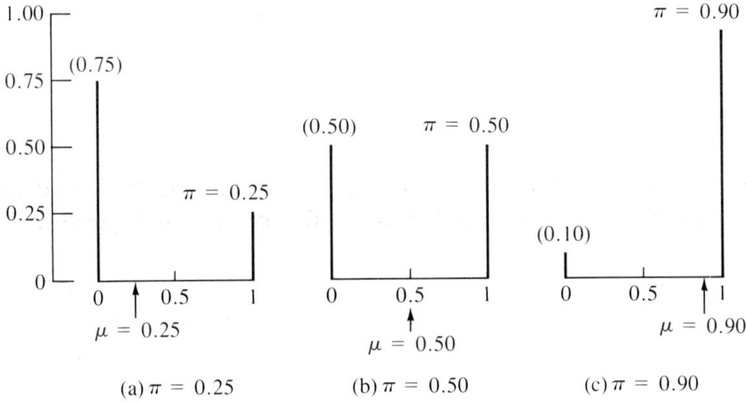

Figure 6.1 *Three binary populations*

in an evenly contested election campaign. The third population shown is one with $\pi = 0.90$, which might represent agreement among college students with the proposition that we should be more concerned about a certain environmental problem. The standard deviations of the three populations are:

$\pi = 0.25$	$\pi = 0.50$	$\pi = 0.90$
$\sigma = \sqrt{\pi(1-\pi)}$ $= \sqrt{0.25(0.75)}$ $= 0.433$	$\sigma = \sqrt{\pi(1-\pi)}$ $= \sqrt{0.50(0.50)}$ $= 0.500$	$\sigma = \sqrt{\pi(1-\pi)}$ $= \sqrt{0.90(0.10)}$ $= 0.300$

Notice that the greatest variability exists for a population evenly divided between absence and presence of an attribute and that the smallest variability is for a population that is very unevenly divided. Preference is most *varied* when 50 percent of a group prefer one candidate and 50 percent prefer the other. A group which is nearly all alike in possessing an attribute represents near uniformity, or lack of variability. **6.1**

The Binomial Sampling Distribution **6.2**

A binary population with a specified value of $\pi = P(X = 1)$ can be shown on the initial branches of a tree diagram as

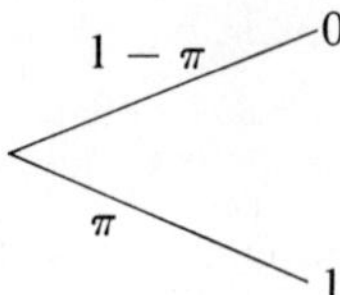

If repeated draws* are made from the binary population, the tree diagram for possible outcomes of the sequence of draws would appear as in Figure 6.2.

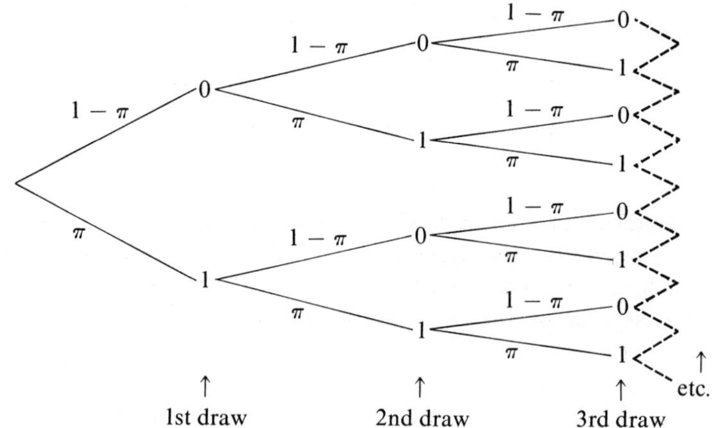

Figure 6.2 *Repeated draws from a binary population*

The Binomial Probability Distribution

We let the number of draws—that is, the size of the sample—be indicated by n. It is clear that, given a value for π, the probability of any sequence of 0s and 1s through the tree can be calculated. Suppose it is known that 25 percent of the registered voters in a town voted in a recent election for mayor. Three registered voters, selected at random, are to be asked if they voted in the election. A yes answer is coded 1 and a no answer is coded 0. Our value of π (the probability of a 1 on any selection) is $\frac{1}{4}$. The possible sequences of outcomes are shown in Figure 6.3.

Our concern is to find the probability distribution of the number (out of three voters) in the sample who voted in the mayorality election. Call this statistic r. Each route through the tree has an associated value of r, shown at the extreme right of Figure 6.3, and the probabilities for each route have been calculated by multiplying the sequence of selection probabilities lying along that route. For example, the sequence (0,1,0) leads to a value of $r = 1$, and has a probability of $\frac{9}{64}$.

*If the universe of units of observation is finite, the draws would be made with replacement.

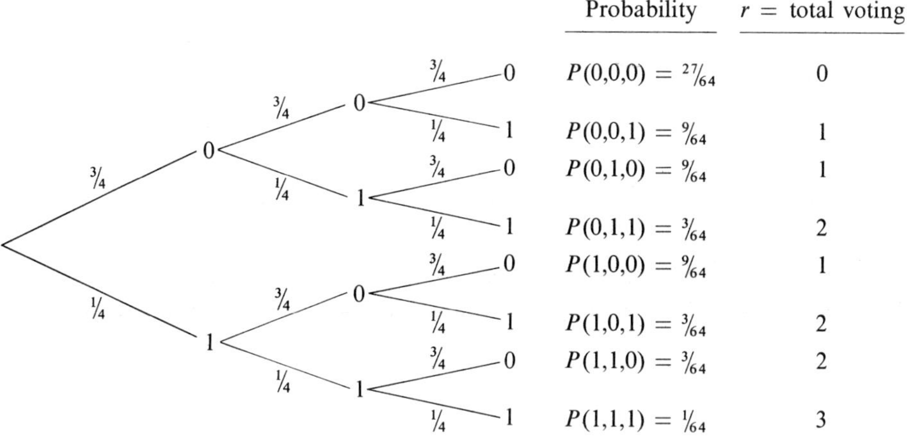

Figure 6.3 *Outcomes for a sample of $n = 3$ and $\pi = 1/4$*

We now sum the probabilities for the different sample sequences leading to zero, one, two, and three persons who voted. The result, shown in Table 6.3, is the probability distribution of number of successes in three independent draws from a binary population with $\pi = 1/4$. "Number of successes" is a general way of saying number of observational units displaying the attribute which was assigned the value of $x = 1$.

Table 6.3 *Binomial Probability Distribution for $n = 3$, $\pi = 1/4$*

Number of Successes (Persons Voting) r		Probability $P(r)$
0		$27/64$
1	$9/64 + 9/64 + 9/64 =$	$27/64$
2	$3/64 + 3/64 + 3/64 =$	$9/64$
3		$1/64$

Definition The **binomial distribution** is the probability distribution of the number of successes (r) encountered in repeated independent draws of size n from a binary population.

The binomial distribution is the sampling distribution of the *number of successes* in a random sample of size n taken from a population with a specified value of π. The statistic is r, the number of sample units displaying the attribute under study. The possible values of r are $0, 1, 2, \cdots, n$. The binomial distribution gives the probabilities of these sample events.

Notice that r, the number of successes, is the sum of the x values drawn in a sample of size n. Since each value drawn is either $x = 0$ or $x = 1$, the sum of the values drawn in a sample is also the *count* of the number of successes, or observational units for which $x = 1$.

Formula and Tables for Binomial Probabilities

A formula is desired for $P(r)$ for $r = 0, 1, 2, \cdots, n$ for any given sample size. The multiplication rule for independent events can be applied to find the probability of any given sequence of 0s and 1s. For example,

$$P(0,1,1,1,0) = (1 - \pi)(\pi)(\pi)(\pi)(1 - \pi)$$

and

$$P(1,0,1,0,1) = (\pi)(1 - \pi)(\pi)(1 - \pi)(\pi).$$

In both cases the factors can be collected to obtain $(\pi)^3(1 - \pi)^2$. In general, any sequence leading to specified r successes out of n trials will have the probability

$$\pi^r(1 - \pi)^{n-r}.$$

Since all sequences leading to a given r out of n have the same probability, we now need a rule for counting the number of equally likely sequences. For example, how many different sequences of 0s and 1s lead to three 1s out of five draws? Here we visualize the five independent draws (in order of draw) as a series of empty boxes

and ask in how many ways three 1s can occupy three of the five boxes. This is the number of combinations of five elements taken three at a time. In general we are interested in n elements (or sample positions) taken r at a time. The general rule for the number of combinations of n elements taken r at a time is

$${}_nC_r = \frac{n!}{(n - r)!r!}.$$

We are now able to write the **general binomial probability formula,**

$$\blacksquare \quad P(r) = {}_nC_r\pi^r(1 - \pi)^{n-r}. \quad \blacksquare \qquad 6.3$$

The parameters we must know before we can apply the binomial probability equation are n, the number of trials, and π, the probability of a success on a single trial. This probability must remain constant over all trials. The trials must be independent of one another.

In the sample of three registered voters from a population with $\pi = \frac{1}{4}$ for the proportion who voted in the mayoralty election, the probability of the sample yielding two who voted is

$$P\left(r = 2 \mid n = 3, \pi = \frac{1}{4}\right) = {}_3C_2\left(\frac{1}{4}\right)^2\left(\frac{3}{4}\right)^1$$

$$= \frac{3(2)(1)}{1 \cdot (2)(1)}\left(\frac{1}{4}\right)^2\left(\frac{3}{4}\right)^1$$

$$= 3\left(\frac{3}{64}\right) = \frac{9}{64}.$$

The three sample sequences that yield two who voted can be traced in Figure 6.3, and the binomial probability of $\frac{9}{64}$ for $r = 2$ appears in Table 6.3.

If six registered voters were selected, the probability of finding four who voted in the mayoralty election is, using decimal fractions,

$$P(r = 4 \mid n = 6, \pi = 0.25) = {}_6C_4(0.25)^4(0.75)^2$$

$$= \frac{6(5)(4)(3)(2)(1)}{2(1) \cdot 4(3)(2)(1)}(0.25)^4(0.75)^2$$

$$= 15(0.00219727) = 0.0330.$$

If you extended the tree diagram of Figure 6.3 to six draws, there would be ${}_6C_4 = 15$ different sample sequences or paths through the tree that would yield two successes, or 1s. Each of these mutually exclusive paths has probability
0.00219727, so the total probability for the event $r = 4$ is $15(0.00219727) = 0.0330$. 6.2

Because the binomial is such a commonly used distribution, tables for binomial
probabilities have been developed to do the work of the binomial formula. Appendix 6.3
E gives such a table. Given specific values of n and π, we read from the table any $P(r)$ desired. Let us check the values we calculated earlier for $P(r)$ for the binomial with $n = 3$ and $\pi = \frac{1}{4}$. The headings of the table give π in decimal proportions, and we are concerned with $\pi = 0.25$. Follow the column down to the row where $n = 3$. There we read

		π
n	r	0.25
⋮	⋮	⋮
3	0	0.4219
	1	0.4219
	2	0.1406
	3	0.0156

These values of $P(r)$ are the decimal equivalents of the fractional probabilities calculated earlier. For example, $P(r = 2) = 9/64 = 0.1406$. A fuller notation for the probability 0.1406 above is

$$P(r = 2 \mid n = 3, \pi = 0.25) = 0.1406.$$

This reads: "The probability of $r = 2$ given $n = 3$ and $\pi = 0.25$ is 0.1406."

If you want probabilities such as the probability that 5 or more chickens among 10 chickens randomly selected from a large flock will be found to have a disease when 15 percent of the entire flock have the disease, you will have to add up the relevant probabilities in Appendix E. For the probability $P(r \geq 5 \mid n = 10, \pi = 0.15)$, you need to add the tabled probabilities for $r = 5$, $r = 6$, $r = 7$, $r = 8$, $r = 9$, and $r = 10$ under $n = 10$ and $\pi = 0.15$. These individual terms are

$$0.0085 + 0.0012 + 0.0001 + 0.0000 + 0.0000 + 0.0000$$

6.3 and their sum is 0.0098, the probability desired.

Summary

Binary populations are a special class of population in which the possible values of a variable are restricted to 0 and 1. They occur in the study of nominal, or attribute, data. The mean and variance of a binary population can be expressed in terms of π, the probability that $X = 1$. The mean is equal to π and the variance is equal to $\pi(1 - \pi)$.

The binomial probability distribution gives the probabilities of $0, 1, 2, \cdots, n$ observations having the attribute under study in a sample of size n from any specified binary population. Thus the binomial is the sampling distribution of the number of successes (r) in a sample of n observations from a population with a specified proportion of successes.

Although you can use a formula to calculate $P(r)$ for any specified number of successes, tables of binomial probabilities are commonly available.

See Self-Correcting Exercises 6A.

Exercise Set A

1. When a fair die is rolled once, what is the probability of a success if we define success as the number turning up

(a) An even number?
(b) A 3?
(c) An odd number greater than 2?
(d) A number in excess of 1?
(e) An even number that exceeds 6?
(f) A number from 2 to 5, inclusive?

2. $P(r = 7 \mid n = 9, \pi = 0.35) = 0.0098$ tells us that the probability of what event is 0.0098? Express this probability in the form of Equation 6.3.

3. Given that $n = 7$ and $\pi = 0.23$, express each of the following probabilities in terms of Equation 6.3. Do not multiply: leave the terms of your answers in factor form.
   (a) $P(r = 0)$
   (b) $P(r > 2)$
   (c) P(number of successes does not exceed 1)
   (d) P(number of failures is 4)
   (e) P(number of failures is less than 5)
   (f) P(number of failures is not less than 5)

4. What is wrong with each of the following as a binomial probability statement? Correct each error that occurs; there may be more than one in a statement.
   (a) $P(r = 7 \mid n = 11, \pi = 0.2) = {}_{11}P_7(0.2)^7(0.8)^{11}$
   (b) $P(r = 2 \mid n = 1, \pi = 0.85) = {}_1C_2(0.85)^2(0.15)^{1-2}$
   (c) $P(r = 4 \mid n = 6, \pi = 0.71) = {}_6C_4(0.71)^6(0.39)^2$
   (d) $P(r = 12 \mid n = 12, \pi = 0.9) = {}_{12}C_{12}(0.9)^{12}(0.1)^0$

5. Use a tree diagram to find the binomial probabilities for $n = 3$, $\pi = 0.20$. Carry your work out in decimals and then check your results with Appendix E.

6. Use the binomial probability formula (Equation 6.3) to find the following probabilities. Check your answers by finding the same probabilities in Appendix E.
   (a) $P(r = 2 \mid n = 3, \pi = 0.60)$
   (b) $P(r = 0 \mid n = 2, \pi = 0.20)$
   (c) $P(r = 4 \mid n = 4, \pi = 0.90)$

7. Find the following binomial probabilities with the aid of Appendix E:
   (a) $P(r \leq 3 \mid n = 8, \pi = 0.20)$
   (b) $P(r > 8 \mid n = 10, \pi = 0.70)$
   (c) $P(r > 2 \mid n = 12, \pi = 0.40)$

8. Why would it be difficult to find the probabilities referred to in Exercise 7 without the aid of Appendix E? Why would Appendix E be of little use to us if we had to compute the probability referred to in Exercise 4(c)?

9. Evaluate ${}_3C_r(0.21)^r(0.79)^{3-r}$ for r equal to
   (a) 0
   (b) 1
   (c) 2
   (d) 3

10. Is $P(r = 10 \mid n = 16, \pi = 0.15)$ equal to zero exactly? Justify your conclusion.

11. Assuming independence and that the probability of the birth of a boy is ½ and that of a girl is ½, determine to which of the following parts, (a) or (b), Equation 6.3 is applicable. Then find the correct answer to each part.
    (a) What is the probability that, in a randomly selected four-child family, the second child is a girl and the other children are boys?

(b) What is the probability that, in a randomly selected four-child family, there are three boys and one girl?

12. A salesman for a computer manufacturer has a history of making successful sales calls one-third of the time. Assuming independence, what is the probability that he will be successful on the first two and fail on the last two of the next four calls?
13. What is the probability that the salesman in Exercise 12 will be successful on two of his next four calls?
14. Parts coming off an assembly line have a probability of 0.15 of being defective. What is the probability that at least 3 of 11 parts randomly selected are defective?

Applications of the Binomial

In finding binomial probabilities, you should have noticed a feature common to our earlier work with sampling distributions. This is the *deductive* nature of the probability statements. We say: *If the population proportion is* π, then the probability of r successes in a sample of n is such and such. The probability for any r out of n is *conditional* on a specified value of π. Our reasoning runs from population to sample. This tends to frustrate people who are anxious to get on to "practical" applications. They ask how a method that seems to require that the user "know" the answer can help anyone who doesn't have the answer to begin with. (It obviously couldn't help anyone who really knew the answer to begin with.) In this section we take up some applications of binomial probabilities that show how deductive probabilities can be useful in decision making.

Our first example is a case of disputed authorship of certain of the *Federalist* papers. Eighty-five papers known to have been written by Alexander Hamilton, John Jay, and James Madison contain important information on the intentions and philosophies of the framers of the Constitution of the United States. Historians agree on the authorship of 70 of the 85 papers: 5 are attributed to Jay, 14 to Madison, and 51 to Hamilton. Moreover, 12 of the remaining 15 are known not to have been authored by Jay. Two statisticians, Frederick Mosteller and David L. Wallace, undertook an analysis of word frequencies in an effort to settle the question whether each of these 12 was written by Hamilton or Madison.* The study was an extensive one, and our example just illustrates their approach. They examined known writings of Hamilton and Madison (94,000 words from Hamilton and 114,000 from Madison) to find their frequency of use of such words as *although, enough, by, also, upon.* Authors differ in their tendencies to use particular "filler" or noncontextual words. For example, the probability that *upon* occurs in 200 words of Hamilton's writing

*See Frederick Mosteller and David L. Wallace, "Deciding Authorship," in *Statistics: A Guide to the Unknown,* edited by Judith M. Tanur (San Francisco: Holden-Day, 1972), pp. 164–175. This book contains examples of statistical work in a variety of fields and is written for beginning students interested in exploring the scope of statistics.

is close to 0.45, and the probability of *upon* appearing in 200 words written by Madison is close to 0.05. These probabilities came from the extensive word study. *Federalist* papers 54 and 55, of disputed authorship, each contain about 2000 words. Thus each contains 10 passages of 200 words each. In paper 54, none of the 10 passages contains an *upon;* in paper 55, 2 of the 10 passages contain an *upon.* So we have the following sample evidence:

$$\text{Paper 54:} \quad n = 10,\ r = 0$$

$$\text{Paper 55:} \quad n = 10,\ r = 2$$

We can ask: What is the probability of zero out of 10 passages from the writing of Hamilton containing an *upon?* The answer is

$$P(r = 0 \mid n = 10, \pi = 0.45) = 0.0025.$$

The probability of zero out of 10 passages from the writing of Madison containing an *upon* is

$$P(r = 0 \mid n = 10, \pi = 0.05) = 0.5987.$$

The evidence in paper 54 strongly supports the view that Madison was the author rather than Hamilton. The absence of *upon* in 10 passages is uncharacteristic of Hamilton but very common in Madison's writing.

On the evidence of $r = 2$ and $n = 10$, paper 55 is a different story. The probability of the sample outcome if the passages were drawn from the writing of Hamilton is

$$P(r = 2 \mid n = 10, \pi = 0.45) = 0.0763,$$

and the probability that 2 out of 10 passages from Madison's writing contain an *upon* is

$$P(r = 2 \mid n = 10, \pi = 0.05) = 0.0746.$$

The sample result of paper 55 would occur with about the same relative frequency in samples of 10 segments from the writings of either author.

The situation for the two papers is shown in Figure 6.4. The result for paper 54 ($r = 0$) is a most typical result for a Madison paper and a very unusual one for a Hamilton paper. The result for paper 55 ($r = 2$) is not especially typical for either author–more segments containing *upon* than is usual for Madison and fewer containing *upon* than is usual for Hamilton. However, the probabilities tell us that two segments with *upon* is a *plausible* result from either author.

Many more words than the one illustrated here were analyzed, and the investigators concluded that Madison was the author of all 12 papers in question. The odds favored Madison overwhelmingly in 11 cases and by a substantial margin in the twelfth (paper 55).

Our second application illustrates the use of the binomial distribution in the inspection of incoming products to assure a level of quality. It is a common industrial

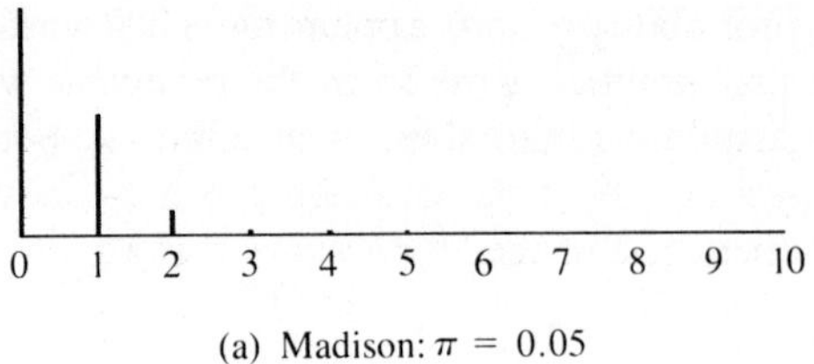

(a) Madison: $\pi = 0.05$

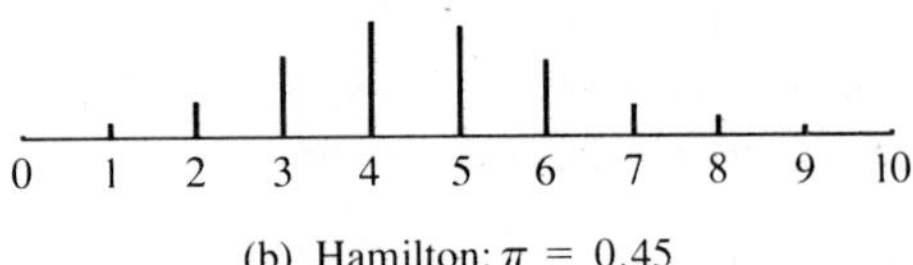

(b) Hamilton: $\pi = 0.45$

Figure 6.4 *Binomial probabilities of r segments out of 10 containing the word* upon *in writings of Madison and Hamilton*

application. Suppose shipments (often called lots) of 1000 small fittings used in assembling a product are received periodically by a manufacturer. A sample of 10 fittings from each shipment is examined. If none of the 10 fittings is defective, the lot is *accepted* as is and the fittings are used in the assembly operation. If *one or more* of the 10 are found defective, the entire lot is inspected and all defective fittings discarded. What can be said about the percentage of defective fittings finally used in the product assembly?

Take, as an example, a shipment that has 50 defective fittings, that is, $\pi = 0.05$. The probability of accepting the lot and moving the 1000 fittings to the production line is $P(r = 0 \mid n = 10, \pi = 0.05) = 0.5987$. The probability of not accepting the lot as is, but proceeding to examine the entire lot, is $P(r \geq 1 \mid n = 10, \pi = 0.05) = 1 - 0.5987 = 0.4013$. If the lot is examined, the 50 defective fittings will be found and removed. There will be 950 good fittings eventually sent to the production line. Thus we have

0.5987 probability of 50 defective and 950 good fittings

0.4013 probability of 0 defective and 950 good fittings

Using Equation 5.1, the expected number of fittings sent to the production line is $0.5987(50 + 950) + 0.4013(0 + 950) = 979.9$, and the expected number of defective fittings sent on from the original lot is $0.5987(50) + 0.4013(0) = 29.9$. The expected percentage of defective fittings used in product assembly is $29.9/979.9$, or 3.05 percent. Similar calculations can be made for shipments with other proportions defective. They are shown in Table 6.4, using the π values found in Appendix E.

What we find in Table 6.4 is that the inspection scheme keeps the average (or expected) outgoing percentage of defective fittings below 4.0 percent—no matter what the lot proportion defective is in incoming lots. When incoming lot quality is poor,

Table 6.4 *Calculation of Average Outgoing Quality (AOQ) for Incoming Lots of 1000 Fittings*

Lot Proportion Defective	Probability of Accepting Lot	Probability of Inspecting Lot	Expected Outgoing Number Defective	Expected Outgoing Number Good	Expected Outgoing Number Total	AOQ % Defective
0.05	0.5987	0.4013	29.9	950	979.9	3.05
0.10	0.3487	0.6513	34.9	900	934.9	3.73
0.15	0.1961	0.8039	29.4	850	879.4	3.34
0.20	0.1074	0.8926	21.5	800	821.5	2.62
0.25	0.0563	0.9437	14.1	750	764.1	1.85
0.30	0.0282	0.9718	8.5	700	708.5	1.20
0.35	0.0135	0.9865	4.7	650	654.7	0.72
0.40	0.0060	0.9940	2.4	600	602.4	0.40

good outgoing quality is achieved by high probabilities of going to 100 percent inspection to remove defectives. When incoming lot quality is good, there is a good chance that the time-consuming and costly process of 100 percent inspection will be avoided.

Of course, quality control savings should be balanced against costs of product failures at the company level or beyond that, at the social level. The binomial distribution has served here to describe the long-run performance of a specified sample inspection plan under a variety of possible conditions.* This would form a necessary input to the cost considerations mentioned. The performance characteristics of a variety of inspection plans are published in manuals used by quality control engineers.

In both the *Federalist* papers and the quality control examples, we did not raise questions about the sample selection process. Perhaps we should do that now. In the *Federalist* papers case, the sample was predetermined—10 passages of 200 words constituting a paper of disputed authorship. The assumption of independent draws needed for proper use of the binomial amounts to a question whether the probability of use of the word *upon* really stays the same over successive passages of 200 words each. If the probability of *upon* occurring in a passage is reduced when *upon* occurs in the preceding passage, independence is violated. Connective or "filler" words would be expected to occur independent of their prior use in passages as long as 200 words and would appear also to occur independent of the subject of passages.

In sampling the fittings in a shipment, it is unlikely that a quality control inspector would number each of 1000 small fittings in order to select 10 in a manner similar to that described in connection with the sales slips in Chapter 5. She might just reach into the shipping carton and select one from here and there until she had 10 fittings to test. She might even take 10 off the top! She would be relying on the stability of the supplier's production and shipping operation to produce an arrangement of the fittings in which order or position in the carton made no difference. This is often true

*Our example covers a range of lot proportion defective that we hope an industrial buyer would not encounter! This was done to fit the example to the scope of Appendix E. In a typical industrial application, the sample size would be larger and the incoming and outgoing percentage defectives generally smaller.

for production situations. However, if there is a greater chance that fittings at the bottom of the box would be damaged in shipment, or if she suspected that a supplier might "salt" the top of the shipping carton with all good fittings, the inspector would surely not sample off the top.

6.4 Mean and Variance of a Binomial Distribution

In the *Federalist* papers example we looked at two binomial distributions. They were graphed in Figure 6.4. We did not need their means and standard deviations in the discussion there, and so we did not stop to get them. However, the mean and variance of a binomial distribution are important in Chapter 10, where we take up statistical inference for classification data, and there is a feature of these measures that should be pointed out now.

The binomial probability distribution for $n = 10$ and $\pi = 0.05$ is listed in Table 6.5 and the mean and variance of the random variable r are calculated. The calculation in Table 6.5 is not really necessary. The mean and variance of a binomial distribution of r can be obtained from

Table 6.5 *Calculation of Mean and Variance of r for Binomial Distribution with n = 10 and π = 0.05*

r	$P(r)$	$r \cdot P(r)$	$(r - \mu_r)^2$	$(r - \mu_r)^2 \cdot P(r)$
0	0.5987	0	0.25	0.1497
1	0.3151	0.3151	0.25	0.0788
2	0.0746	0.1492	2.25	0.1678
3	0.0105	0.0315	6.25	0.0656
4	0.0010	0.0040	12.25	0.0122
5	0.0001	0.0005	20.25	0.0020
⋮	⋮	⋮	⋮	⋮
10				
		$\mu_r = 0.500$		$\sigma_r^2 = 0.476$

$$\mu_r = n\pi \qquad (6.4)$$

and

$$\sigma_r^2 = n\pi(1 - \pi). \qquad (6.5)$$

Remember that π is the mean of the underlying binary population and $\pi(1-\pi)$ is the variance of the underlying binary population. The mean and variance of the number of successes in a sample of n independent observations are just n times the mean and variance of the underlying binary population. We review the formulas by showing the calculations for the binary population with $\pi = 0.05$ and the related binomial with $n = 10$:

Binary population: $P(X = 1) = 0.05$

Mean	$\pi = P(X = 1) = 0.05$
Variance	$\sigma^2 = \pi(1-\pi) = 0.05(0.95) = 0.0475$
Standard deviation	$\sigma = \sqrt{\pi(1-\pi)} = \sqrt{0.0475} = 0.218$

Binomial distribution of r with $n = 10$ and $\pi = 0.05$

Mean	$\mu_r = n\pi = 10(0.05) = 0.50$
Variance	$\sigma_r^2 = n\pi(1-\pi) = 10(0.05)(0.95) = 0.475$
Standard deviation	$\sigma_r = \sqrt{n\pi(1-\pi)} = \sqrt{0.475} = 0.689$

The mean of the binomial distribution above implies that the average number of successes out of 10 in repeated samples of that size from a population with 5 percent successes is 0.5 successes per sample. These figures apply to the Madison case in Figure 6.4(a). Notice where 0.5 lies on the variable (horizontal) axis for the Madison case. Also note that we just found the standard deviation of this binomial to be 0.689. For the Hamilton case ($\pi = 0.45$), the mean of the binomial for $n = 10$ is $10(0.45) = 4.5$. This is the expected number of passages containing *upon* in samples of 10 passages from a population in which 45 percent of the passages contain *upon.* The variance of the binomial with $n = 10$ and $\pi = 0.45$ is $n\pi(1-\pi) = 10(0.45)\cdot(0.55) = 2.475$. This leads to a standard deviation of $\sqrt{2.475} = 1.573$. A feeling for this magnitude compared to 0.689 for the standard deviation of the binomial with $n = 10$ and $\pi = 0.05$ can be gained by looking at the spread of the two distributions in Figure 6.4.

Sampling Distribution of the Proportion

Should we want to present the information of Table 6.5 in terms of the *proportion* of passages containing the word *upon,* all that would be required is that we divide the number of passages containing *upon* (r) by the sample size of ten passages (n). The probabilities are unchanged. The result is the probability distribution of the sample proportion–also called the **sampling distribution of the proportion**. The letter p is used for sample proportion.

Table 6.6 Sampling Distribution of the Proportion for $n = 10$ and $\pi = 0.05$

$p = {}^r/_n$	0.0	0.1	0.2	0.3	0.4	0.5	. . .	1.0
$P(p)$	0.5987	0.3151	0.0746	0.0105	0.0010	0.0001	. . .	0.0000

The mean and variance of the sampling distribution of the proportion are related to those of the underlying binary population by

$$\blacksquare \quad \mu_p = \pi \quad \blacksquare \qquad \textbf{6.6}$$

and

$$\blacksquare \quad \sigma_p^2 = \frac{\pi(1-\pi)}{n}. \quad \blacksquare \qquad \textbf{6.7}$$

The mean and variance of the sampling distribution of the proportion in Table 6.6 are

$$\mu_p = \pi = 0.05$$

and

$$\sigma_p^2 = \frac{\pi(1-\pi)}{n} = \frac{0.05(0.95)}{10} = 0.00475.$$

The positive square root of σ_p^2 is called the standard error of the proportion.

$$\blacksquare \quad \sigma_p = \sqrt{\frac{\pi(1-\pi)}{n}} \quad \blacksquare \qquad \textbf{6.8}$$

In our example,

$$\sigma_p = \sqrt{\frac{0.05(0.95)}{10}} = 0.0689.$$

The relations between means and standard deviations of the binomial distribution and the sampling distribution of the proportion can be seen below. The binomial expectation of 0.5 passages out of ten becomes a proportion of 0.05, or 5 percent (of 10 passages) containing the word *upon*. The binomial standard deviation of 0.689 passages becomes 0.0689. The proportion 0.0689 is the same as 0.689 out of 10 (the binomial variable), or 6.89 percentage points.

	Binary Population	Binomial Distribution	Sampling Distribution of the Proportion
Variable	X	$r = \sum X$	$p = r/n$
Mean	$\pi = 0.05$	$n\pi = 0.5$	$\pi = 0.05$
Standard deviation	$\sqrt{\pi(1-\pi)} = 0.218$	$\sqrt{n\pi(1-\pi)} = 0.689$	$\sqrt{\frac{\pi(1-\pi)}{n}} = 0.0689$

6.4

Summary

Two examples of applications of the binomial distribution have been presented. The objective of the examples has been to suggest how deductive probabilities can be turned to a practical purpose. The *Federalist* papers example showed that sample evidence can support one of two contending views (about authorship) because it is more consistent with one contention (Madison) than the other (Hamilton). The acceptance sampling example showed how the performance of a decision rule based on sample outcomes can be traced out by considering the consequences of applying the rule over a range of possible values for the population parameter. The binomial was used to gain answers to "if–then" probability questions that were appropriate to each case. We will find this use of sampling distributions an important feature of a method of inference called hypothesis testing, which is introduced in Chapter 8.

See Self-Correcting Exercises 6B.

Exercise Set B

1. Referring to the quality control of fittings example and Table 6.4, verify, showing all your computations, that if the proportion of defectives in the lot is 0.25, then the percentage of defective fittings leaving the plant is 1.85.
2. Compute the correct value for each entry in a line that would be entered in Table 6.4 for a lot in which the proportion of defectives is 0.50.
3. Referring to Table 6.4 we find that the greater the proportion of defectives in the lot, the lower the percentage of defective fittings leaving the plant. Why is this so?
4. An anonymous 13-page typewritten report alleging misconduct on the part of a department manager is sent to the president of the firm. The president questions the manager about the charges made in this report, which by its content reveals that it was almost certainly submitted by one of the manager's three secretaries. The manager vows to determine the author of the report and to challenge the writer, with the president present, to prove the allegations. For a recent efficiency study, the typing efficiency of the firm's secretaries was tested on a lengthy report. Each page of each secretary's report was rated acceptable if there were two or fewer errors on that page and unacceptable otherwise. The manager's secretaries scored as follows on this test:

Name	Proportion of Unacceptable Pages
Ms. Fitch	0.15
Mr. Joubert	0.20
Mr. Sanchez	0.10

(a) If in the report alleging managerial misconduct 2 of the 13 pages are found to contain more than two errors, what is the probability that it was typed by (i) Ms. Fitch? (ii) Mr. Joubert? (iii) Mr. Sanchez?

(b) From your results in part (a), if the manager were to base his decision on the probable authorship entirely on this criterion, whom would he most likely confront? Would his decision be clear-cut, or are two of the secretaries almost equally likely to be the author?

5. With reference to Exercise 4, whom would the manager confront as a likely writer of the report under each of the following circumstances? Would his decision be clear-cut in each case?
   (a) Three of the 13 pages contain two or more errors.
   (b) The report has only 12 full pages, 2 of which contain two or more errors. The thirteenth page has only a few lines of type on it and is ignored by the manager in his hunt for clues.

6. Find the mean and standard deviation of the following binomial distributions:
   (a) $\pi = 0.20, n = 30$
   (b) $\pi = 0.08, n = 100$
   (c) $\pi = 0.50, n = 2$
   (d) $\pi = 0.95, n = 14$

7. For $n = 16$, determine the mean and standard deviation of the binomial distribution having
   (a) $\pi = 0.10$
   (b) $\pi = 0.30$
   (c) $\pi = 0.50$
   (d) $\pi = 0.70$
   (e) $\pi = 0.90$

8. In Exercise 7, what happens to the mean as π increases? What happens to the standard deviation?

9. For $\pi = 0.40$, determine the mean and standard deviation of the binomial distribution for
   (a) $n = 1$
   (b) $n = 9$
   (c) $n = 49$
   (d) $n = 100$

10. In Exercise 9, how are the mean and standard deviation affected as n increases?

11. Twenty percent of the consumers in an urban market *prefer* Low-Cal softdrink, while 50 percent of all consumers have *tried* Low-Cal. Find the mean and variance of each of the binary populations mentioned.

12. Four consumers are to be selected at random in the situation of Exercise 11. Find the mean and variance of the probability distribution of the number of consumers who *prefer* the product, and the mean and variance of the probability distribution of the number who have *tried* the product.

13. Find the mean and variance of the sampling distribution of the *proportion* of consumers (a) who prefer and (b) who have tried the product in the situation of Exercise 12.

Overview

1. For a binomial distribution with $\pi = 0.02$ and $n = 2$, determine exactly
   (a) $P(r = 0)$
   (b) $P(r = 1)$

(c) $P(r = 2)$
(d) $P(r = 3)$
(e) $P(r \geq 2)$
(f) $P(r \leq 2)$
(g) $P(r \neq 0)$
(h) $P(r$ does not exceed 1)

2. Find the probability that, in four rolls of a fair die,
   (a) No 6s turn up.
   (b) Exactly two 6s turn up.
   (c) At least two 6s turn up.
   (d) Five 5s turn up.
   (e) Only even numbers turn up.
   (f) No 1s or 2s turn up.
   (g) Only 1s and 2s turn up.
   (h) The sum of the numbers that turn up is 24.

3. A student of French, unable to attend class on the day of a promised true-false quiz, prevails upon her friend to attend the class and take the quiz for her. The friend knows no French, but he learns readily the French words for "true" and "false." Thus armed, he figures his probability of answering any one question correctly is about 0.50. If there are 10 questions on the quiz and each one is worth 10 points, what is the probability that he
   (a) Gets fewer than half right?
   (b) Passes with a grade of 60 or better?
   (c) Scores 100?

4. A student prepares for an exam by studying a set of 10 problems. He can work 7 of the 10. If the professor chooses randomly 5 of the 10 problems for the exam, what is the probability that the student can work at least 4 of them? (The binomial is not applicable. Why?)

5. In a random sample of 15 students, 10 are found to favor a certain candidate for public office. What is the probability of this sample outcome if
   (a) 65 percent of all students favor the candidate?
   (b) 45 percent of all students favor the candidate?

6. As part of quality control efforts at a gasket factory, 12 gaskets are randomly selected from each day's production for thorough inspection and testing. If no more than 1 of the 12 is found to be defective, the production lot is approved for packaging; otherwise, the entire lot is inspected. Find the probability of the rejection of a day's production lot if, in actuality, unknown to the quality control team, the proportion of defectives in that lot is (a) 0.05; (b) 0.25; (c) 0.45.

7. An acceptance sampling plan calls for lots to be acceptable if a random sample of 10 items yields 2 or fewer defectives. What is the probability of
   (a) Rejecting a lot that has 5 percent defectives?
   (b) Accepting a lot that has 40 percent defectives?

8. If the sample size in Exercise 7 is increased to 15, is there a maximum acceptance number that will result in decreasing the probabilities in (a) and (b)? If so, what is it?

9. Find the mean and standard deviation of the binomial distribution with
   (a) $n = 100$ and $\pi = 0.27$
   (b) $n = 27$ and $\pi = 0.81$
   (c) $n = 16$ and $\pi = 0.49$

10. What is the most likely number of successes in 16 independent attempts if the probability of a success on any one trial is (a) 0.95? (b) 0.90? (c) 0.50?

11. Using Appendix E, determine each of the following probabilities:
    (a) $P(r \geq 15 \mid n = 16, \pi = 0.25)$

(b) $P(r \leq 3 \mid n = 14, \pi = 0.85)$
(c) $P(r = 7 \mid n = 15, \pi = 0.90)$

12. What does your result for part (a) of Exercise 11 tell you about the probability of 15 or more successes in 16 independent trials if the probability of a success on any one trial is ¼?

13. Should obtaining 15 heads and 1 tail on 16 flips of a coin make you doubt the fairness of the coin? Why?

Glossary of Equations

6.1 $\mu = \pi = P(X = 1)$ — The mean of a binary {0,1} population is the probability assigned to the value $x = 1$.

6.2 $\sigma^2 = \pi(1 - \pi)$ — The variance of a binary {0,1} population is the product of the mean and its complement.

6.3 $P(r) = {}_nC_r\pi^r(1 - \pi)^{n-r}$ — The binomial probability formula for r successes in n independent trials from a binary population with probability of success equal to π on any single draw.

6.4 $\mu_r = n\pi$ — The mean (or expected value) of the binomial sampling distribution of r successes in a sample of n independent draws from a binary population is n times the mean of the underlying binary population.

6.5 $\sigma_r^2 = n\pi(1 - \pi)$ — The variance of the binomial sampling distribution of r is n times the variance of the underlying binary population.

6.6 $\mu_p = \pi$ — The mean of the sampling distribution of the proportion is equal to the mean of the underlying binary population.

6.7 $\sigma_p^2 = \dfrac{\pi(1 - \pi)}{n}$ — The variance of the sampling distribution of the proportion is equal to the variance of the underlying binary population divided by n, the sample size.

6.8 $\sigma_p = \sqrt{\dfrac{\pi(1 - \pi)}{n}}$ — The standard deviation of the sampling distribution of the proportion is equal to the square root of the variance of the sampling distribution of the proportion.

7

Continuous Variables and the Normal Distribution

This chapter introduces continuous probability models and a particular model, the normal distribution, that is very important in statistical work. The first half of the chapter develops the measure of probability that is appropriate to continuous variables—the area measure. Following this, we introduce the normal distribution and explain how the table of areas of the standard normal distribution is used in solving probability problems. The second half of the chapter deals with normal probability distributions of sample statistics. The essential concept of a sampling distribution developed in earlier chapters dealing with discrete variables carries over into the discussion of continuous variables. With the discussion of the normal distribution as a sampling distribution, we will have completed our presentation of probability as a foundation for statistical inference.

Continuous Random Variables 7.1

In the previous two chapters we treated probability as a function of values of a random variable. However, we considered only discrete random variables. Their values were capable of being listed and their frequencies of occurrence could be counted. For example, consider a draw from a card deck consisting of the 13 spades. The sample space is $\{A,2,3,4,\cdots,9,10,J,Q,K\}$. In many card games an ace has the value 1 and a jack, queen, or king counts as 10. If the elements in the sample space are equally likely, the probability distribution of the random variable X, where X is the customary point value associated with the cards, is

x	1	2	3	4	5	6	7	8	9	10
$P(x)$	$1/13$	$1/13$	$1/13$	$1/13$	$1/13$	$1/13$	$1/13$	$1/13$	$1/13$	$4/13$

Consider now a slightly different kind of problem. A finely balanced spinner is mounted in the center of a board. Around the board is a scale of numbers from 1 to 12 arranged like the hours on a clock. Finer decimal divisions are indicated as well as the exact "hours." If the spinner is spun, what is the probability that it will come to rest between 2.0 and 4.0?

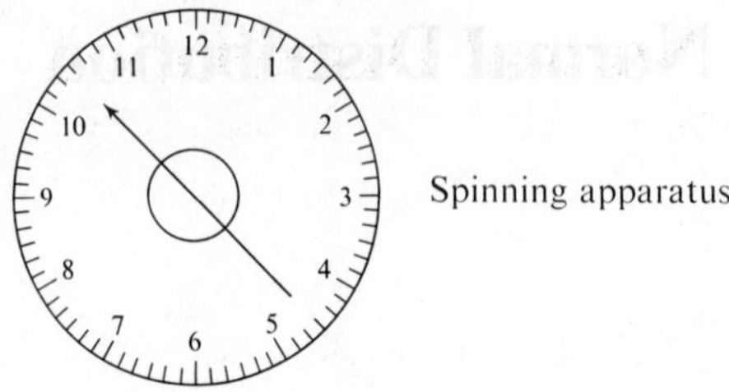

Most students answer $2/12$ with little hesitation. Without counting elements in the sample space, they assume that the spinner is as likely to come to rest at any one point as another and that the interval between 2.0 and 4.0 contains two-twelfths of the total points in the sample space. The points are, of course, not countable, because we are dealing with a continuous random variable. The spinner may come to rest anywhere along the continuum from 0.0 to 12.0. The possible values of the variable are restricted only by the practical limitations of our measurement apparatus.

Figure 7.1 graphs the discrete probability distribution of card values and the continuous probability distribution of pointer values. For the discrete card values, ordinates are drawn up to a vertical height corresponding to the probability of each

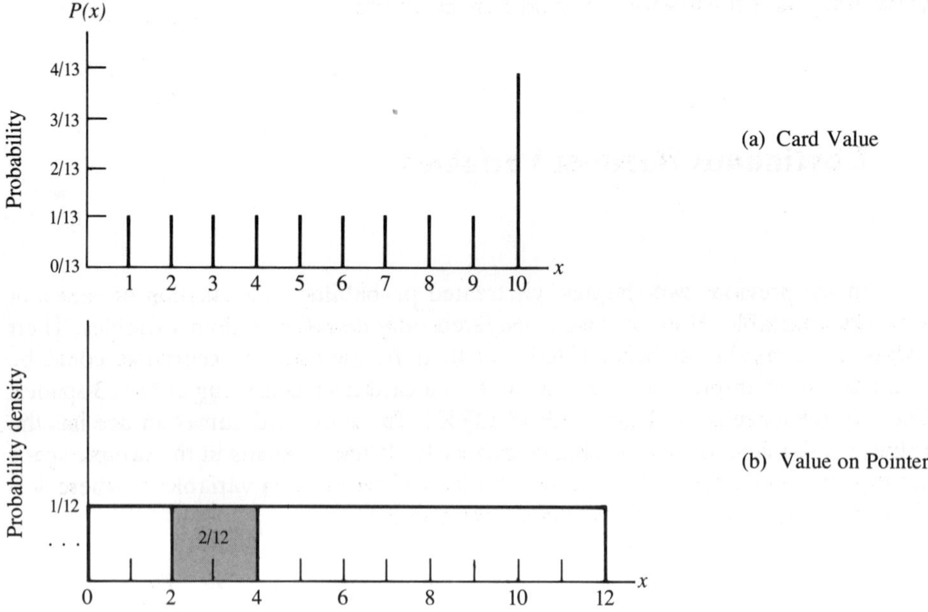

Figure 7.1 *Discrete and continuous probability distributions*

value. The sum of all these vertical ordinates equals 1.0, the total probability for the sample space. In the pointer value representation, we cannot draw vertical ordinates at discrete x values whose heights total 1.0, because probability must be distributed over the *continuum* of values of the random variable. The appropriate representation is a figure whose total enclosed area is 1.0. Now the probability associated with any *interval* of the continuous variable is the *proportion* of the area in the figure between the endpoints of the interval. The rectangular shape of the figure for pointer values reflects the equally likely assumption about the pointer coming to rest within any interval of equal width. The shaded area represents the probability that the pointer comes to rest between 2.0 and 4.0. This area is $\frac{2}{12}$, or 0.1667 of the total area of the rectangle.

The base of the rectangle in Figure 7.1(b) is the range of pointer values, $12 - 0 = 12$. For the area under the rectangle to total 1.0, the height of the rectangle must be $\frac{1}{12}$. Then the total area (base times height) will be $12(\frac{1}{12}) = 1.0$. The height (at any value of x) of the figure representing a continuous probability distribution is called the **probability density**. The figure itself shows the relation between probability density and values of the random variable, and is called the graph of the **probability density function**.

In the spinner example, the rectangular probability density function was arrived at directly through an equally likely sample points assumption. However, we can imagine an empirical process that would approach the same result. We can imagine repeated trials of spinning the spinner and being able to read the point at which the spinner comes to rest with more and more accuracy. If we were able to read and record results to the nearest tenth of a point, the sample space would have $12 \times 10 = 120$ points. With a more accurate gauge we can imagine reading points to the nearest hundredth, and the sample space would contain $12 \times 100 = 1200$ points. In each of these cases, the probability distribution over the sample points would be approximated by the relative frequencies of outcomes of repeated trials. The continuous probability density function is a model of the relative frequency distribution that would result when the resolution of the measurements gets finer and finer and the number of trials gets larger and larger. The continuous model represents the limit of that process.

An actual measurement apparatus is, of course, not infinitely accurate. Nevertheless, the continuous model may be useful on one of two counts. First, the possible outcomes may be so numerous that the continuous model provides a reasonable representation. Second, the continuous model is much more convenient than the discrete alternative and is applied in the same manner regardless of how fine the actual gradations in measurements.

Cumulative Distribution Function

Figure 7.2(a) presents a probability density function which might describe the processing times for a certain class of jobs submitted to a computer. The distribution

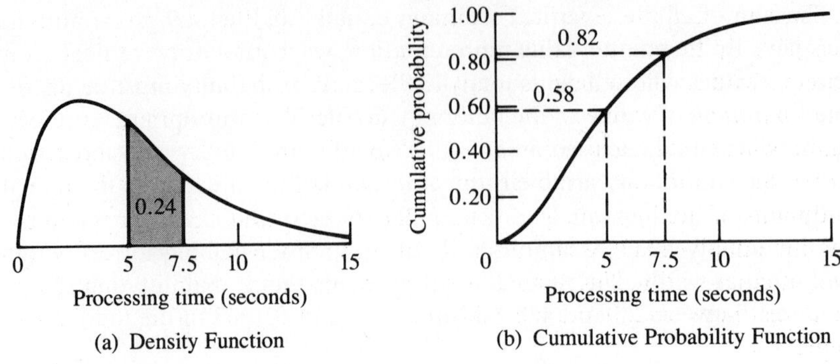

Figure 7.2 Probability density and cumulative probability functions

has considerable positive skewness. The maximum probability density appears to occur around 3 seconds, which would be called the modal, or most typical, processing time. Some jobs take 10 and more seconds, and the mean of the distribution is actually at 5 seconds.

Suppose we wanted to find the probability that a randomly selected job would be found to take between 5.0 and 7.5 seconds to process. The cumulative distribution of processing times is shown in Figure 7.2(b). This is the continuous equivalent of the relative frequency ogive from Chapter 2. The rapid early rise in the cumulative curve reflects the high frequency densities at low to moderate processing times, and the flattening out of the cumulative curve at high processing times reflects the long right tail of the density function. If we had an equation for the cumulative distribution, we could find the height of the curve at $x = 7.5$ seconds. This would give us the probability of a processing time of 7.5 seconds or less. Then the height of the cumulative curve at 5.0 seconds would give the probability of a processing time of 5.0 seconds or less. Subtracting the cumulative probability for 5.0 seconds from the cumulative probability for 7.5 seconds would give us the probability of a processing time between 5.0 and 7.5 seconds. From the graph we can approximate these probabilities:

$$\begin{array}{r} P(X \leq 7.5) \approx 0.82 \\ P(X \leq 5.0) \approx 0.58 \\ \hline P(5.0 \leq X \leq 7.5) \approx 0.24 \end{array}$$

The continuous probability distributions that are encountered most in statistical work have been tabulated in cumulative form so that calculation of cumulative probabilities from equations is not necessary.

In many uses of continuous distributions, the mean and variance are known beforehand. In the case of the processing times, for example, the continuous distribution does not appear out of nowhere. A computer center employee could have studied a large collection of job processing times, computing their mean and variance and graphing the frequency distribution in the manner of Chapters 1 and 2. Then a continuous distribution model would have been selected to match the characteristics of the actual distribution—particularly the mean, variance, and shape. Thus the continuous model is a generalization about actual (or in some cases potential) real data. In either case, the mean and variance of the distribution are *inputs* to selection of the continuous model. 7.1

Mean and Variance of Sample Sums and Means 7.2

The distribution of processing times has a mean of 5.0 seconds and a variance of 10.0. Suppose now that four processing times are drawn independently from the population of processing times for individual jobs. As with the sampling situations in the preceding chapters, the four draws constitute a repeatable experiment and a statistical result from the experiment is a random variable. One result of interest is the total of the four processing times.

Statement

For n independent observations of a continuous random variable with mean μ and variance σ^2, the probability distribution of $\sum X$ has the following mean and variance:

$$\blacksquare \quad \mu_{\Sigma X} = n\mu \quad \blacksquare \tag{7.1}$$

and

$$\blacksquare \quad \sigma^2_{\Sigma X} = n\sigma^2. \quad \blacksquare \tag{7.2}$$

In our example, the mean and variance of the sample sum of four processing times would be

$$\mu_{\Sigma X} = n\mu = 4(5.0) = 20.0 \text{ seconds}$$

and

$$\sigma^2_{\Sigma X} = n\sigma^2 = 4(10.0) = 40.0.$$

The standard deviation of the probability distribution of total processing time for four jobs would be $\sqrt{40.0} = 6.32$ seconds.

Suppose 40 jobs of the class represented by the original job-time distribution are awaiting execution. Assuming that their processing times are a random sample of the population, the expected total processing time for the 40 jobs is $40(5.0) = 200$ seconds. The standard deviation of the probability distribution of total processing time is $\sqrt{40(10.0)} = \sqrt{400} = 20$ seconds. The empirical rule used before in connection with other distributions still applies. No matter what the shape of a distribution, the probability is at least 0.90 that an observation lies within three standard deviations of the mean. Chances are very good that the 40 jobs will not take less than 140 seconds or more than 260 seconds [that is, 200 ± 3(20)].

We might have been concerned in our examples with the average time for the 4 or 40 jobs. This you will recognize as the sample mean, which is the sample sum divided by n, the number of observations composing the sample. The formulas for mean and variance of the sampling distribution of a mean are the same for continuous as for discrete underlying variables.

Statement

For n independent observations of a continuous random variable with mean μ and variance σ^2, the probability distribution of $\overline{X}$ has the following mean and variance:

$$\blacksquare \quad \mu_{\overline{X}} = \mu \quad \blacksquare \qquad 7.3$$

and

$$\blacksquare \quad \sigma_{\overline{X}}^2 = \frac{\sigma^2}{n}. \quad \blacksquare \qquad 7.4$$

For 40 jobs representing independent draws from the original population of processing times, the expected sample mean is

$$\mu_{\overline{X}} = \mu = 5.0 \text{ seconds,}$$

and the variance of the probability distribution of the sample mean is

$$\sigma_{\overline{X}}^2 = \frac{\sigma^2}{n} = \frac{10.0}{40} = 0.25.$$

The standard deviation of the probability distribution of the sample mean is $\sqrt{0.25} = 0.50$. Thus, the probability is at least 0.90 that the mean for 40 jobs will
7.2 lie within $3(0.50) = 1.50$ seconds of the population mean of 5.0 seconds.

7.3 The Normal Distribution

A continuous distribution model of crucial importance in statistical theory and practice is the **normal probability distribution**. We encounter the normal distribution in two situations. In the first, we may know that the distribution of a particular X population follows the normal distribution form. The second situation involves sampling distributions such as the probability distribution of the sample mean. Here statistical theory tells us that in certain situations the sampling distribution follows the normal probability model. The first situation is taken up now and the second is dealt with in the latter half of the chapter. In both situations we use the normal distribution model for obtaining probabilities.

Figure 7.3 shows a normal distribution having a mean of 24 and a standard deviation of 2. As you can see, the distribution here is symmetrical about its mean. This is a feature of all normal distributions. Because of this symmetry, all the measures of central location that we have studied—mean, median, mode—have the same value in a normal distribution. A further consequence of symmetry is shown by the shaded areas left and right of the mean. Since areas under a continuous distribution are proportional to probabilities, we can see in the normal distribution in Figure 7.3 that the probability of X in the interval 24 to 26 is the same as the probability of X in the interval 22 to 24. In symbols, $P(24 \leq X \leq 26) = P(22 \leq X \leq 24)$. In general for a normal distribution, the probability in the interval from μ to any distance above μ is equal to the probability in the interval from μ to the same distance below μ.

The normal density curve is bell-shaped with symmetrically decreasing density as one considers x values that deviate from the mean in either direction. At large distances from the mean, the curve approaches but never reaches the x axis. The equation for the normal density function is

$$f(x) = \frac{1}{\sqrt{2\pi}\sigma} e^{(-1/2)[(x-\mu)/\sigma]^2},$$

where π is the constant $3.14159 \cdots$ and e is the constant $2.71828 \cdots$. The random variable is x and μ and σ are the mean and standard deviation. As we stated before, the mean and standard deviation are values that would be known, or specified, beforehand. Given the constants π and e, and $\mu = 24$ and $\sigma = 2$, substitution of different values for x in the normal density function would produce the curve of Figure 7.3. For the main use of the normal distribution model—obtaining probabilities—this is not necessary because tables are provided.

Figure 7.4 shows five normal distributions with different means and standard deviations. As with any distribution, the mean is a measure of central location and the standard deviation is a measure of dispersion. The normal distribution with $\mu = 36$ and $\sigma = 3$ is more dispersed about a greater mean than the one with $\mu = 24$ and $\sigma = 2$. On the other hand, the normal distribution with $\mu = 17$ and $\sigma = 1$ is

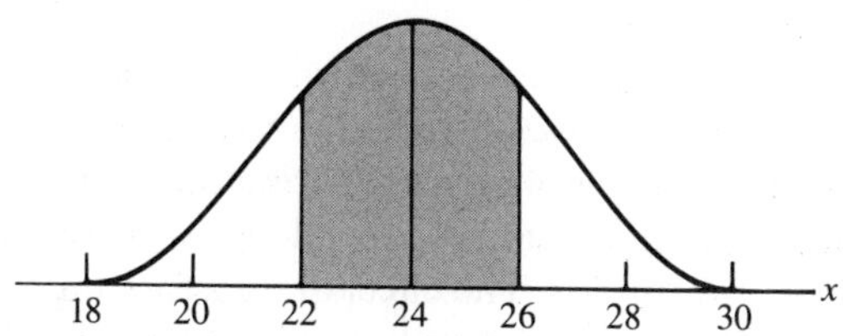

Figure 7.3 *Normal distribution with $\mu = 24$ and $\sigma = 2$*

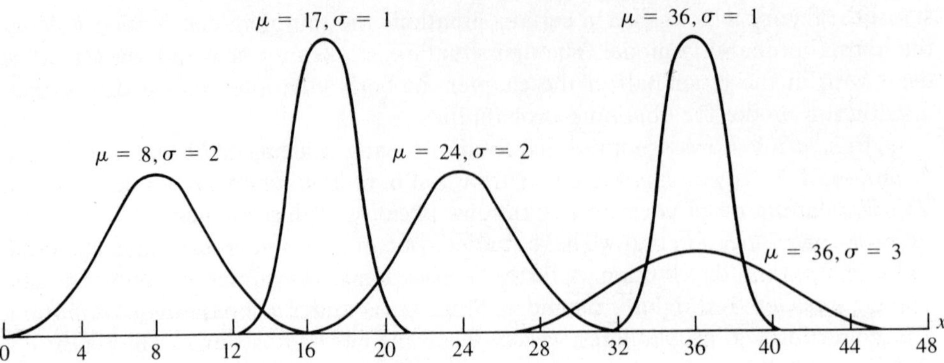

Figure 7.4 Five normal probability distributions

more tightly concentrated about a smaller mean than the normal distribution with $\mu = 24$ and $\sigma = 2$. While normal distributions differ in these two important characteristics, they all have similar properties which we can now summarize:

1. A normal distribution is symmetrical about its mean and has a bell-shaped density curve.
2. While the normal distribution curve approaches but never reaches the horizontal axis, the total area under the curve is equal to 1.0. Thus it can serve as a probability distribution.
3. The area under the curve for a fixed interval immediately to the left of the mean is the same as the area under the curve for an equal interval immediately to the right of the mean.

Examination of the equation for a normal density function reveals that the density depends on the value of $(x - \mu)/\sigma$, which appears in the exponent in the equation. We have seen this term before:

$$z = \frac{x - \mu}{\sigma} \qquad 7.5$$

is the expression of x in standard deviation units from the mean. The variable z was called *standard score* in Chapter 1.

When converted to standard scores, any variable has a mean of zero and a standard deviation of 1.0. The *normal* distribution with a mean of zero and standard deviation of 1.0 is called the **standard normal distribution**. Conversion of x values for any normally distributed variable into z scores allows us to use the standard normal distribution for finding probabilities. The effect of transforming two different normal distributions into the *standard* normal distribution is illustrated graphically in Figure 7.5. The figure shows two of the normal distributions from Figure 7.4 plotted first on

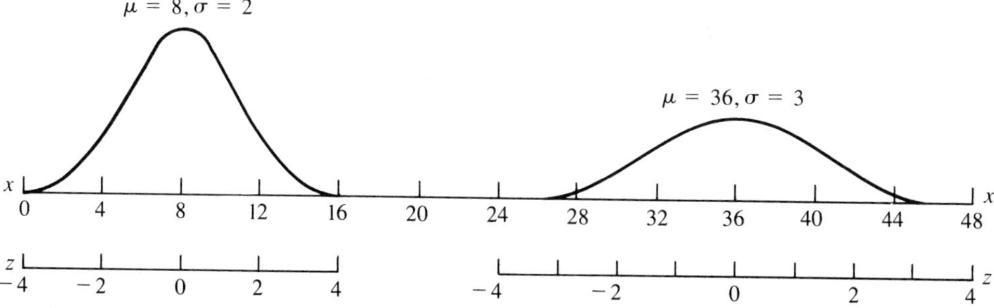

(a) Scales of x Given Same Spacing

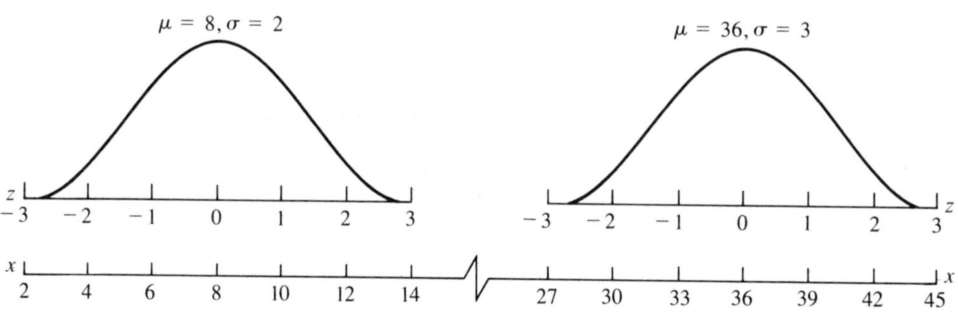

(b) Scales of z Given Same Spacing

Figure 7.5 *Two normal distributions converted to the standard normal distribution*

the scale of x. Their shapes reflect their different standard deviations. When the distributions are plotted with equal z-score spacing, however, their shapes appear the same. In fact, they could now be plotted right on top of each other because both means are now zero. Thus they have been converted to the same (standard normal) distribution.

Areas and Percentiles of the Standard Normal Distribution

Appendix A contains cumulative probabilities for the standard normal distribution. For any value of z the table gives the cumulative probability up to that level of z. We call this a *lower-tail* probability. For example, at $z = -1.50$ we read 0.0668. Thus

$$P(Z \leq -1.50) = 0.0668.$$

Graphically, we have found the proportionate area shown in Figure 7.6. Note that if we wanted $P(Z < -1.50)$ the answer would be the same, because there is zero probability (area) at any exact value of a continuous variable; that is, $P(Z = 1.50) = 0$.

You might want the probability that z exceeds some value. We call this an *upper-tail* probability. As an example consider $P(Z > 0.26)$. The table gives

$$P(Z \leq 0.26) = 0.6026.$$

Since the total probability under the curve is 1.0, the probability you want is the complement of this:

$$P(Z > 0.26) = 1 - P(Z \leq 0.26) = 1 - 0.6026 = 0.3974.$$

This is shown graphically in Figure 7.7.

A third situation is a bounded interval. An example is to find the probability that z lies in the interval from -0.50 to 2.00. Here we find the cumulative probability associated with the upper limit of the interval and subtract from it the cumulative probability associated with the lower limit. In symbols,

$$\begin{aligned} P(-0.50 \leq Z \leq 2.00) &= P(Z \leq 2.00) - P(Z \leq -0.50) \\ &= 0.9772 - 0.3085 = 0.6687. \end{aligned}$$

This is shown graphically in Figure 7.8.

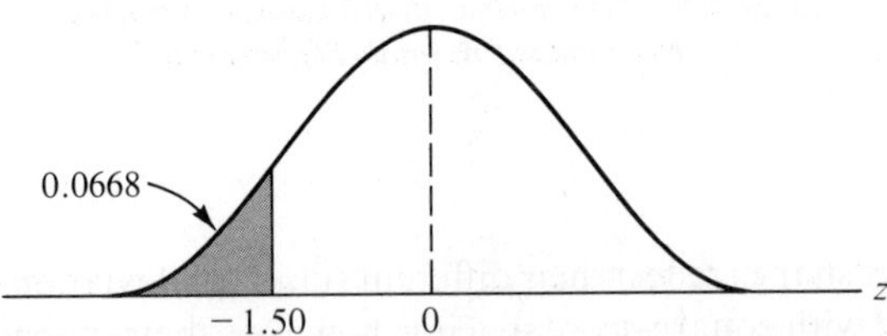

Figure 7.6 A lower-tail standard normal probability

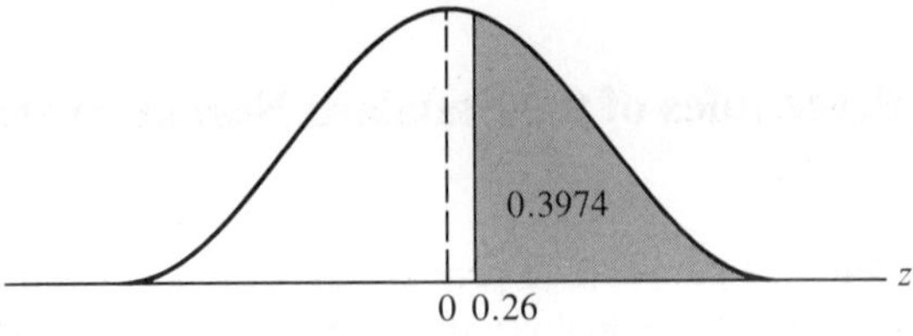

Figure 7.7 An upper-tail standard normal probability

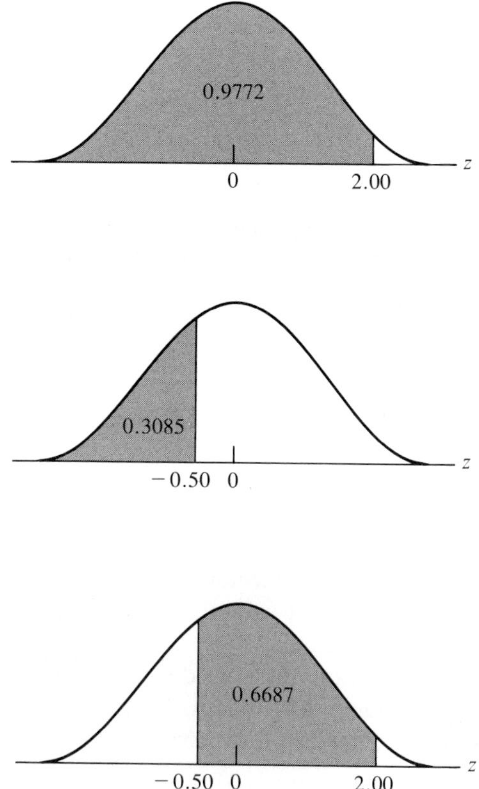

Figure 7.8 *Finding the standard normal probability for a bounded interval*

Sometimes we need to find the value of z that has an associated cumulative probability. For example, find the value of z with an associated cumulative probability of 0.10. Here you must find the cumulative probability 0.10 (or the closest value to 0.10) in the body of the table and read the associated value of z from the margin of the table. Looking through the table we find the cumulative probability 0.1003 at $z = -1.28$. The value -1.28 is the 10th percentile of z, or the value of z which has a probability of 0.10 of not being exceeded and a probability of 0.90 of being exceeded. A common way to refer to such a cutoff value is to write $z_{0.10} = -1.28$.

Suppose we want the value of z that is exceeded with probability 0.05. First we recognize that this is $z_{0.95}$. Looking for the cumulative probability 0.95 in the table, we find it eventually at 1.64, that is, $z_{0.95} = 1.64$.

Finally, suppose we want a value of z such that the total probability associated with values less than $-z$ or greater than z is, say, 0.05. Since the normal curve is symmetrical, we recognize that these cutoff values of z are $z_{0.025}$ and $z_{0.975}$. Consulting the table, we find $z_{0.025} = -1.96$ and $z_{0.975} = 1.96$. This is illustrated graphically in Figure 7.9. 7.3

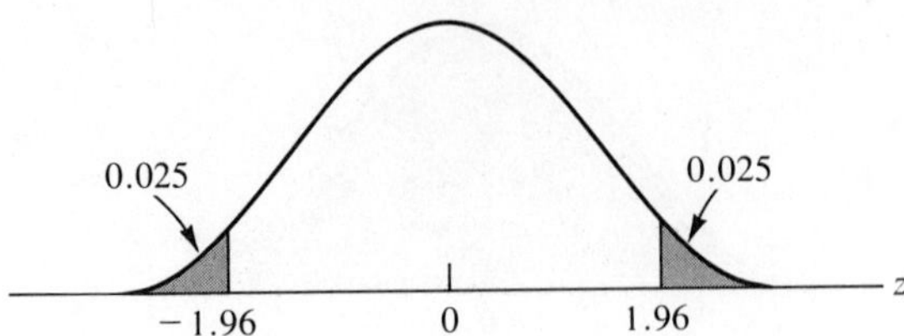

Figure 7.9 *The 5 percent most extreme values of z under the standard normal distribution*

7.4 Probabilities under a Normal Distribution of X

To find probabilities under a normal distribution other than the standard normal, we have to convert the particular normal distribution to the standard form. We do this by converting the relevant values (x) to the standard form through the expression $z = (x - \mu)/\sigma$. This amounts to transforming the scale of the x variable to standard deviation units. For example, consider a normal distribution with a mean of 20.0 and a standard deviation of 4.0. What is the probability that X is equal to or less than 17.0? The steps in obtaining the answer are:

1. Problem: $P(X \leq 17.0) = ?$
2. Convert x to z:

$$z = \frac{x - \mu}{\sigma}$$

$$= \frac{17.0 - 20.0}{4.0} = -0.75.$$

3. Solve equivalent standard normal probability problem:

$$P(Z \leq -0.75) = 0.2266.$$

4. State answer in terms of x:

$$P(X \leq 17.0) = P(Z \leq -0.75) = 0.2266.$$

Graphically, these steps represent a substitution of the z scale for the x scale in order to use the table of probabilities for the standard normal distribution. Having done this, we then convert back to the original x scale of the problem. (See Figure

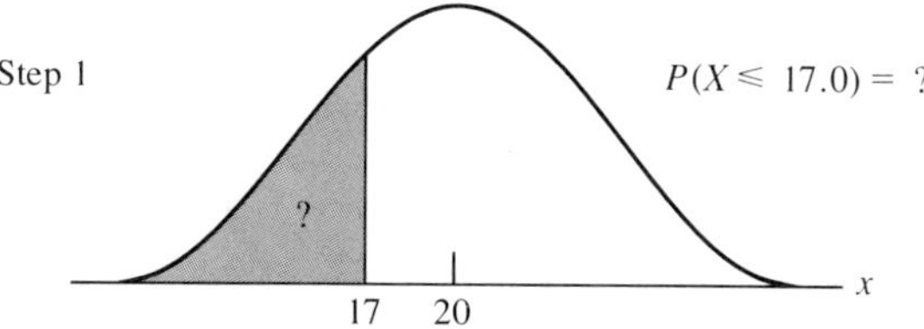

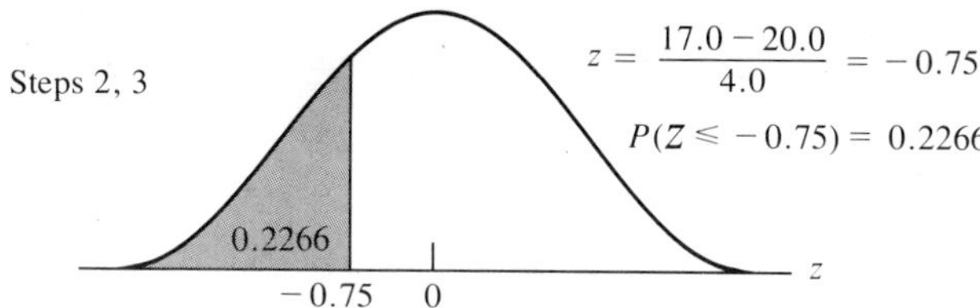

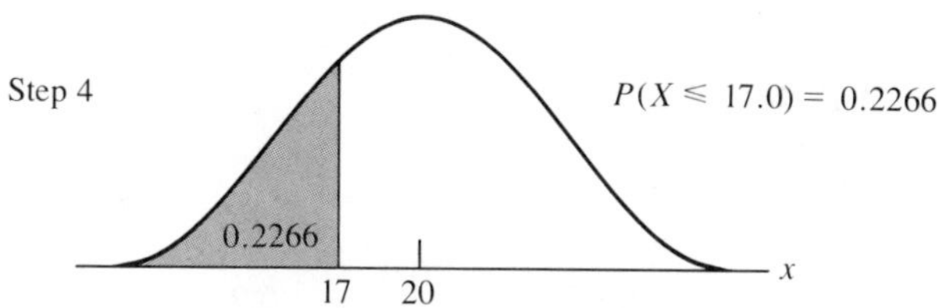

Figure 7.10 *Graphic representation of normal probability problem and its solution*

7.10.) You can visualize the procedure in one graph with the z scale shown below the x scale as in Figure 7.11.

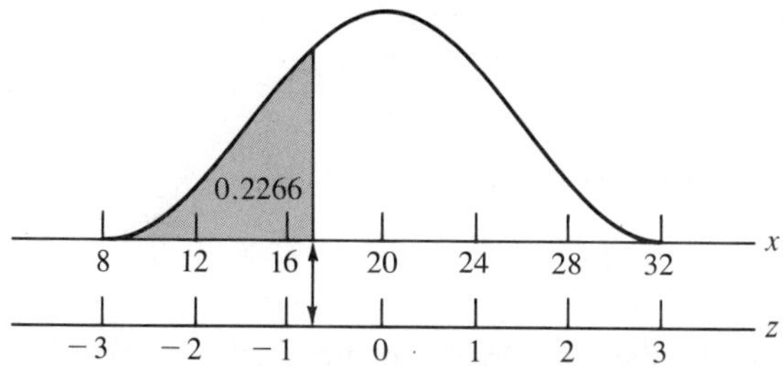

Figure 7.11 *Representation of normal probability problem using x and z scales*

For an upper-tail problem let us find the probability that X exceeds 18.0 in a normal population with a mean of 20.0 and a standard deviation of 4.0.

1. Problem: $P(X > 18.0) = ?$

2. Convert x to z:

$$z = \frac{x - \mu}{\sigma}$$

$$= \frac{18.0 - 20.0}{4.0} = -0.50.$$

3. Solve equivalent standard normal probability problem:

$$P(Z > -0.50) = 1 - P(Z \leq -0.50) = 1 - 0.3085 = 0.6915.$$

4. State answer in terms of x:

$$P(X > 18.0) = 0.6915.$$

For a bounded-interval problem let us find the probability that x lies in the interval from 19.0 to 26.0, given a normal distribution with $\mu = 20.0$ and $\sigma = 4.0$.

1. Problem: $P(19.0 \leq X \leq 26.0) = ?$
2. Convert x to z:

$$z = \frac{19.0 - 20.0}{4.0} = -0.25$$

$$z = \frac{26.0 - 20.0}{4.0} = 1.50.$$

3. Solve equivalent standard normal probability problem:

$$P(-0.25 \leq Z \leq 1.50) = P(Z \leq 1.50) - P(Z \leq -0.25)$$
$$= 0.9332 - 0.4013 = 0.5319.$$

4. State answer in terms of x:

$$P(19.0 \leq X \leq 26.0) = 0.5319.$$

To find a value of x having a specified cumulative probability, you must follow the same process of converting to terms of z, solving the problem in terms of z, and converting back again. In a normal distribution with $\mu = 20.0$ and $\sigma = 4.0$, let us find the value of x that is not exceeded with probability 0.80 (the 80th percentile of x):

1. Problem: $x_{0.80} = ?$
2. Convert x to z: $z_{0.80} = ?$
3. Solve equivalent standard normal probability problem:

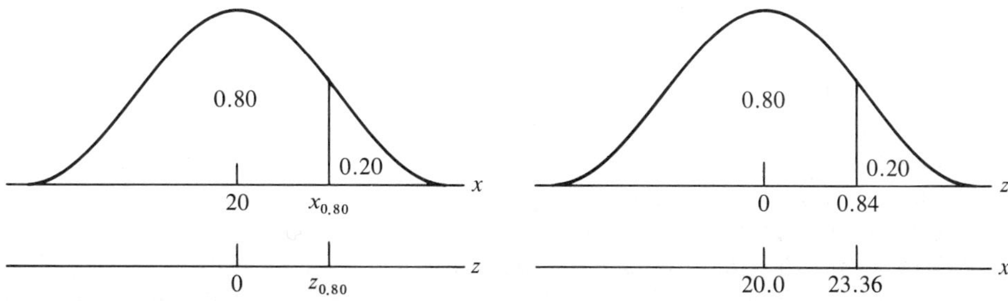

Figure 7.12 *Graphic representation of procedure for finding a percentile of a normal distribution*

$$z_{0.80} = 0.84.$$

4. State answer in terms of x. To do this it is convenient to restate the standard score expression as an equation for x:

$$z = \frac{x - \mu}{\sigma}$$

$$x = \mu + z\sigma.$$

For our problem we would write

$$\begin{aligned} x_{0.80} &= \mu + z_{0.80}\sigma \\ &= 20.0 + 0.84(4.0) = 23.36. \end{aligned}$$

We know μ and σ because they are given as the parameters of the normal distribution of interest to us. The value of z that we want, $z_{0.80}$ in the problem above, comes from the specification of the cutoff point. We find $z_{0.80}$ by looking for the cumulative probability 0.80 in the standard normal probability table and noting at what z value this cumulative probability appears. Then we find the corresponding x value, as illustrated in Figure 7.12. 7.4

Summary

A probability distribution of a continuous random variable is a model for outcomes of an experiment that incorporates an abstraction of infinite accuracy of measurement. The model is useful when relatively many values of a variable are produced by applying an interval or ratio scale to units of observation. An event is an interval of the variable, and the probability of an event is the proportion of area under the density function within the interval.

In applied work, the mean and variance of a population of continuous measures are usually known from the situation that the model has been selected to represent. If repeated independent draws are conceived from such a population, the sample sum and the sample mean are random variables whose mean and variance are related to the mean and variance of the population.

The normal distribution is a particular form of continuous distribution that you will encounter as a population distribution and as a sampling distribution. The table of areas of the standard normal distribution can be applied to normally distributed populations. The key to using the standard normal table to find probabilities under a particular normal distribution is conversion of the x scale of the variable to the standard z scale. This conversion involves the mean and standard deviation of the population through $z = (x - \mu)/\sigma$. You may want to find the probability associated with a particular interval of x, or you may need to find a specified percentile of x. Both kinds of problems were carried out.

See Self-Correcting Exercises 7A.

Exercise Set A

1. If a rectangular probability density function is defined for x values ranging from 7.2 to 11.7:
   (a) What is the area of the rectangle above the interval from 7.2 to 11.7?
   (b) What is the height of the density function above any possible value of x?
   (c) Is this distribution symmetrical about the mean? What is the mean?
2. Referring to the probability density function of Exercise 1, determine numerical values for
   (a) $P(X < 8.6)$
   (b) $P(X = 9.7)$
   (c) $P(X < 8 \text{ or } X > 10)$
   (d) $P(9 \leq X \leq 11)$
   (e) $P(7.2 \leq X \leq 11.7)$
   (f) $P(X < 9.645)$
3. A machine cuts wood blanks used in making pencils. The minimum length of the blanks is 8.00 inches. The mean blank length is 8.15 inches. If all blank lengths between the minimum and the maximum length are equally probable, what is the maximum length of wood blanks for this machine?
4. In the situation of Exercise 3, what is the probability that a blank will measure (a) 8.10 inches to the nearest 0.005 inch? (b) 8 inches exactly?
5. What must the mean of the population be if the standard deviation is 2 and a z score of 1.25 is obtained for an x value of (a) 12? (b) 9? (c) 14.6?

6. What must the standard deviation of the population be if the mean is 8 and a z score of -1.8 is obtained for an x value of (a) 6? (b) 4? (c) 2.4?

7. A high school student receives, on a standardized examination in Russian, a score which exceeds that of 68 percent of all students taking the exam.
   (a) What percentile of x does the student's score represent?
   (b) If you know that the student's raw or x score is 432, and the standard deviation of all scores is known to be 20, what must the mean score be?
   (c) If you know that the test scores are normally distributed, what is the numerical value of this student's z score?

8. Convert the following x values to z scores. Assume $\mu = 6.2$ and $\sigma = 2.1$.
   (a) $x = 8.3$
   (b) $x = 6.5$
   (c) $x = 3.94$
   (d) $x = -2$

9. If the variable x in Exercise 8 is normally distributed, determine:
   (a) $P(X \leq 8.3)$
   (b) $P(X > 6.5)$
   (c) $P(X < 3.94)$
   (d) $P(-2 < X \leq 3.94)$

10. State whether each of the following is true or false. Assume z has the standard normal distribution.
   (a) $P(Z = 0.8) = 0.7881$
   (b) $P(Z \geq 0.8) = 0.2119$
   (c) $P(Z \neq 0.8) = 1$
   (d) $P(-0.8 < Z < 0.8) = 0.5762$
   (e) $P(Z < -0.8 \text{ or } Z > 0.8) = 0.4238$
   (f) $P(Z < -0.8 \text{ and } Z > 0.8) = 0$
   (g) $P(Z > -0.8 \text{ or } Z < 0.8) = 1$
   (h) $P(Z > -0.8 \text{ and } Z < 0.8) = 1.5762$
   (i) $P(Z \text{ does not exceed } 0.8) = 0.7881$

11. Provide correct numerical values, without consulting Appendix A, for each of the following:
   (a) $P(Z > z_{0.10})$
   (b) $P(Z \leq z_{0.85})$
   (c) $P(z_{0.35} \leq Z \leq z_{0.85})$
   (d) $P(Z \text{ is not less than } z_{0.90})$
   (e) $P(Z < z_{0.12} \text{ and } Z < z_{0.04})$
   (f) $P(Z < z_{0.12} \text{ or } Z < z_{0.04})$
   (g) $P(Z \text{ does not exceed } z_{0.50})$

12. Before you can find $P(1.2 < z < 1.7)$ for a normally distributed variable x, what else must you know? Why?

13. For a normal distribution with $\mu = 10$ and $\sigma = 2$, convert each of the following to z scores:
   (a) $x = 3.7$
   (b) $x = -4$
   (c) $x = 10$
   (d) $x = 11$
   (e) $x = 9.8$

14. Find the probability, correct to four places, that each of the values in Exercise 13 is exceeded.

15. Referring to a normally distributed variable X, state under what conditions, if any, each of the following is negative:
   (a) x values
   (b) z values
   (c) F' values in symbols of the form $z_{F'}$
   (d) Probabilities read from Appendix A

16. The length of the interval under the standard normal curve from $z = 1.2$ to $z = 1.4$ is the same as that from $z = 1.4$ to $z = 1.6$. Yet $P(1.2 \leq Z \leq 1.4)$ exceeds $P(1.4 \leq Z \leq 1.6)$. Why is this so?

17. Without referring to Appendix A, determine which of the following is larger for a normal distribution and justify your conclusion:

$$P(-1.2 < Z < -0.8) \quad \text{and} \quad P(0.9 \le Z \le 1.3).$$

18. Which, if any, of the following are negative in a normal distribution?

(a) $z_{0.9}$
(b) $z_{0.10}$
(c) $z_{0.01}$
(d) $z_{0.50}$
(e) $z_{0.495}$
(f) $z_{0.338}$
(g) $z_{0.975}$

19. From Appendix A, find numerical values for each term in Exercise 18.

20. Can you find a value for z such that the probability of a standard score falling below that value is 1? If so, what is it? If not, why not?

21. Our rule of thumb that "at least 90 percent of the data are within three standard deviations of the mean" applies to data distributions in general. If the distribution of concern to us is in fact normal, what stronger statement can we make about the proportion of data falling within this range?

22. Determine the 10 percent most extreme values of z under the standard normal distribution.

23. For a normal distribution with $\mu = 12$ and $\sigma = 4$, use the formula $z = (x - \mu)/\sigma$ and your results from Exercise 22 to find the 10 percent most extreme values of the distribution. These values could be represented in the form $z_{F'}$ for what two values of F', respectively?

24. In a normal distribution with $\mu = 38.4$ and $\sigma = 4$,

(a) What are the 10 percent most extreme values of x?
(b) Determine $x_{0.10}$ and $x_{0.90}$.

25. Completion times for an assembly operation among a group of workers had a mean of 80 seconds and a standard deviation of 10 seconds. Can you count on a completion time of 100 seconds being in the extreme 10 percent of all completion times (a) if the distribution of times is normal? (b) If nothing is known about the shape of the distribution?

26. A machine fills bottles of wine with an average of 0.200 gallon. The standard deviation of fills is 0.008 gallon. If the fills are normally distributed, below what value will the shortest 10 percent of the fills lie?

27. If the wine bottler wants to guarantee that no more than 1 percent of the fills will be short of 0.200 gallon and he cannot change the standard deviation, what mean fill will he have to provide?

28. An improved filling machine has a standard deviation for fills of 0.005 gallon. What percentage saving will this secure for the bottler in meeting the guarantee in Exercise 27?

7.5 Sampling from Normal Populations

In the first half of this chapter we found that the sample space represented by a truly continuous variable is infinite. The values of x composing the population cannot be listed. Nevertheless, the probability distribution for the population can be portrayed by an equation or by the graph of the probability density function.

Let us return to our normal population with a mean of 20.0 and standard deviation of 4.0. Suppose this population represents the lengths of life (in hours) of a specified type of flashlight bulb. Suppose now that four bulbs are selected at random. What is the probability distribution of the total length of life of the four bulbs?

Equations 7.1 and 7.2 give us the mean and variance of the sampling distribution of $\sum X$, the total length of life of the four bulbs:

$$\mu_{\Sigma X} = n\mu = 4(20.0) = 80.0 \text{ hours}$$

and

$$\sigma^2_{\Sigma X} = n\sigma^2 = 4(4.0)^2 = 4(16) = 64.$$

The square root of the variance gives us the standard deviation of the probability distribution of the sample sum:

$$\sigma_{\Sigma X} = \sqrt{\sigma^2_{\Sigma X}} = \sqrt{64} = 8.0 \text{ hours.}$$

What about the shape of the probability distribution of a sample sum? Here we rely on a rule which can be proved mathematically: Linear functions of normally distributed random variables are also normally distributed. The sample sum is a linear function of n identical normally distributed random variables. That is, the probability distribution of the first x drawn is the same as the second x drawn, and so on to the nth x drawn, and

$$\sum_{i=1}^{n} x_i = x_1 + x_2 + x_3 + \cdots + x_n.$$

The probability distribution of the total length of life of four randomly selected bulbs is then completely identified by saying:

1. The probability distribution is normal.
2. Its mean is 80.0 ($\mu_{\Sigma X} = 80$).
3. Its standard deviation is 8.0 ($\sigma_{\Sigma X} = \sqrt{64.0}$).

Since the mean is $\mu_{\Sigma X}$, and the standard deviation is $\sigma_{\Sigma X}$, the standard score for $\sum x$ is

$$■ \quad z = \frac{\sum x - \mu_{\Sigma X}}{\sigma_{\Sigma X}}. \quad ■ \qquad 7.6$$

With this information we can answer a question such as this: What is the probability that four successive bulbs (independently selected) last less than a total of 70.0 hours? Following our previous four-step method, we have:

1. Problem: $P(\sum X < 70.0) = ?$
2. Convert $\sum x$ to z:

$$z = \frac{\sum x - \mu_{\Sigma X}}{\sigma_{\Sigma X}}$$

$$= \frac{70.0 - 80.0}{8.0} = -1.25.$$

3. Solve equivalent standard normal probability problem:

$$P(Z < -1.25) = 0.1056.$$

4. State answer in terms of $\sum x$:

$$P\left(\sum X < 70.0\right) = 0.1056.$$

Sampling Distribution of $\overline{X}$ from a Normal Population

Sometimes, as above, the sum of x for n independent draws will be a statistic of interest. If we take four bulbs along on a camping trip, we are interested in how much light (in hours) we may have. More generally, in statistical work the sample mean $\sum x/n$ will be of paramount interest.

Statement If n independent draws are made from a normally distributed population with mean μ and standard deviation σ, the probability distribution of the sample mean is a normal distribution.

We know from our previous work that $\mu_{\overline{X}} = \mu$ and $\sigma^2_{\overline{X}} = \sigma^2/n$. For our experiment, where $\mu = 20.0$, $\sigma = 4.0$, and $n = 4$, we would have

$$\mu_{\overline{X}} = \mu = 20.0,$$

$$\sigma^2_{\overline{X}} = \frac{\sigma^2}{n} = \frac{(4.0)^2}{4} = 4.0.$$

We can now identify the sampling distribution of mean length of life of four independently selected bulbs by saying:

1. The probability distribution of $\bar{X}$ is normal.
2. Its mean is 20.0 ($\mu_{\bar{X}} = 20.0$).
3. Its standard deviation is 2.0 ($\sigma_{\bar{X}} = \sqrt{4.0}$).

Figure 7.13 shows our original population along with the two sampling distributions we have developed. Here we show all three against a scale of hours. The sampling distribution of $\sum X$ for four draws has a wider spread ($\sigma_{\sum X} = 8.0$) than the original population ($\sigma = 4.0$). The sampling distribution of $\bar{X}$ for four draws is centered about the mean of the population ($\mu_{\bar{X}} = \mu$) but has a smaller spread ($\sigma_{\bar{X}} = 2.0$). The spread of the sampling distribution of $\bar{X}$ depends on the size of sample, n. More formally, $\sigma_{\bar{X}}$ is called the **standard error of the mean.**

$$\blacksquare \quad \sigma_{\bar{X}} = \frac{\sigma}{\sqrt{n}} \quad \blacksquare \qquad 7.7$$

With our knowledge about the sampling distribution of the mean from a normal population, we can answer a question such as this: What is the probability that the mean life of a sample of 100 bulbs will be within 1.0 hour of the population mean life? We know that the sampling distribution of the mean is normal with

$$\mu_{\bar{X}} = \mu = 20.0,$$

$$\sigma_{\bar{X}} = \frac{\sigma}{\sqrt{n}} = \frac{4.0}{\sqrt{100}} = 0.40.$$

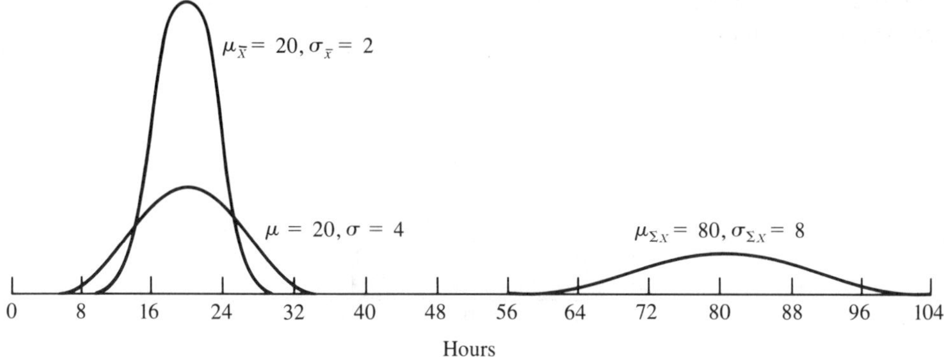

Figure 7.13 *A probability distribution of X along with sampling distribution of $\sum X$ and sampling distribution of $\bar{X}$ for $n = 4$*

The standard score for $\bar{x}$ is

$$■ \quad z = \frac{\bar{x} - \mu}{\sigma_{\bar{X}}}. \quad ■ \qquad 7.8$$

Sample means within 1.0 hour of the population mean are in the interval 19.0 to 21.0 for $\bar{x}$. Following our previous steps for normal probability problems, we have:

1. Problem: $P(19.0 \leq \bar{X} \leq 21.0) = ?$
2. Convert $\bar{x}$ to z:

For $\bar{x} = 19.0$: $$z = \frac{\bar{x} - \mu}{\sigma_{\bar{X}}} = \frac{19.0 - 20.0}{0.40} = -2.5.$$

For $\bar{x} = 21.0$: $$z = \frac{\bar{x} - \mu}{\sigma_{\bar{X}}} = \frac{21.0 - 20.0}{0.40} = 2.5.$$

3. Solve equivalent standard normal probability problem:

$$\begin{aligned} P(-2.5 \leq Z \leq 2.5) &= P(Z \leq 2.5) - P(Z \leq -2.5) \\ &= 0.9938 - 0.0062 = 0.9876. \end{aligned}$$

4. State answer in terms of $\bar{x}$:

$$P(19.0 \leq \bar{X} \leq 21.0) = 0.9876.$$

The two-scaled representation of our problem is shown in Figure 7.14.

Finding a particular cutoff value of $\bar{x}$ parallels the percentile problem shown previously for a normal probability distribution of X. Let us find the 75th percentile of the mean length of life of 100 bulbs. This is a value of $\bar{x}$ so located that the probability of its not being exceeded is 0.75. The key to solving for $\bar{x}$ is the expression derived from Equation 7.8:

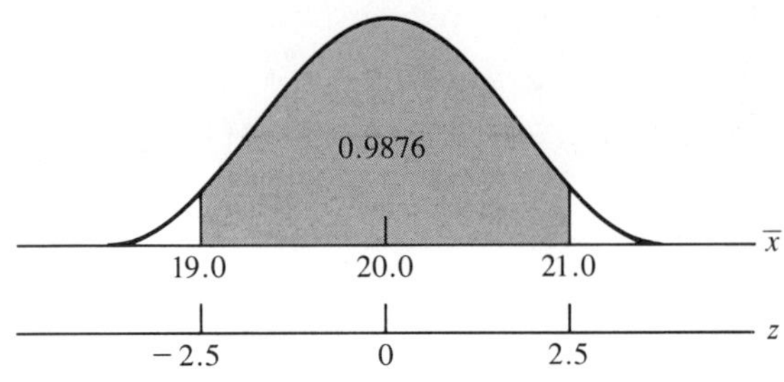

Figure 7.14 Representation of normal sampling distribution problem using $\bar{x}$ and z scales

$$\blacksquare \qquad \bar{x} = \mu + z\sigma_{\bar{X}}. \qquad \blacksquare \tag{7.9}$$

Applying our four-step procedure, we have:

1. Problem: $\bar{x}_{0.75} = ?$
2. Convert $\bar{x}$ to z:

$$z_{0.75} = ?$$

3. Solve equivalent standard normal probability problem:

$$z_{0.75} = 0.67.$$

4. State answer in terms of $\bar{x}$:

$$\begin{aligned} \bar{x} &= \mu + z\sigma_{\bar{X}} \\ \bar{x}_{0.75} &= 20.0 + 0.67(.40) \\ &= 20.268. \end{aligned}$$

We have dealt in this section with the normal probability distribution of $\sum X$ and the normal probability distribution of $\bar{X}$. Earlier we solved some probability problems relating to the distribution of an underlying population. The key to solving any normal probability problem is the z expression for the distribution involved. All z expressions represent the deviation of a value of a random variable from its expected value expressed in standard deviation units. The three expressions for the distributions we have studied are given below. Also given are the equivalent expressions for a value of the random variable. We used the second form in solving for percentiles of the random variable.

x	$\sum x$	$\bar{x}$
$z = \dfrac{x - \mu}{\sigma}$	$z = \dfrac{\sum x - \mu_{\Sigma X}}{\sigma_{\Sigma X}}$	$z = \dfrac{\bar{x} - \mu}{\sigma_{\bar{X}}}$
$x = \mu + z\sigma$	$\sum x = \mu_{\Sigma X} + z\sigma_{\Sigma X}$	$\bar{x} = \mu + z\sigma_{\bar{X}}$

7.5

Sampling from Nonnormal Populations 7.6

Some populations encountered in statistical work are normally distributed or very nearly so. For example, physical characteristics of species of animals and plants tend to be normally distributed. So do the measured characteristics of outputs of manufacturing processes that are operating under a more or less constant set of controlled conditions. Aptitude and achievement measures in education are sometimes normally distributed and sometimes not. Skewness to the right is common in variables that reflect economic or social power, such as incomes of families or measures of personal influence. Further, distributions that are symmetrical are not necessarily normal in shape. They may have a central clustering that is more (or less) peaked (or bunched up in the middle) than a normal distribution.

To illustrate sampling from a nonnormal population, we begin with a continuous population of incomes–shown in Figure 7.15. The distribution has a mean of 10 (thousand dollars) and a standard deviation of 1.732 (thousand dollars). The distribution has considerable skewness, as evidenced by the tailing off of the density curve toward the high values and the fact that the population mean of $10,000 exceeds the most typical, or modal, income.

With the aid of a computer, we could simulate the drawing of repeated samples of any desired size from this population. Of course, we cannot simulate an infinite number of samples of a particular size, but we can make the number of samples large enough to gain a good indication of the characteristics of the probability distribution of the sample mean. We elected to draw 400 samples of size $n = 3$, 400 samples of size $n = 9$, 400 samples of size $n = 15$, and 400 samples of size $n = 30$. It is not possible or desirable to show the results of each sample, but the first, second, and four-hundredth samples of size $n = 3$ are given in Table 7.1.

Figure 7.16 shows the frequency distributions of the sample mean for each sample size. The distribution of means from samples of size $n = 3$ is obviously skewed to the right, but the skewness diminishes in the distributions for samples of size 9, 15, and 30. Actually, both the skewness and the periodic irregularities in the distribution pattern on the long tail are characteristic of sampling distributions from skewed populations. As the sample size is increased, however, both these features tend to disappear. The skewness decreases and the periodic irregularities get smoothed out.

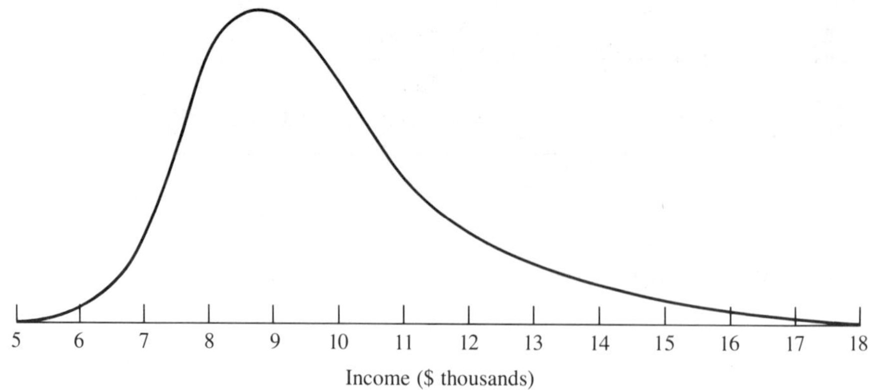

Figure 7.15 A nonnormal population of incomes

Table 7.1 Results of Simulated Draws of 400 Random Samples of Three Observations Each

	Sample 1	Sample 2	...	Sample 400
x_1	12.8266	9.3202	...	10.8615
x_2	8.9128	10.3434	...	10.3293
x_3	10.6368	8.6066	...	12.0986
Mean	10.7920	9.4234		11.0964

We see here the operation of a very remarkable and important theorem in statistics—the central limit theorem. Even though our simulation was limited to 400 trials, the distribution of means of samples of size $n = 15$ and $n = 30$ from our highly skewed population is highly suggestive of the normal curve shape.

Statement The **central limit theorem** states that, as the sample size is increased, the sampling distribution of the mean of a continuous variable with a finite variance approaches the normal distribution form regardless of the distribution form of the underlying population.

The central limit theorem does not tell us how large the sample size must be before the sampling distribution of the mean is reasonably close to the normal form. This depends on the shape of the population distribution: the approach to a normal sampling distribution of the mean is more rapid for populations with some central clustering than for those without. Experienced statisticians feel safe in assuming essentially normal sampling distributions for means of samples of size 30 and larger.

The existence of the central limit theorem greatly extends the applicability of normal sampling distribution theory. Consider our incomes population. If you had been asked to estimate the mean income from the mean of a sample of 100, how good might your estimate have been? You know that

$$\mu_{\bar{X}} = \mu = 10.0 \qquad \text{and} \qquad \sigma_{\bar{X}} = \frac{\sigma}{\sqrt{n}} = \frac{1.732}{\sqrt{100}} = 0.173.$$

You can be assured that the probability distribution of $\bar{X}$ for $n = 100$ is essentially normal. What is the probability that the mean of a sample of 100 incomes will differ from the population mean by no more than 0.50, that is, $500?

1. Problem: $P(9.5 \leq \bar{X} \leq 10.5) = ?$
2. Convert $\bar{x}$ to z:

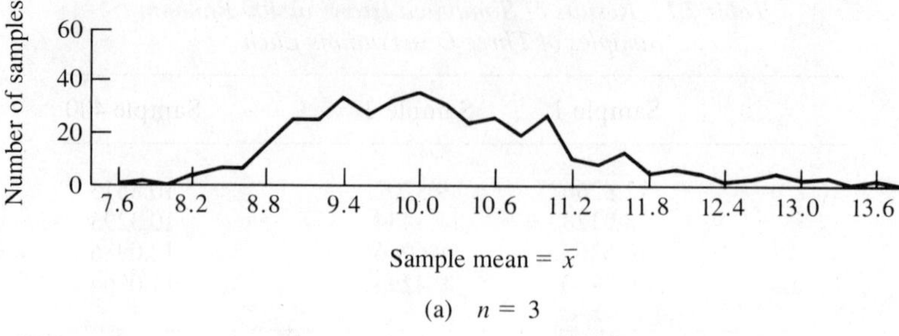

(a) $n = 3$

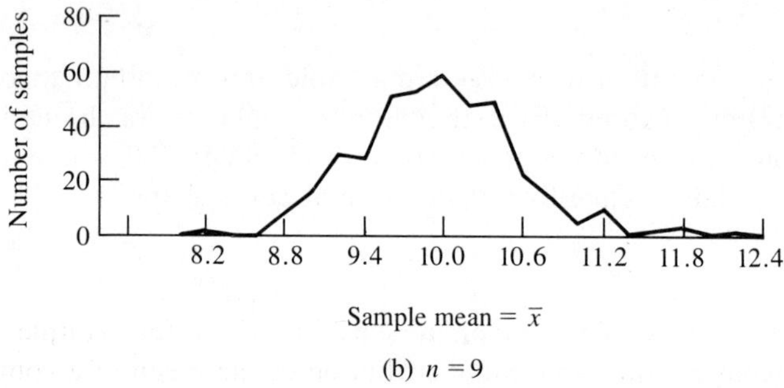

(b) $n = 9$

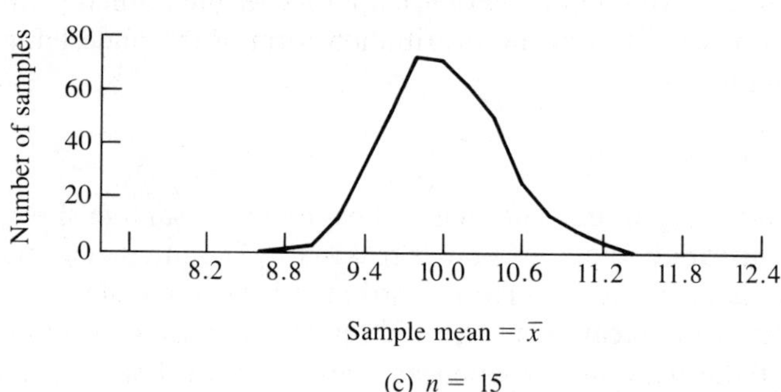

(c) $n = 15$

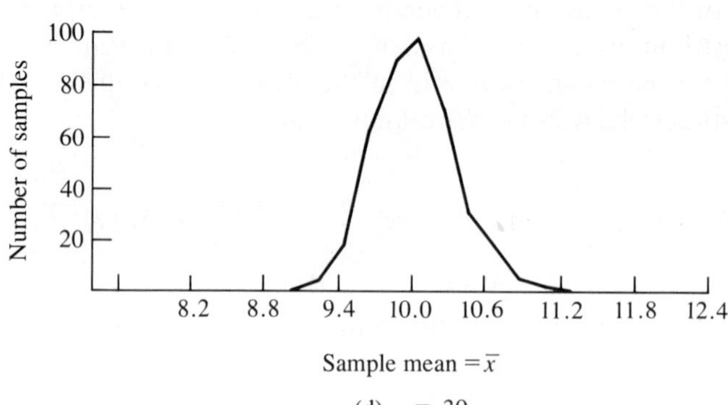

(d) $n = 30$

Figure 7.16 *Simulated sampling distributions of the mean for samples of different size from population of incomes*

$$z = \frac{\bar{x} - \mu}{\sigma_{\bar{X}}}.$$

For $\bar{x} = 9.5$:
$$z = \frac{9.5 - 10.0}{0.173} = -2.89.$$

For $\bar{x} = 10.5$:
$$z = \frac{10.5 - 10.0}{0.173} = 2.89.$$

3. Solve equivalent standard normal probability problem:

$$P(-2.89 \leq Z \leq 2.89) = P(Z \leq 2.89) - P(Z \leq -2.89)$$
$$= 0.9981 - 0.0019 = 0.9962.$$

4. State answer in terms of $\bar{x}$:

$$P(9.5 \leq \bar{X} \leq 10.5) = 0.9962.$$

The chances are better than 99 out of 100 that a sample of 100 incomes would have yielded a sample mean within $500 of the true population mean. This may be sufficient accuracy for some purposes but not for others. That is another question. Knowledge of sampling distribution theory has at least answered the first necessary question. 7.6

Summary

The sampling distribution of the mean for samples of any given size drawn from a normally distributed population is a normal distribution.

The central limit theorem assures us that when the sample size is at least moderate ($n \geq 30$), the sampling distribution of the mean will be a normal distribution (approximately) even when the underlying population is not normally distributed. This theorem greatly extends the applicability of normal sampling distribution theory, because many populations encountered in applied statistics are not normally distributed.

The mean and standard deviation of the sampling distribution of a mean are known from formulas involving the mean and standard deviation of the underlying X population. If, in addition, the sampling distribution is normal, then probability questions about the sample mean can be answered by using the table of areas under the normal distribution.

See Self-Correcting Exercises 7B.

Exercise Set B

1. Given a normally distributed population with $\mu = 80$ and $\sigma = 10$ and an experiment of four independent draws from the population, what is the probability that $\sum X$
   (a) Exceeds 350?
   (b) Fails to exceed 275?
   (c) Lies between 300 and 350?
2. A secretary has found that the time the boss takes for appointments is normally distributed with a mean of 13 minutes and a standard deviation of 3 minutes. What is the probability that four randomly selected appointments will take less than 1 hour?
3. If the secretary wants to be 99 percent sure that four appointments will not exceed 1 hour, how much reduction in average time taken for appointments is required?
4. Over a period, the price changes of a universe of 500 common stocks were normally distributed with a mean of +4.0 percent and a standard deviation of 8.0 percentage points. If four stocks had been randomly selected, what is the probability that their average price change would have been negative? Answer the same question for a random selection of 25 stocks.
5. Part of a manual dexterity test administered individually to preschool children consists of nine tasks, each of which has a mean performance time of 1 minute and 40 seconds and a standard deviation of 20 seconds. Determine the mean and standard deviation of the distribution of performance times for the entire test.
6. The average cost of feeding a teenage boy per week at a summer camp is determined to be \$21.56. The distribution of such costs has a standard deviation of \$2. Eighty-one boys are due to arrive for a 2-week stay.
   (a) How much money should the food services director expect to spend on feeding the boys for this period?
   (b) The director should expect with at least 90 percent certainty that her expenditures for feeding the boys should not fall below or exceed what values?
7. A youngster, concerned with the waste of his valuable childhood washing the dinner dishes each evening, decides to support his argument against child labor with some pertinent statistics. Estimating an average of 300 at-home dinners per year and a daily mean dishwashing time of 20 minutes with a standard deviation of 2 minutes, the child concludes that over a period of 8 years, the total time spent washing dishes would have a distribution with a mean of 800 hours and a standard deviation of 20 hours. Why should the child recalculate his figures before presenting his argument to his parents?

Exercises 8 through 12 refer to bulb lifetimes from a normally distributed population (of such lifetimes) with $\mu = 20$ hours and $\sigma = 5$ hours.

8. What is the probability that nine independently selected bulbs last a total of:
   (a) More than 178 hours?
   (b) Less than 170 hours?
   (c) At least 184 hours?
   (d) Less than 180 hours?
9. What is the probability that nine randomly selected bulbs will have a mean lifetime of:
   (a) At least 20 hours?
   (b) Between 12 and 28 hours?
   (c) Less than 21.5 hours?
   (d) Not less than 18 hours?

10. For 16 randomly selected bulbs, find:
    (a) $P(\sum X \geq 318)$
    (b) $P(321 \leq \sum X \leq 324)$
11. For an independently selected sample of size 25 from this population, determine:
    (a) $P(19 \leq \bar{X} \leq 20)$
    (b) $P(18.5 \leq \bar{X} \leq 19.7)$
12. What is the probability that the mean life of a sample of 25 randomly selected bulbs will be within 2 hours of the population mean lifetime?

Exercises 13 through 17 refer to the following: The lengths of yarn in 4-ounce skeins of wool produced by a certain spinning machine are found to be normally distributed with a mean length of 104.8 feet and a standard deviation of 1 foot.

13. Determine the probability of the sample mean exceeding 104.5 feet in a randomly selected sample of size
    (a) 1
    (b) 4
    (c) 25
    (d) 100
14. Determine, for each sample size in Exercise 13, the probability that the sample mean does not
    (a) Exceed the population mean by more than 6 inches.
    (b) Differ from the population mean by more than 6 inches.
15. State what event is referred to in each of the following expressions, and find the numerical value of the probability of that event for a sample size of 4:
    (a) $P(\bar{X} - 104.8 < 1)$
    (b) $P(\bar{X} - 104.8 > -1)$
    (c) $P(-1 < \bar{X} - \mu < 1)$
    (d) $P(\bar{X} - \mu > 1)$
16. Find the probability that the sample mean of yarn lengths of a sample of size 10 differs from the population by more than (a) 6 inches and (b) two standard errors.
17. For random samples of size 1, 4, 25, and 100 from the population of your yarn lengths, find: (a) $\bar{x}_{0.50}$; (b) $\bar{x}_{0.75}$; (c) $\bar{x}_{0.01}$.
18. How does changing the sample size affect the numerical value of (a) $\bar{x}_{0.50}$? (b) $\bar{x}_{0.75}$? (c) $\bar{x}_{0.01}$?
19. Express the probability that a sample mean is within three standard deviations of the mean of a normally distributed population from which the sample is taken by completing the following probability statement correctly. Insert an algebraic expression in the blank at left and the appropriate numerical value in the blank at right:

$$P(\_\_\_\_\_ \leq \bar{X} \leq \mu + 3\sigma_{\bar{X}}) = \_\_\_\_\_.$$

20. When is the term *standard error of the mean* applied to a standard deviation? How is it related to the standard deviation of the population being sampled? Can it ever exceed the standard deviation of the population being sampled? Justify your conclusion.

Overview

1. Find the following probabilities under the standard normal distribution:
    (a) $P(0 \leq Z \leq 1.25)$
    (b) $P(-0.15 \leq Z \leq 0.36)$
    (c) $P(Z \geq 2.50)$
    (d) $P(Z < -0.52)$

2. For a normal distribution with $\mu = 25.0$ and $\sigma = 4.0$, find:
   (a) $P(X \leq 27.0)$
   (b) $P(22.0 \leq X \leq 28.0)$
   (c) $P(X > 35.0)$

3. Find the following "percentiles" for the standard normal distribution:
   (a) $z_{0.10}$
   (b) $z_{0.60}$
   (c) $z_{0.90}$
   (d) $z_{0.99}$
   (e) $z_{0.995}$

4. After a household survey in a particular community, a researcher announced that the cumulative probability distribution of the amount of money desired for a "guaranteed annual income" could be expressed by $F'(x) = x^2/16$ where $0 \leq x \leq 4$, and x is income in thousands of dollars. Find the probability for the following intervals:
   (a) $0 \leq X \leq 1$
   (b) $1 \leq X \leq 2$
   (c) $2 \leq X \leq 3$
   (d) $3 \leq X \leq 4$

5. For distributions of what sort of data do we use the following symbols for the mean and standard deviation:
   (a) $\mu_{\Sigma X}$ and $\sigma_{\Sigma X}$?
   (b) μ and σ, or μ_X and σ_X?
   (c) $\mu_{\bar{X}}$ and $\sigma_{\bar{X}}$?

6. Determine each value of x for which the following standard scores are obtained if μ and σ for the original population are 4 and 6, respectively:
   (a) $z = 1$
   (b) $z = -2$
   (c) $z = 1.2$
   (d) $z = -4.62$
   (e) $z = 0$
   (f) $z = 7$

7. For a population with $\mu = 10$ and $\sigma = 4$, complete the following probability statements by entering the correct numerical value in the blank provided:
   (a) $P(X < 7) = P(Z <$ ______$)$
   (b) $P(X \geq 12) = P(Z \geq$ ______$)$
   (c) $P(X <$ ______$) = P(Z < 1.2)$
   (d) $P(8 < X < 12) = P($______ $< Z <$ ______$)$
   (e) $P($______ $< X <$ ______$) = P(-1.6 < Z < 1.6)$
   (f) $P(4 < X < 13) = P($______ $< Z <$ ______$)$

8. If, in addition to the information given in Exercise 7, you know that the population is normal, determine the correct numerical value of each probability referred to in that exercise.

9. Without referring to Appendix A for the standard normal distribution, state which (if either) of the following probability pairs is the larger. Justify your conclusions.
   (a) $P(Z = 0)$ and $P(Z = 1.2)$
   (b) $P(Z < 0)$ and $P(Z \leq 0)$
   (c) $P(Z < 0)$ and $P(Z > 0)$
   (d) $P(1.2 \leq Z \leq 1.7)$ and $P(-1.2 \leq Z \leq 1.7)$
   (e) $P(-0.6 \leq Z \leq 0.6)$ and $P(-1.6 \leq Z \leq 1.6)$
   (f) $P(Z < 1.5)$ and $P(Z > 1.5)$

10. An attitude test was given by a large firm to job applicants to determine their suitability for employment. Scores are distributed with a slight negative skewness and have a mean of 38 and a standard deviation of 4.
    (a) If a random sample of size 36 is taken from the file of test results, what is the probability that the mean score for that sample does not exceed 37?

(b) If a random sample of size 50 is taken, determine $P(38 \leq \bar{X} \leq 39.5)$.

(c) If a random sample of 64 test results is considered, what is the probability that its mean differs from the true population mean by at least 1.5?

11. As the size n of samples from a normal population increases, the sampling distribution of $\bar{X}$ remains centered at the mean of the original population but the shape of the bell curve changes. In what way does it change and why?

12. If the variable z represents the standard score, what is the mean of the distribution z? What is the standard deviation of this distribution? Do your answers depend in any way on the distribution of the original x values?

13. Without consulting Appendix A, determine correct answers to the following. The area under the standard normal curve to the left of each value given below is equal to the area under the standard normal curve to the right of what value? Express your answers in the form $z_{F'}$, where F' stands for cumulative (less than) probability.

(a) z_0
(b) $z_{0.10}$
(c) $z_{0.95}$
(d) $z_{0.875}$

14. State lottery ticket sales in a certain city result in a mean weekly revenue of $48,000 to the state, with a standard deviation of $2000.

(a) Find the total expected revenue to the state for the next 10-week period from lottery ticket sales in this city.

(b) The state can expect with at least 90 percent certainty that its revenue from such sales over the next 10 weeks will fall within what range of dollar values?

(c) If you know that the dollar revenues from such weekly sales are normally distributed, with what probability are you assured that the total revenue over the next 10 weeks will fall within the dollar range you obtained in part (b)?

(d) Comparing your results from parts (b) and (c), how does your knowledge that a population is normal affect the probability that a sample sum will fall within a certain distance of $\mu_{\Sigma X}$?

15. The amount of additives in a certain food product has a right triangular distribution over a range of 0 to 12 percent by weight. Such a distribution has a mean of 4 and a variance of 8. For samples of size $n = 50$ from this population, what is the probability that the sample mean will lie within two percentage points of the population mean?

16. Consider a triangular probability density function, as shown in the illustration, such that the height above any value of x is $x/32$.

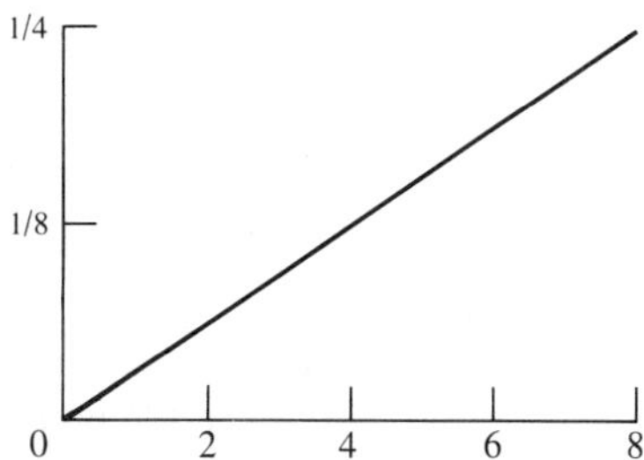

(a) What is the height of the function above

(i) $x = 0$?
(ii) $x = 2$?
(iii) $x = 3$?
(iv) $x = 8$?

(b) If the area of a triangle is given by the formula *area* = ½ *times the base times the height*, what is the area of the triangle shown?

(c) Bearing in mind the formula given in part (b), determine numerical values for the probability that

(i) $X < 2$ (iv) $X = 2$

(ii) X exceeds 2 (v) $4 \leq X \leq 6$

(iii) $X \neq 2$ (vi) X is not less than 8

(d) If the triangle were such that values of x range from 0 to 8 but the height over any value of x is $x/3$, why could it not possibly represent a probability density function?

Glossary of Equations

7.1 $\mu_{\Sigma X} = n\mu$ — The expected value of ΣX in n independent draws from a population with mean μ is $n\mu$.

7.2 $\sigma^2_{\Sigma X} = n\sigma^2$ — The variance of ΣX in n independent draws from a population with variance σ^2 is $n\sigma^2$.

7.3 $\mu_{\bar{X}} = \mu$ — The expected value of $\bar{X}$ in n independent draws from a population with mean μ is equal to the population mean.

7.4 $\sigma^2_{\bar{X}} = \dfrac{\sigma^2}{n}$ — The variance of $\bar{X}$ in n independent draws from a population whose variance is σ^2 is the population variance divided by the sample size n.

7.5 $z = \dfrac{x - \mu}{\sigma}$ — The standard score (z score) of an observation x is its deviation from the population mean divided by the standard deviation.

7.6 $z = \dfrac{\sum x - \mu_{\Sigma X}}{\sigma_{\Sigma X}}$ — The standard score for a statistic defined as a sample sum.

7.7 $\sigma_{\bar{X}} = \dfrac{\sigma}{\sqrt{n}}$ — The standard deviation of the sampling distribution of $\bar{X}$, also called the standard error of the (sample) mean.

7.8 $z = \dfrac{\bar{x} - \mu}{\sigma_{\bar{X}}}$ — The standard score for a statistic defined as a sample mean $\bar{x}$.

7.9 $\bar{x} = \mu + z\sigma_{\bar{X}}$ — The sample mean expressed as a function of the standard z score and the mean and standard deviation of the sampling distribution of the mean.

Part Three
Basic
Statistical Inference

8

Inference for Means – Variance Known

In the section of the book immediately prior to this we typically began our discussions with a completely defined statistical population. We knew the shape of the distribution and the values of its parameters—the mean μ and standard deviation σ. We used this information and sampling theory to find important characteristics of the sampling distribution of some statistic such as the sample mean $\bar{X}$. In this and in many of the following chapters we will reverse the direction of thought. We will assume that we have begun by finding the value of some sample statistic such as the sample mean $\bar{x}$. Our objective will be to use this information to make an inference about the value of some related population parameter such as the population mean μ. This type of procedure is called **statistical inference**.

Two major classes of activity in statistical inference are **estimation** and **testing hypotheses**. Both begin with the selection of a random sample from the statistical population under study. Both are concerned with the value of some parameter of that population, such as the mean μ. Statistical estimation begins with no preconception as to the value of the parameter. The objective is to use the information in the sample to find a reasonable value, whatever it is. On the other hand, hypothesis testing begins with a definite statement as to what the value of the parameter is and then compares the sample evidence with that value.

Estimation

Point Estimation

Suppose a market survey researcher must estimate the current mean weekly income per household (μ) in a certain city. Having selected a random sample of 80 households, the researcher obtains the observations of weekly income that constitute the data. How can the value of the parameter best be estimated? Should the sample mean or the sample median or the sample mode be calculated to mention only a few of the many possible statistics that might be calculated? Suppose the researcher arbitrarily elects to calculate the sample mean $\bar{x}$, finds the value to be \$104, and takes this value as the estimate of the value of the parameter sought (μ). In statistical language, the researcher has chosen to use the sample mean $\bar{X}$ as the **estimator** of the population mean μ, and the **point estimate** of the population mean is \$104. It is called a point estimate because it is only a single value or point on the real number line.

In the example above, the selection of the sample mean as the estimator of the population mean was arbitrary. It would be better to have a sound theoretical basis for selecting an estimator, and such a basis is provided by estimation theory.

Estimation theory has been developed by studying the behavior of different sample estimators with respect to the parameter being estimated. The goal is to determine how well each estimator conforms to several desirable characteristics for estimators. One such characteristic is *absence of bias.* Under random sampling conditions, the sampling distribution of values of an unbiased estimator will be centered on the parameter being estimated. For example we have learned that the various values of the sample mean $\bar{x}$ from random samples of n observations each are centered on the mean of the population μ from which the samples came. If we use a single value of $\bar{x}$ as an estimator of an unknown value of μ, we know that the selected value of $\bar{x}$ will very rarely equal μ but we can expect it to be correct on the average.

Definition An **unbiased estimator** is a statistic having an expected value the same as that of the parameter being estimated. In other words, an unbiased estimator is correct *on the average.*

The sample mean is an unbiased estimator of the population mean for any statistical population whatever. This general absence of bias does not hold for all estimators, as we will see in the next chapter.

Interval Estimation

All point estimators have one major shortcoming. In any given instance, there is nothing to tell us how much reliance to place on the estimate obtained. The probability that the mean of a particular random sample will be exactly the same as the population mean, for example, is usually very small and cannot be determined in most sampling situations. An interval estimate, on the other hand, carries with it a measure of the confidence we can place in it. Such an estimate consists of a range of values, rather than just one, for our estimate of the underlying parameter.

Three things determine how we must proceed when we want to make an interval estimate of the population mean: our knowledge about the shape of the population distribution, our knowledge about the value of the population variance σ^2, and the number of observations in our random sample. In this chapter we will assume that the population variance σ^2 is known. Often in practice we don't know the population variance, but we will see in the next chapter that this void can be filled.

Normally Distributed Population A food packing plant has a machine that fills No. 303 cans with cream style corn. The operator sets a dial to control the amount of corn deposited in each can, but experience has shown that, regardless of the dial setting, machines of this type fill cans with a standard deviation σ of 0.80 ounce. In other words, the statistical population of fill weights made at any fixed dial setting has a standard deviation of 0.80 ounce. In addition, the fill weights at any one setting are assumed to be normally distributed, because experience with many large samples from this type of machine has shown this to be the appropriate shape. **8.1**

Suppose upon coming to work one morning the operator finds the weight dial on the filling machine has fallen off. After replacing and tightening the dial and setting it at a particular reading, the operator does not know what true fill weight this reading indicates. A sample of fill weights must be taken and the sample used to estimate the mean of the population of fill weights at this setting. The operator will then loosen the dial, move it to this estimate without changing the amount of fill delivered by the machine, and retighten it.

The operator arbitrarily decides to fill 25 cans at the fixed setting and weigh the contents of each can. Then the sample mean $\bar{x}$ is calculated and found to be 17.60 ounces. If the 25 content weights are treated as independent random observations, 17.60 ounces can be used as an unbiased point estimate of the population mean μ. On the other hand, the sampling theory covered in Chapter 7 can be used to form a different type of estimate of the population mean—an interval estimate. We will describe how to make such an estimate and the rationale behind it.

We have the mean of a single random sample that we assume came from a normally distributed population of fill weights. The sample mean is 17.60 ounces and there are 25 observations in the sample. We also know that the standard deviation of the population is 0.80 ounce.

We can now refer to the statement on page 186 and the discussion following it. We conclude from this material that our sample mean of 17.60 belongs to a distribution of all possible means that could arise from random samples of 25 observations each. Furthermore, we can state three characteristics of this distribution of sample means:

1. It is normally distributed.
2. It has the same mean as does our objective, the mean of the parent population, even though we don't know the numerical value of either mean, that is, $\mu_{\bar{X}} = \mu$.

3. It has a standard deviation $\sigma_{\bar{X}}$, called the **standard error of the mean**, the value of which can be found from the standard deviation of the parent population ($\sigma = 0.80$ ounce) and the sample size ($n = 25$). Specifically,

$$\sigma_{\bar{X}} = \frac{\sigma}{\sqrt{n}} \quad \text{or} \quad \sigma_{\bar{X}} = \frac{0.80}{\sqrt{25}}, \text{ or } 0.16;$$

that is, the standard error is 0.16 ounce.

We know that we have one observation ($\bar{x}$) belonging to a distribution for which we know two of the three determining characteristics. We know the shape of the distribution (normal) and we know its dispersion ($\sigma_{\bar{X}} = 0.16$ ounce). But we do not know where our observation lies with respect to the distribution's mean ($\mu_{\bar{X}} = \mu$) because we do not know the value of that mean.

The single sample mean we have observed ($\bar{x}$) could be someplace well below the distribution's mean $\mu_{\bar{X}}$, very near the population mean, or well above it. From our work with normal probabilities, however, we can be confident that our observation lies within, say, three standard errors of the population mean. This circumstance is illustrated in Figure 8.1.

In Figure 8.1, the distribution to the left is built on the assumption that our sample mean $\bar{x}$ of 17.60 ounces lies exactly three standard errors above the population mean $\mu_{\bar{X}}$. To be in this position the standard normal deviate (z) must be $+3$ in the expression

$$z = \frac{\bar{x} - \mu_{\bar{X}}}{\sigma_{\bar{X}}}.$$

We already know that $\bar{x}$ is 17.60 and $\sigma_{\bar{X}}$ is 0.16 in this expression. Consequently we can state that

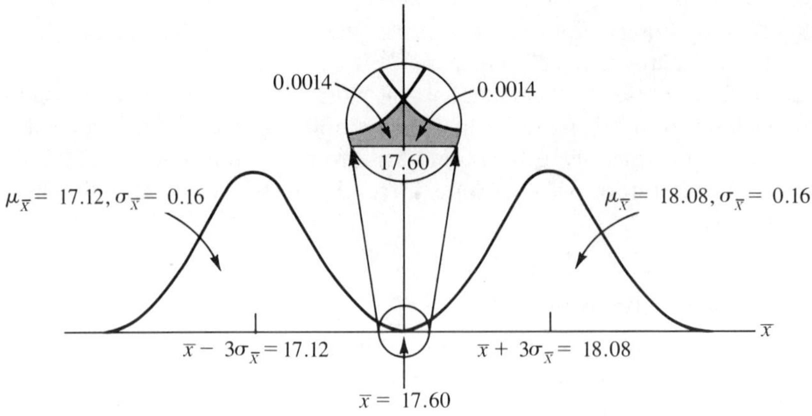

Figure 8.1 *Two possible sampling distributions of the mean*

$$3 = \frac{17.60 - \mu_{\bar{X}}}{0.16},$$

from which

$$\mu_{\bar{X}} = 17.60 - 3(0.16)$$

or

$$\mu_{\bar{X}} = 17.12 \text{ ounces.}$$

If the mean of the population of fill weights is 17.12 ounces, the mean of the random sample we selected (17.60) is three standard errors greater. It is not very plausible that a sample mean of 17.60 could have resulted from a population with a mean as small 17.12. Similarly, for our sample mean to lie three standard errors below the population mean, $\mu_{\bar{X}}$ must be 18.08 ounces, 17.60 + 3(0.16). We can designate 18.08 and 17.12 as the limits of our interval estimate of the population mean. Alternatively we can reason that if it is likely that $\bar{x}$ is within three standard errors of $\mu_{\bar{X}}$ (and of μ) it is also likely that μ is within three standard errors of $\bar{x}$. In other words, it is likely that μ is somewhere between $\bar{x} - 3\sigma_{\bar{X}}$ and $\bar{x} + 3\sigma_{\bar{X}}$. Hence for our example it is likely that μ is between 17.12 and 18.08 ounces.

By using normal curve theory we can get a measure of how much confidence we can place in this estimate. In Appendix A, the probability of a randomly selected observation falling three standard deviations or more above the mean is seen to be 0.0014. Hence only 14 times in 10,000 would we expect to get a sample mean in the tail of the distribution above 17.60 ounces in Figure 8.1 when the population mean is 17.12 ounces. Similarly, only 14 times in 10,000 would we expect to get a sample mean in the tail of the distribution below 17.60 ounces when the population mean is 18.08 ounces. Now 17.12 is 0.48 ounce (0.5 to one decimal) below 17.60, and 18.08 is half an ounce above it. As a result the operator can change the dial to 17.60 ounces and be virtually certain that it is reading within a half ounce of the true mean fill at the machine's current setting. He can be highly confident that the population mean lies in the interval which has 17.12 ounces as its lower limit and 18.08 ounces as its upper limit. This interval centered on 17.60 ounces ($\bar{x}$) and running from three standard errors below $\bar{x}$ to three standard errors above it constitutes an **interval estimate of the population mean**.

Next consider Figure 8.1 in the light of what would happen were we to set the lower limit of our interval estimate even farther to the left than 17.12 and the upper limit even farther to the right than 18.08. Then the sum of the two normal tail areas would come to less than 0.0028 and we could place even greater confidence in this wider interval estimate. Conversely, bringing the two limits in so that they are less than 0.48 ounce away from 17.60 will establish a narrower interval in which we would place less confidence.

Using the reasoning just developed, we can construct a measure of our degree of confidence in a particular interval estimate. We call this measure the **confidence coefficient**. As with probabilities, a confidence coefficient can range from 1 to 0. When we are certain that an interval contains the population mean, we have complete confidence in that estimate and assign to it a confidence coefficient of 1. When we are certain an interval does *not* contain the population mean, we assign to it a confidence coefficient of zero. Between these extremes we find the confidence coefficient by subtracting the sum of the tail areas, such as those illustrated in Figure 8.1, from 1. Hence, for the situation shown in that figure, the confidence coefficient is 1 − 0.0028, or 0.9972. Compare this coefficient with the value we would get by basing the interval on two standard errors rather than three. If we did so, the limits for our interval estimate would be at 17.28 and 17.92 ounces. The sum of the tail areas would increase to 0.0456, and the confidence coefficient would drop to 0.9544 for this narrower interval.

Before formally summarizing our discussion of interval estimates, we must consider another major influence—the effect of variations in the number of sample observations on an interval estimate of the population mean. Instead of 25 observations, suppose the operator in our example had chosen a random sample of 100 observations. Further suppose that the mean $\bar{x}$ of this larger sample turned out to be 17.30 ounces rather than 17.60. In this case, the standard error of the mean $\sigma_{\bar{X}}$ would be only 0.08 ounce because

$$\sigma_{\bar{X}} = \frac{\sigma}{\sqrt{n}} = \frac{0.80}{\sqrt{100}}, \text{ or } 0.08.$$

An interval with limits three standard errors above and below the new sample mean would have a lower limit (μ_L) of

$$\mu_L = \bar{x} - 3\sigma_{\bar{X}} = 17.30 - 3(0.08), \text{ or } 17.06 \text{ ounces}$$

and an upper limit (μ_U) of

$$\mu_U = \bar{x} + 3\sigma_{\bar{X}} + 17.30 + 3(0.08), \text{ or } 17.54 \text{ ounces.}$$

For this larger sample of 100 observations, the length of the interval estimate is only 0.48 ounce (17.54 − 17.06) as compared with an interval length of 0.96 ounce (18.08 − 17.12) based on the sample of 25 observations. Nonetheless, the confidence coefficients are 0.9972 for both intervals, because both sets of limits are three standard errors from their respective sample means. This example shows an important point: By increasing the sample size, we can maintain a given degree of confidence while narrowing the length of an interval estimate. More information allows us to narrow the interval without loss of confidence. Alternatively, given a larger sample we could elect to maintain the same length of interval and increase the value of the confidence coefficient. Changes in the opposite direction would accompany a decision to select a smaller, rather than a larger, sample.

In the situation just discussed, we are given the values of a sample mean $\bar{x}$, the number of observations n on which it is based, and the standard deviation σ of the population from which it came. We also have reason to assume that the population is normally distributed. When the population sampled is normally distributed and its standard deviation is known we can form an interval estimate by executing the steps in the following procedure:

Rules

1. Select a relatively large value (usually 0.90 or greater) for the confidence coefficient.* Set this value equal to $1 - \alpha$ and solve for α, where α is the sum of the areas in the tails of the sampling distributions corresponding to those shown in Figure 8.1.
2. From Appendix A find the magnitude of the standard normal deviate (z) that cuts off a tail area equal to $\alpha/2$. Call this magnitude $z_{\alpha/2}$ and disregard its algebraic sign.
3. Find the standard error of the mean $\sigma_{\bar{X}}$ by dividing the population standard deviation by the square root of the sample size:

$$\sigma_{\bar{X}} = \frac{\sigma}{\sqrt{n}}.$$

4. Find the lower (μ_L) and upper (μ_U) limits of the interval estimate by subtracting $(z_{\alpha/2} \cdot \sigma_{\bar{X}})$ from and adding $(z_{\alpha/2} \cdot \sigma_{\bar{X}})$ to the sample mean $\bar{x}$:

$$\blacksquare \quad \mu_L = \bar{x} - z_{\alpha/2} \cdot \sigma_{\bar{X}} \quad \blacksquare \qquad \textit{8.1(a)}$$

and

$$\blacksquare \quad \mu_U = \bar{x} + z_{\alpha/2} \cdot \sigma_{\bar{X}} \quad \blacksquare \qquad \textit{8.1(b)}$$

5. Alternatively, we can say that the $(1 - \alpha)$ **confidence interval estimate** of the population mean μ is defined symbolically as

$$\bar{x} \pm z_{\alpha/2} \cdot \sigma_{\bar{X}}. \qquad \textit{8.2}$$

Suppose in our can-filling example we are beginning with the sample mean of 17.60 ounces, based on 25 observations from a normal population with a standard deviation of 0.80 ounce. Furthermore, suppose we choose a confidence coefficient of 0.95. Then α is 0.05 and $\alpha/2$ is 0.025. In Appendix A, we see that a tail area of 0.025 is cut off by a standard normal deviate (z) of 1.96. Hence, disregarding sign, $z_{\alpha/2}$ is 1.96. In addition, the standard error $\sigma_{\bar{X}}$ is 0.16 ounce, as before. It follows from Equation 8.2 that the 0.95 confidence interval estimate of the population mean is $17.60 \pm 1.96(0.16)$, or 17.60 ± 0.31 ounces. From this we obtain a lower limit of 17.29 ounces and an upper limit of 17.91 ounces for the 0.95 confidence interval estimate of the population mean.

Table 8.1 shows some values commonly used in setting confidence interval estimates when the situation is as described in this section. Disregarding sign, the value of $z_{\alpha/2}$ from Appendix A associated with each value of $(1 - \alpha)$ is also listed.

*The coefficient must be large enough to represent a high level of confidence in the estimate yet not be so high that the cost of sampling is excessive. Judgment and trial and error can be used to find a satisfactory value.

Table 8.1 Values Commonly Used for Confidence Interval Estimates

$1 - \alpha$	$z_{\alpha/2}$
0.90	1.645
0.95	1.96
0.9544	2.00
0.99	2.58
0.9972	3.00

When the population is normally distributed, the sampling distribution of the mean is normally distributed for any sample size whatsoever. Hence Equations 8.1
8.1 and 8.2 and the rules which accompany them apply to *any* sample size.

Unspecified Population Distribution We turn now to the case for which we are given the sample mean $\bar{x}$, the sample size n, and the population standard deviation σ as before. But we have no reason to believe that the population is normally distributed. In this case we can make use of the central limit theorem–provided that the number of observations in the sample is at least 30. For all but extremely skewed populations, this theorem and a minimum sample size of 30 assure us that the sampling distribution of the mean will be essentially normally distributed, even though the parent population is not.

A look at the rules stated in conjunction with Equation 8.1 shows they require only that the sampling distribution of the mean, and not the parent population, be normally distributed. If the sampling distribution of the mean is normally distributed, then we can use Appendix A to find the value of $z_{\alpha/2}$ in Equation 8.2. Since the central limit theorem provides the assurance that the distribution is essentially normal, we can also apply the steps of the rule in the previous section to the case being considered here.

Rule When the population standard deviation σ is known but there is no reason to believe that the population is normally distributed, the procedure leading to Equations 8.1 should be used to set a $(1 - \alpha)$ confidence interval estimate of the population mean μ provided that the sample size n is at least 30.

When the population distribution is not specified, the possibility of selecting a random sample of less than 30 observations sometimes arises. The case of parent populations characterized by considerable skewness is especially troublesome. Sampling distributions of the mean for small samples for such populations will also be skewed, though not to the same extent as the populations. This rules out using the procedures just discussed for making interval estimates of the population mean. When the population is not known to be normally distributed, it is usually most effective to increase the sample size to at least 30.

Determining Sample Size

In the previous section, the operator of the can-filling machine arbitrarily chose 25 as the number of observations to be selected for his random sample. We saw that this permitted him to set the dial on his machine at 17.60 ounces with virtual certainty that this is within a half ounce of the mean fill. The 0.9972 confidence interval estimate of the population mean runs from 17.12 to 18.08 ounces. We also saw that increasing the sample size to 100 observations made the 0.9972 confidence interval half as long as it was with 25 observations. It is practically certain that setting the dial to correspond with the mean of such a sample would result in a mean fill that would be in error by a maximum of only a quarter ounce. We can see that the error associated with an interval estimate depends on the number of sample observations. Finding this relationship will enable us to specify the sample size needed to achieve a desired expected maximum error before the sample is selected.

Equations 8.1 tell us what the lower (μ_L) and upper (μ_U) limits are for the population mean at specified confidence level ($1 - \alpha$). Alternatively, we can state that the population mean falls in the interval

$$\bar{x} \pm z_{\alpha/2} \cdot \sigma_{\bar{X}}$$

with the specified level of confidence. We can think of $z_{\alpha/2} \cdot \sigma_{\bar{X}}$ as being the maximum distance the population mean (μ) may reasonably be expected to lie above or below the sample mean. Hence $z_{\alpha/2} \cdot \sigma_{\bar{X}}$ is the maximum expected error between $\bar{x}$ and μ when $1 - \alpha$ is the level of assurance. Let's assign this maximum error the symbol E. Then, disregarding algebraic sign,

$$E = z_{\alpha/2}\sigma_{\bar{X}}.$$

We know that the standard error of the mean ($\sigma_{\bar{X}}$) can be stated as a function of sample size; that is, $\sigma_{\bar{X}} = \sigma/\sqrt{n}$. Substituting this in the preceding equation for error E will give us a statement of maximum error as a function of sample size n:

$$E = z_{\alpha/2} \cdot \frac{\sigma}{\sqrt{n}}. \qquad 8.3$$

Therefore: With probability ($1 - \alpha$), the expected maximum error E *in the estimate of the population mean is equal to the product of the associated standard normal* *deviate and the standard error of the mean.* In Equation 8.3, σ is a constant and so is $z_{\alpha/2}$ for a specified value of α. As a result the value of E varies only with sample size n. The larger the sample the smaller the expected maximum error.

Equation 8.3 gives us the value of E for a given sample size n. Alternatively we often want to find the size of sample that will be required to limit the expected maximum error to a specified value for E. Solving Equation 8.3 for sample size, we get the following:

Statement

The required number of observations n to limit the expected maximum error in an interval estimate of the population mean μ to a specified quantity E with probability $(1 - \alpha)$ is

$$n = \left(\frac{z_{\alpha/2} \cdot \sigma}{E}\right)^2 \qquad 8.4$$

where $z_{\alpha/2}$ is the standard normal deviate and σ is the population standard deviation.

The filling machine operator described earlier is beginning another production run of several thousand cans. He wants to fill a different size can such that the mean fill will be 7.6 ounces with a maximum error of 0.2 ounce. He wants the probability to be 0.95 that 0.2 ounce is the maximum error. He moves the control dial to a reading of 7.6 ounces. How many cans should he fill to check this reading?

We already know that the standard deviation σ of fill weights is 0.80 ounce for any setting of this machine. We also know that the distribution is normal. From Table 8.1 we find that the standard normal deviate $(z_{\alpha/2})$ associated with 0.95 is 1.96. The expected maximum error E is to be 0.2 ounce. From Equation 8.4,

$$n = \left[\frac{1.96(0.80)}{0.2}\right]^2$$
$$= (7.84)^2, \text{ or } 61.47.$$

Since the number of observations must be an integer, a sample of 62 observations will provide the desired protection.

The operator fills 62 cans, weighs them, and finds that the sample mean $\bar{x}$ is 7.72 ounces rather than 7.6. Being careful not to change the fill delivered by the machine, he loosens the control knob on its shaft, moves it to a reading of 7.72, and retightens it. Then he changes the reading to 7.6, which also changes the fill setting to this quantity. He can now begin his production run with 0.95 confidence that the mean fill will be within 0.2 ounce of 7.6 ounces.

Note that in Equation 8.3 the expected maximum error E is a function of the sample size n but *not of the size of the population.* Most people find this strange when they first encounter it. Intuitively it seems that sampling error should be strongly influenced by the size of the population. We will see later in the chapter on sampling techniques that population size does influence sampling error. But the influence is negligible unless the number of observations in the sample is more than about 10 percent of the population.

Summary

Estimation is a very important type of statistical inference. There are many instances when an estimate of a population mean is needed and only random sample observations are available. The sample mean is an unbiased point estimator of the population mean, but there is no way to assess the degree of reliance to place on a point estimate. An interval estimate does not have this shortcoming. Adding and subtracting an error term to and from the sample mean establishes an interval estimate of the population mean. For random samples of any size from normal populations and 30 or more observations from other population distributions, the error term is the product of a selected standard normal deviate and the standard error of the mean. A confidence coefficient associated with the selected normal deviate is a measure of the reliance to be placed in the estimate. The same theory can be used to provide a way for choosing the number of observations in a sample to limit the expected maximum sampling error to a specified value with a known probability.

See Self-Correcting Exercises 8A.

Exercise Set A

1. To what two characteristics is it desirable for a sample estimator to conform?
2. Does the fact that the sample mean is an unbiased estimate for the mean of a population depend in any way on special characteristics of the population or the size of the sample? Explain.
3. Why is an interval estimate preferable to a point estimate?
4. Without changing the sample size, does increasing the confidence level of an interval estimate result in the narrowing or the widening of the interval? How does this fit in with the statement that "the more certain we are, the less there is of which to be certain"?

5. From a normal distribution with $\sigma = 16$, a sample of size 36 is taken and its mean is found to be 9. We wish to establish an interval estimate of the population mean with its lower and upper limits 2.5 standard errors from the sample mean.
   (a) What is the confidence coefficient for this interval?
   (b) Determine: (i) μ_L and (ii) μ_U.

6. Determine the confidence coefficients for interval estimates of μ for a normal population for which μ_L and μ_U are the following numbers of standard errors from the sample mean:
   (a) 1.64
   (b) 1.96
   (c) 2.33
   (d) 2.58
   (e) 3.00

7. Do your answers to the preceding exercises depend in any way on the sample size for which $\bar{x}$ is obtained? Do they depend in any way on the standard deviation of the population from which the sample is taken?

8. From a normal population with standard deviation 5, a sample is taken and $\bar{x}$ is found to be 30. Determine an interval estimate for μ having μ_L and μ_U two standard errors from the sample mean if the sample size is:
   (a) 4
   (b) 25
   (c) 50
   (d) 100

9. We can narrow the interval estimate and still maintain the same level of confidence by doing what to the sample size?

10. What happens to $5/\sqrt{n}$ as n increases? Why? As the sample size increases, what happens to the standard error of the mean?

11. Determine, using Equation 8.2, the following confidence interval estimates for μ if the sampled population has a standard deviation of 4 and a sample of size 100 has a mean of 0:
    (a) 0.90
    (b) 0.95
    (c) 0.98

12. From a normal population having a standard deviation of 16, a sample is taken and found to have a mean of 16. Find a 95 percent confidence interval estimate of the population mean if the sample size is:
    (a) 4
    (b) 16
    (c) 100
    (d) 400

13. From a skewed population having a standard deviation of 8, a sample is taken and its mean is found to be 112. If the size of the sample is 16, Equation 8.2 does not apply. Why not? What is one way to avoid this difficulty?

14. From a population with slight negative skewness and a variance of 36, a sample of size 50 has a mean of 68.6. Does Equation 8.2 apply to interval estimation of μ here? If not, why not? If so, find the 0.99 confidence interval estimate of μ.

15. For a fixed level of confidence, how can we make the expected maximum error E as small as we wish? In practical application, why might we have to settle on a happy medium between expected maximum error and sample size?

16. Use Equation 8.4 to determine how large a sample must be taken from a population of sawmill-cut board lengths having a standard deviation of 13 millimeters if we wish to

be assured, at a level of confidence of 0.95, that the true mean length differs from the sample by:

(a) At most 4 millimeters.

(b) No more than 2 millimeters.

(c) Less than 1 millimeter.

17. Referring to the population of Exercise 16, answer parts (a), (b), and (c) for a level of confidence of 0.98.

18. A statewide study has shown that weights of 12-year-old boys are normally distributed with a standard deviation of 10 pounds for any major city in the state. A random sample of 225 such weights in one of these cities has a mean of 85 pounds. Find the 0.9972 confidence interval estimate of the mean weight of all 12-year-old boys in the city.

19. Balances for commercial accounts in a branch bank are known to be distributed with moderate positive skewness. The standard deviation of the distribution has remained essentially constant at $210 for the past 3 years, and there is no reason to assume it has changed this year. A random sample of 49 accounts has a mean balance of $832 today. What is the 0.95 interval estimate of the mean balance for all accounts?

Tests of Hypotheses

In the first part of this chapter we considered how to use the data from a sample to estimate the value of the mean of the population from which the sample came. We had no preconception as to the value of the population mean. We simply wanted the best estimate of its value that could be obtained from a random sample.

As discussed here hypothesis testing begins with a statement about the value of the population mean. We have some basis on which to form a preconception, or hypothesis, about what the value is. We then test the hypothesis. The test consists of combining the data from a random sample with relevant sampling theory to make a decision about the hypothesis. The alternatives are to accept the statement about the value of the population mean or to reject it because some other values are much more plausible.

In hypothesis testing, as in estimation, drawing a conclusion about the value of the population mean from a sample that makes up only part of the population always entails a risk. Estimation presents the risk that our sample estimate of the parameter is more in error than we have specified. In hypothesis testing we run the risk that the sample evidence leads us to the wrong alternative concerning the statement being tested.

Two-Tailed Tests

In a two-tailed test, we want to guard against a bad decision in either direction. If we hypothesize that the population mean is a certain value, we want a test which is highly likely to lead us to reject that hypothesis when the true mean lies substantially above or below that value.

8.2 **Normally Distributed Population** A cement plant has just installed a new machine to fill 100-pound sacks. Although the manufacturer of the filling machine sets machines for the correct fill, vibration in shipment sometimes introduces a fixed amount of error into the dial which controls fill weight. Therefore, the cement manufacturer is advised to check. A two-tailed test protects him from overfilling bags and from underfilling them. On the one hand he won't be giving cement away and on the other he won't get into trouble with state inspectors for shorting customers.

The population standard deviation of fill weights σ for this type of machine is 0.90 pound, regardless of the setting. Similarly, any machine of this type can be assumed to deliver a normally distributed population of fill weights at any single setting. The cement manufacturer thinks that his new filling machine has not been disturbed in shipment and will deliver a mean fill weight of 100 pounds. Hence he decides to state that the population mean μ is 100 pounds and to treat this as a hypothesis to be tested.

To test the hypothesis that the population mean μ is 100 pounds, the manufacturer will fill 4 sacks and treat their content weights as a random sample of four observations. If the mean $\bar{x}$ of these four weights is far greater or far less than 100 pounds, he will reject his original hypothesis. If the mean is close to 100 pounds he will not reject it.

We can use sampling theory to establish a range for the sample mean $\bar{x}$. If the sample mean falls inside this range, we will accept the hypothesis. If the mean falls outside the range, we will reject it. We will have a single random sample of 4 observations that we assume to be from a normally distributed population with a standard deviation σ of 0.90 pound. Furthermore, if the manufacturer's hypothesis happens to be correct, the population mean fill weight μ will be 100 pounds. We begin with the premise that this hypothesis is true.

The statement on page 186 and the discussion following it now come into play. The mean $\bar{x}$ of the four observations that the manufacturer will select belongs to a distribution of sample means from all possible random samples of four observations each. If the hypothesis and the foregoing assumptions are true, then the distribution of sample means has three characteristics:

1. It is normally distributed.
2. It has the same mean ($\mu_{\bar{X}}$) as does the parent population (μ). This is 100 pounds, by hypothesis.
3. It has a standard deviation ($\sigma_{\bar{X}}$), called the standard error of the sample mean, that can be found from the standard deviation of the parent population ($\sigma = 0.90$ pound) and the sample size ($n = 4$). Specifically,

$$\sigma_{\bar{X}} = \frac{\sigma}{\sqrt{n}} = \frac{0.90}{\sqrt{4}}, \text{ or } 0.45 \text{ pounds.}$$

If the assumptions and the hypothesis are true, then the sample mean $\bar{x}$ of the four observations made by the manufacturer will be a member of the sampling distribution just described. Since this distribution of sample means is normal and we know its mean and standard deviation, we can use Appendix A to assess the probability—when the hypothesis is true—that $\bar{x}$ will lie more than a given number of standard errors ($\sigma_{\bar{X}}$) from the hypothesized mean. For example, we find that the probability is only 0.01 (0.0049 in either tail) that a sample mean selected randomly from the distribution of sample means will lie more than 2.58 standard errors above or below 100 pounds. Should the sample mean selected by the manufacturer be farther from 100 pounds than 2.58 standard errors, he may consider this too rare an event to accept and, instead, prefer to conclude that the hypothesized mean is wrong,

Suppose the manufacturer decides to reject the hypothesized value of 100 pounds if the sample mean turns out to be more than 2.58 standard errors away. Then we say he has selected a **level of significance** of 0.01 for his test and we give the level of significance the symbol α. The level of significance (α) is the probability of getting a sample result farther from the hypothesized value than the associated number of standard errors when the hypothesis is true.

To convert 2.58 standard errors to pounds, we multiply 0.45 pounds (the value of one standard error) by 2.58 to get 1.16 pounds. Hence, if the hypothesis is true, the sample mean will fall inside the interval from 98.84 to 101.16 (that is, 100 ± 1.16) pounds with probability 0.99 and outside the interval with probability 0.01. The values at the ends of this interval are called **critical values** and the interval itself is called the **acceptance region**. The situation is illustrated in Figure 8.2.

In Figure 8.2 the sample size and the assumptions determine the shape of the

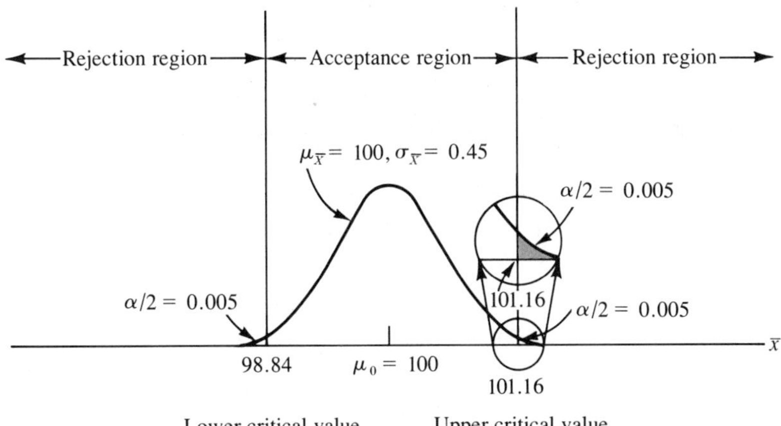

Null hypothesis: $\mu = 100$
Alternative hypothesis: $\mu < 100, \mu > 100$
Level of significance: $\alpha = 0.01$

Figure 8.2 *Hypothesized distribution of $\bar{X}$ and test criteria: two-tailed test*

sampling distribution of the mean and the value of its standard error. The hypothesized value of the population mean determines the sampling distribution's central location. We will give the hypothesized value of the mean the symbol μ_0. The statement that μ_0 is equal to 100 pounds is called the **null hypothesis**. It is the statement to be tested. A null hypothesis must always be stated in such a way that we know what to expect if it is true. Figure 8.2 illustrates our complete knowledge of the sampling distribution assuming the truth of the null hypothesis.

Now suppose the cement manufacturer selects a random sample of 4 filled sacks, weighs them, and finds the mean $\bar{x}$ to be 98.3 pounds. This value falls below 98.84 pounds. Consequently, it is in the **rejection region** and we must reject the null hypothesis. The evidence is that the machine is out of adjustment. The population mean fill weight appears to be considerably less than 100 pounds.

When the hypothesis test just described is compared with the interval estimation procedure described earlier in this chapter, two major differences are apparent. In the first place, an interval estimate is centered on the sample mean $\bar{x}$ but an acceptance region for a two-tailed hypothesis test is centered on the hypothesized value of the population mean (μ_0). The situation is as illustrated for our cement sack example in Figure 8.3. The upper portion of the figure shows the 0.99 confidence interval estimate of the population mean μ centered on the observed sample mean of 98.3 the manufacturer found. The lower portion shows the acceptance region for the hypothesis test for which the level of significance is 0.01.

The second difference between the two procedures involves the meaning of α. In the case of interval estimates, α is the extent to which we lack confidence in the

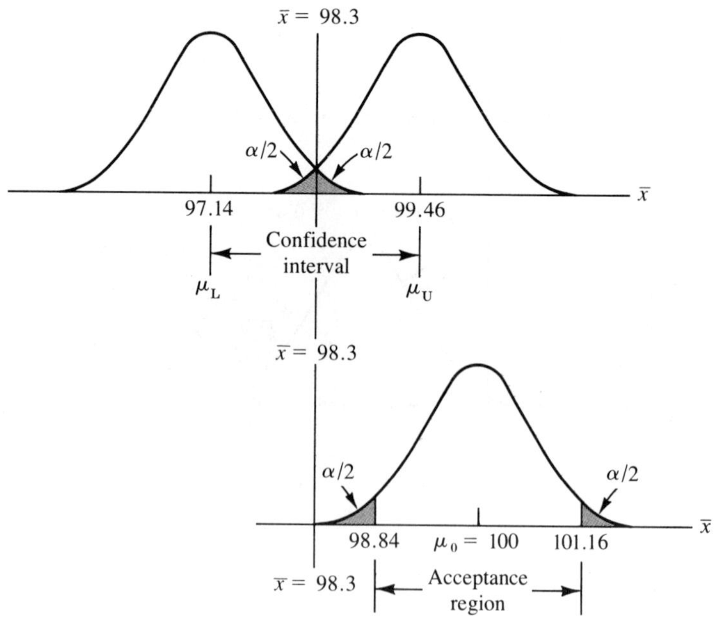

Figure 8.3 *Confidence interval and acceptance region compared*

estimate; in hypothesis tests, α is the probability of rejecting the null hypothesis when it is true.

The purposes of interval estimation and hypothesis testing are also different. Interval estimation is meant to provide a plausible range of values for the population mean μ based on the value of the sample mean $\bar{x}$. The purpose of hypothesis testing is to test the credibility of a presupposed value of μ against the sample result.

Now that we have discussed the rationale of testing a hypothesis about the mean of a statistical population, we can introduce a more generally accepted approach for performing this same test. Instead of defining $\bar{x}$ as the variable, we put everything in terms of standard scores, or z scores. Now

$$z = \frac{\bar{x} - \mu_0}{\sigma_{\bar{x}}} \qquad \textbf{8.5}$$

and

$$\sigma_{\bar{x}} = \frac{\sigma}{\sqrt{n}},$$

since our observations are sample means. Hence a value of z tells us how many standard errors $\sigma_{\bar{x}}$ a given value of $\bar{x}$ is away from the hypothesized value of the population mean μ_0. Many statisticians would say that we have changed the **test statistic** from $\bar{x}$ to z.

Suppose that, as above, we choose a level of significance α of 0.01 for the two-tailed test of the null hypothesis that the population mean is 100 pounds. Then in terms of z the *lower critical value* will be $z_{\alpha/2}$, or $z_{0.005} = -2.58$, and the *upper critical value* will be $z_{1-\alpha/2}$, or $z_{0.995} = +2.58$, as found directly from Appendix A. Hence the acceptance region will be $-2.58 \leq z \leq +2.58$ and the rejection regions will lie below -2.58 and above $+2.58$. This assumes that the sampling distribution of $\bar{x}$ is normal with a mean of 100 pounds and a standard error of 0.45 pound, or equivalently, that the sampling distribution of z in Equation 8.5 is normally distributed with a mean of zero and a standard deviation of 1.

Since the random sample we selected has a mean of 98.3 pounds, we find from Equation 8.5 that

$$z = \frac{98.3 - 100}{0.45}, \text{ or } -3.78.$$

The sample mean of 98.3 is 3.78 standard errors less than the hypothesized mean of 100 pounds. This value of z lies in the lower rejection region, and we arrive at the same conclusion as before. This is a direct parallel of our earlier procedure in which the $\bar{x}$ value of 98.3 fell below the acceptance region. When $\bar{x}$ is the statistic, the acceptance region is $98.84 \leq \bar{x} \leq 101.16$. When z is the statistic the logically equivalent acceptance region is $-2.58 \leq z \leq +2.58$.

We can now summarize the discussion to this point. Our objective is to perform a two-tailed test of a hypothesis about the value of a population mean μ. The test is based on three assumptions: independent random sampling, a normally distributed population, and a known value for σ, the population standard deviation. In addition, the sample size n to be used in the test has already been decided. We can perform the two-tailed test as follows:

Rules

1. State the null hypothesis and the alternative to it (by assigning, in this case, a numerical value to μ_0 in the statements

$$H_0: \quad \mu = \mu_0,$$
$$H_A: \quad \mu \neq \mu_0,$$

where H_0 is the null hypothesis and H_A is the alternative statement to be accepted if H_0 is rejected).

2. Select a value for α, the level of significance.
3. Select the test statistic (z in this case).
4. From the assumptions, determine the appropriate sampling distribution for the test statistic (z) and use that distribution (Appendix A) to find the critical values of the statistic ($z_{\alpha/2}$ and $z_{1-\alpha/2}$).
5. Find the sample value of the test statistic (from the sample mean $\bar{x}$ and Equation 8.5) and accept the null hypothesis if the value of the test statistic lies within the acceptance region ($z_{\alpha/2} \leq z \leq z_{1-\alpha/2}$). Otherwise accept the alternative hypothesis.

So long as the parent population is normally distributed and the value of its standard deviation σ is known, the sampling distribution of $\bar{X}$ will be precisely normal for any specified value of n, the sample size. As a result, the procedure just
8.2 described applies to both large and small samples—to any sample size whatsoever.

Unspecified Population Distribution Although we do not assume that the parent population is normally distributed in this case, we do know the value of its standard deviation σ. Furthermore, as in interval estimation for this case, a sample of 30 or more observations and the central limit theorem provide reasonable assurance that we are working with a sampling distribution of $\bar{X}$ that is virtually normal in shape. Consequently, we have everything we need to make the procedure described in the previous section applicable. Hence the test statistic is again z from Equation 8.5 and Appendix A gives its sampling distribution.

In interval estimation we pointed out that usually the most effective procedure is to increase the sample size to at least 30 when the population is not known to be normally distributed. For hypothesis testing in the same situation, the same advice applies. Increase the sample size to at least 30 observations when the shape of the population distribution is not known but the population standard deviation is known.

One-Tailed Tests

8.3

In the preceding section the cement manufacturer decided to use a two-tailed test. A two-tailed test protected him from either seriously overfilling or underfilling on the average.

Now let's consider this same situation from the viewpoint of a state inspector of weights and measures. The inspector's job in this instance is to see that customers get at least 100 pounds of cement per sack on the average. If the mean fill is exactly 100 pounds, then half the sacks will contain less than 100 pounds. This is unacceptable to the inspector. He wants the mean fill weight to be somewhat greater than 100 pounds so the majority of sacks will weigh at least 100 pounds. The manufacturer, aware of the slight variability in fill weights, is willing to set the machine to deliver a mean fill slightly greater than 100 pounds to satisfy the inspector. Now the inspector wants to check the resulting process to see that it is satisfactory.

The inspector decides to select a random sample of 36 fill weights and find the sample mean $\bar{x}$ as a check on the process mean μ. Will he conclude that the process mean is greater than 100 pounds if the sample mean turns out to be less than 100 pounds? No. Such a sample result would indicate that if the process mean is not 100 pounds then it is *less* than 100 pounds. A sample mean of 98 pounds, for example, certainly will not be accepted as evidence that sacks usually are *heavier* than 100 pounds. Will he conclude that the process mean is greater than 100 pounds if the sample mean is greater than 100 pounds? The decision depends on how much greater than 100 pounds his sample mean is. We know about the sampling distribution of the sample mean $\bar{X}$ when the process mean is exactly 100 pounds. If the sample mean is enough greater than 100 pounds to be in the far upper tail of this sampling distribution, he would be inclined to reject the notion that the process mean is 100 pounds in favor of the notion that it is some value greater than 100 pounds. This type of evidence would support the conclusion that the majority of sacks weigh more than 100 pounds.

Formally, the inspector states the null hypothesis that the process mean μ is exactly 100 pounds because he can define a suitable sampling distribution for this condition. He knows the filling machine delivers a normal distribution of fill weights with a standard deviation σ of 0.90 pound. He has selected a sample size of 36 observations ($n = 36$). Given a process mean of 100 pounds, the sampling distribution of the sample mean $\bar{X}$ will be normally distributed around 100 pounds with a standard error of 0.15 pound ($\sigma/\sqrt{n} = 0.90/\sqrt{36}$). Furthermore, the probability is only 0.01, for example, that the sample mean $\bar{X}$ will be more than 2.33 standard errors greater than the hypothesized process mean of 100 pounds, and 2.33 standard errors is 0.35 pound (2.33×0.15). Symbolically $\mu_0 + z_\alpha \cdot \sigma_{\bar{X}}$ is $100 + (2.33)(0.15)$, or 100.35 pounds. Hence, at the 0.01 level of significance, he will reject the null hypothesis that

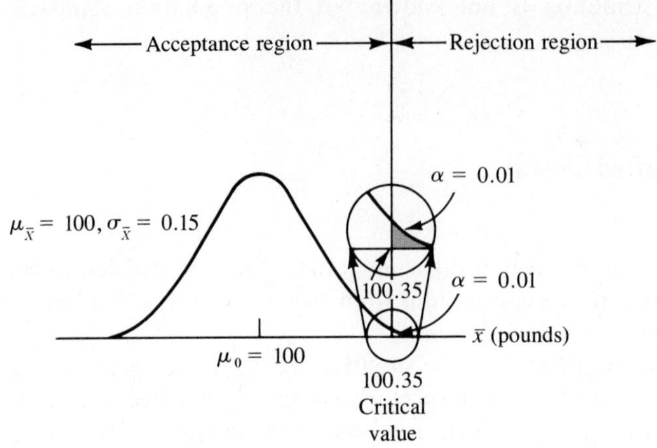

Figure 8.4 Hypothesized distribution of $\bar{X}$ for upper one-tailed test

the process mean μ is 100 pounds if the sample mean exceeds 100.35 pounds. Instead he will accept the alternative hypothesis that the process mean is greater than 100 pounds.

Figure 8.4 illustrates the one-tailed test just described. To guard against a low setting the inspector requires the sample mean to exceed a single critical value (100.35) greater than the desired minimum mean fill of 100 pounds. The entire probability area associated with the level of significance is placed in the upper tail. The result is called an *upper* one-tail test.

For our example, one form in which the hypothesis statement can be put is

$$H_0: \quad \mu = 100 \text{ pounds};$$

$$H_A: \quad \mu > 100 \text{ pounds}.$$

Here, in contrast to the two-tailed test, the inspector wants to reject the null hypothesis (H_0) in favor of the alternative (H_A). This is typical of most one-tailed tests. The desired outcome usually appears as the alternative to the null hypothesis.

Our example required an upper one-tailed test. In other circumstances, *lower* one-tailed tests are called for. For instance, at the same time the inspector is carrying out the upper one-tailed test just described, the cement manufacturer could be carrying out a lower one-tailed test to guard against excessive overfilling. A suitable hypothesis statement for such a test is

$$H_0: \quad \mu = 101.5 \text{ pounds};$$

$$H_A: \quad \mu < 101.5 \text{ pounds}.$$

Here the manufacturer wants a process mean just enough larger than 100 pounds to avoid both frequent underfilling and excessive overfilling.

For the same level of significance (0.01) and sample size ($n = 36$), the critical value would be located 0.35 pound *below* 101.5 pounds, at 101.15 pounds. The cement manufacturer wants the sample mean to be enough less than 101.5 pounds to convince him that he is not overfilling.

To be complete we need to alter the statement of the null hypothesis because all possible values of μ are not covered by the statements given above. For example, in the hypothesis statement for an upper one-tailed test (H_0: $\mu = 100$ pounds; H_A: $\mu > 100$ pounds), cases in which μ is less than 100 pounds are not covered. To overcome this objection, it is customary to put the hypothesis statement for upper one-tailed tests in the form

$$H_0: \quad \mu \leq \mu_0$$
$$H_A: \quad \mu > \mu_0$$

and to put the hypothesis statement for lower one-tailed tests in the form

$$H_0: \quad \mu \geq \mu_0$$
$$H_A: \quad \mu < \mu_0.$$

The general procedure outlined in the rules on page 218 also applies to one-tailed tests. Of course, the hypothesis statement in the first step must be changed to the correct one-tailed form and the critical values in the last two steps must be changed to $z_{1-\alpha}$ for an upper tail test or to z_α for a lower tail test. Table 8.2 shows some critical values commonly used in one-tailed tests.

Table 8.2 *Commonly Used Critical Values for One-tailed Hypothesis Tests*

$1-\alpha$	z_α (lower tail)	$z_{1-\alpha}$ (upper tail)
0.90	−1.28	+1.28
0.95	−1.645	+1.645
0.99	−2.33	+2.33

8.3

Types of Error

Decisions about the design of hypothesis tests can be improved by considering the types of error that are possible and the probabilities of making each type. We can determine the types of error by enumerating the possible sequences of events which can occur in a hypothesis test. The test begins with the statement of a null hypothesis (H_0) and an alternative hypothesis (H_A). One of these statements is true and the other is false, but we don't know which is which. In the upper one-tailed test discussed above, the statements were "H_0: $\mu \leq 100$ pounds" and "H_A: $\mu > 100$ pounds."

The second step in the test is to gather evidence by conducting a suitable random sampling experiment. In the upper one-tailed test, the mean of a sample of 36 observations will be compared with a critical value of 100.35 pounds. If the sample mean is less than this critical value, H_0 will be accepted; if the sample mean is greater than 100.35 pounds, H_A will be accepted. The crucial thing to realize is this: The sample mean can fall on either side of the critical value regardless of which hypothesis statement is true. If the true process mean μ is 99.8 pounds, for instance, random sampling fluctuation can produce a sample mean $\bar{x}$ on either side of 100.35 pounds. The same result is possible if the process mean is greater than 100 pounds.

If the null hypothesis is true (if the process mean is 100 pounds or less) and the sample mean happens to fall below the critical value, the null hypothesis will be accepted and no error will have occurred. But if the null hypothesis is true and the sample mean happens to fall above the critical value, the null hypothesis will be rejected in favor of the alternative hypothesis (H_A: $\mu > 100$ pounds) even though the null hypothesis is, in fact, correct. If we reject the null hypothesis when it is correct, we make a **Type I error**.

If the alternative hypothesis is true (if the process mean is some value greater than 100 pounds) and the sample mean happens to fall above the critical value, the alternative hypothesis will be accepted and no error will have occurred. But if the alternative hypothesis is true and the sample mean happens to fall below the critical value, the alternative hypothesis will be rejected in favor of the null hypothesis. If we reject the alternative hypothesis when it is correct, we make a **Type II error**.

Now look at Figure 8.5. The top half of the figure shows a situation when the null hypothesis is true. Note that 1 percent of the sampling distribution of $\bar{X}$ lies above 100.35 pounds, the critical value. One percent of the time the sample mean can be expected to fall above the critical value and thereby cause a Type I error. Not only is α the level of significance for a hypothesis test: it is also the probability of making a Type I error.

In the bottom half of Figure 8.5 we have a situation in which the null hypothesis is false and the alternative is true. Here the process mean is 100.50 pounds. For a random sample of 36 observations, the critical value (100.35 pounds) is 1.00 standard error less than the process mean to two decimal places. The percentage of area under the sampling distribution lying to the left of the critical value is 15.87. The probability is 0.1587 that the sample mean will fall below the critical value and cause a Type II error. In general, the probability of a Type II error is designated with the symbol β (beta). Note that the critical value, 100.35, stems directly from the null hypothesis. As such the same critical value remains the reference point for calculating beta for any value of μ not included in the null hypothesis. The four possible results of a hypothesis test are summarized in Table 8.3.

We would prefer to reduce the probability of either type of error as much as possible. We see in the lower half of Figure 8.5 that we can reduce the probability of

Table 8.3 Possible Outcomes of a Hypothesis Test

Decision	H_0 True	H_A True
Accept H_0	Correct decision	Type II error Probability: β
Accept H_A	Type I error Probability: α	Correct decision

a Type II error by decreasing the critical value below 100.35 pounds. But, at the same time, we see in the upper half of the figure that this will only produce an increase in the probability of a Type I error.

On the other hand, consider what would happen if we decided to take a random sample of 225 observations instead of 36. Intuitively, we might hope that more

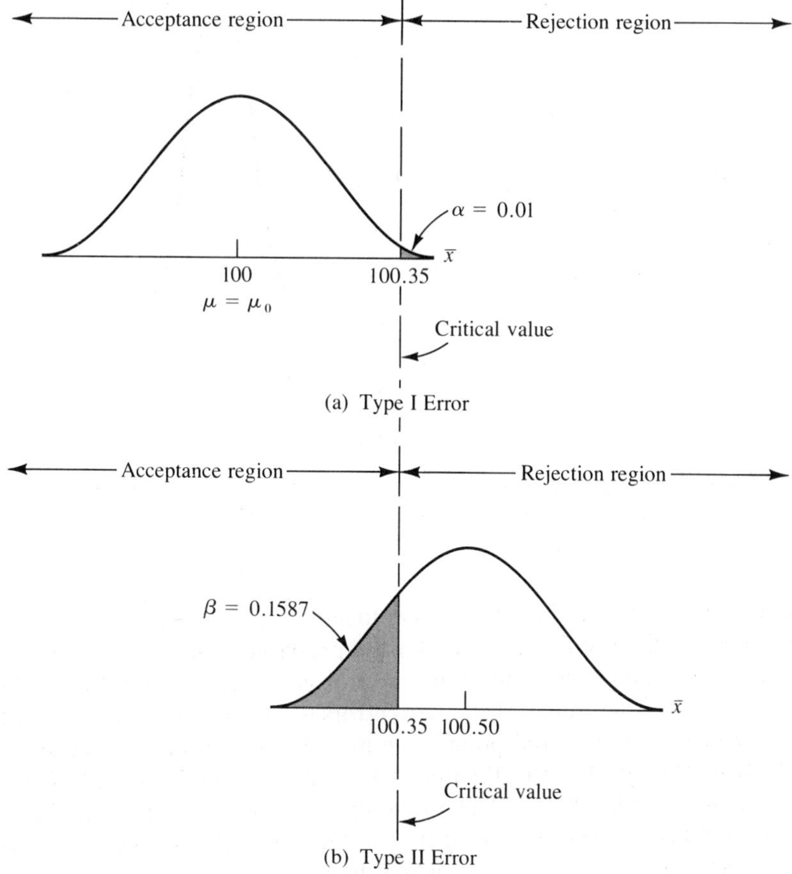

Figure 8.5 *Types of error and their probabilities for $n = 36$ observations (H_0: $\mu \leq 100$; H_A: $\mu > 100$)*

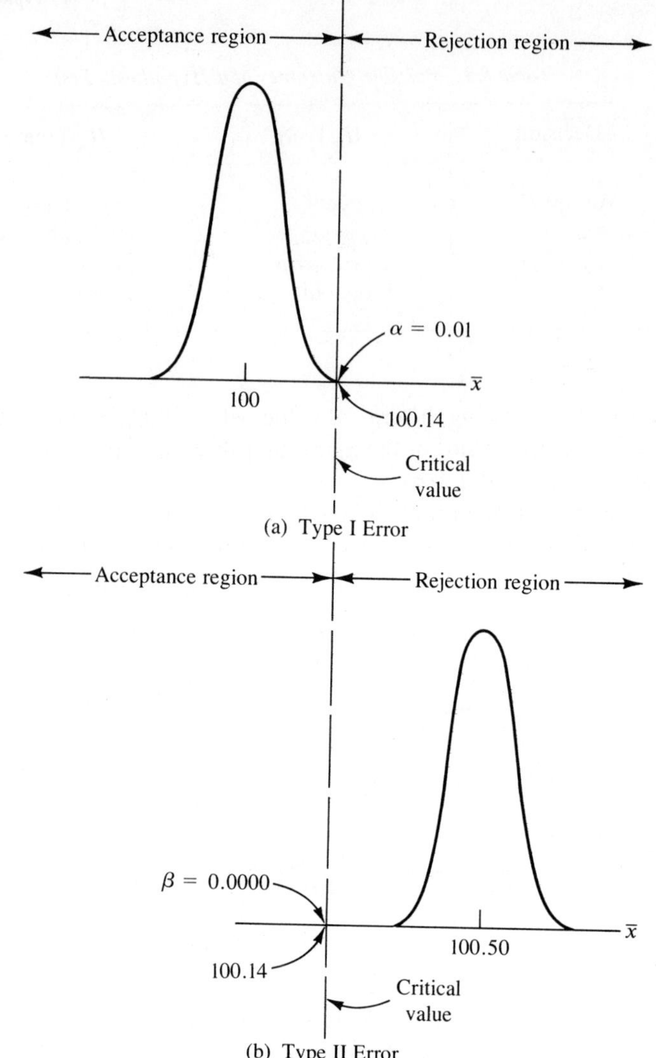

Figure 8.6 *Types of error and their probabilities for $n = 225$ observations (H_0: $\mu \leq 100$; H_A: $\mu > 100$)*

information could be used in some manner to reduce the probability of error. Suppose we decide that 0.01 is still acceptable for the probability of a Type I error. Because of the increased sample size the standard error is reduced from 0.15 to 0.06 pound ($\sigma/\sqrt{n} = 0.90/\sqrt{225}$), and 2.33 standard errors is 0.14 pound to two decimal places. By adding 0.14 pound to 100 pounds we find the new critical value to be 100.14 pounds. Even though the critical value is now 100.14 pounds instead of 100.35, the level of significance is still 0.01. The probability of making a Type I error is still 0.01.

The new test is illustrated in Figure 8.6. The top half of the figure illustrates the narrower sampling distribution of $\overline{X}$ when the process mean is 100 pounds and n is 225 rather than 36. This was described in greater detail in the preceding paragraph. The bottom half of Figure 8.6 illustrates the sampling distribution of $\overline{X}$ when the process mean is 100.50 pounds. The critical value, 100.14 pounds, is now 6.0 standard errors below the mean. The portion of area below the mean is 0.0000 to 4 decimal places. Hence the probability of making a Type II error is now 0.0000 instead of 0.1587 as it was with 36 observations.

We used all the added information in the sample of 225 observations to reduce probabilities of Type II errors. Probabilities of Type I errors were left unchanged. Alternatively, we could have increased the critical value to some value slightly larger than 100.14 pounds. This would have resulted in reductions of probabilities of both types of error, but the reductions in probabilities for Type II errors would not have been as great as they are with a critical value of 100.14 pounds. By fixing sample size at larger and larger values, we can reduce the probabilities of Type I and Type II errors for specified values of the process mean μ to any desired value. Of course, larger samples are more expensive and we must balance the benefits of lower probabilities of error against increased sampling costs.

Summary

Testing statements about the mean value of a statistical population, like estimation, is an application of the characteristics of the distribution of sample means discussed in Chapter 7. Both one-tailed and two-tailed hypothesis tests have been developed. It is assumed that the population variance and, hence, the population standard deviation are known. The procedures for both one-tailed and two-tailed tests apply to random samples of any number of observations from normally distributed populations. By reason of the central limit theorem, these same procedures apply to random samples of 30 or more observations from populations not normally distributed.

Two types of error are possible when a hypothesis is being tested. If the null hypothesis is true and you reject it, a Type I error occurs. If you reject the alternative hypothesis when it is true, a Type II error occurs. Increasing the number of observations in the random sample can reduce the probabilities of either or both types of error.

See Self-Correcting Exercises 8B.

Exercise Set B

1. State whether a two-tailed test, an upper one-tailed test, or a lower one-tailed test is most appropriate in each of the following studies. Justify your conclusion.

(a) A consumer advocacy group is concerned with the advertised claim that a certain deodorant "protects for twenty-five hours."
(b) The Food and Drug Administration is testing whether a prescription drug is of the potency claimed.
(c) The FDA is checking whether the amount of butylated hydroxytoluene in a frozen food product is really "less than .1%" as stated on the package.
(d) The university food services manager is concerned that her 4-ounce-patty-making machine might be forming unduly large hamburger patties.

2. Determine the number of standard errors associated with each of the following levels of significance for normal two-tailed tests:
(a) 10 percent
(b) 5 percent
(c) 2 percent
(d) 1 percent

3. Determine the number of standard errors associated with each of the following levels of significance for normal one-tailed tests:
(a) 10 percent
(b) 5 percent
(c) 2 percent
(d) 1 percent

4. When we assume a level of significance of 5 percent in hypothesis testing, what type of error do we risk with a probability of 0.05?

5. The hypotheses

$$H_0: \quad \mu = 16.8$$
$$H_A: \quad \mu \neq 16.8$$

are to be tested at a 5 percent level of significance using a sample of size 16 from a normal population with standard deviation of 12.
(a) Give the numerical value of $\sigma_{\bar{x}}$.
(b) What is the lower critical value for $\bar{x}$?
(c) What is the upper critical value for $\bar{x}$?
(d) Define the limits of the acceptance region.
(e) If the sample mean is found to be 8, should H_0 be rejected? Why or why not?
(f) If the sample mean is found to be 16, should H_0 be rejected? Why or why not?

6. Using the conversion formula

$$z = \frac{\bar{x} - \mu}{\sigma_{\bar{x}}},$$

restate the null and alternative hypotheses of the preceding exercise, find the lower and upper critical values for z, define the limits of the acceptance region as z scores, and convert the sample mean to z scores. Then run the hypothesis test of parts (e) and (f) from Exercise 5 at a level of significance of 0.05. What are your conclusions?

7. Perform the hypothesis testing of Exercise 5 at a level of significance of 0.02.

8. Perform the hypothesis testing of Exercise 5 assuming a sample size of 100 rather than 16.

9. For a normal distribution with known variance 16, the hypotheses

$$H_0: \quad \mu \geq 12$$
$$H_A: \quad \mu < 12$$

are tested at a 2 percent level of significance using the mean from a sample of size 64. What is

(a) α?

(b) β if the true mean is actually 10.2?

(c) β if the true mean is actually 10.8?

(d) β if the true mean is actually 11.4?

(e) β if the true mean is 12.0?

(f) β if the true mean is 12.6?

10. How does increasing the sample size to 400 affect the answer to each part of Exercise 9?

11. How does reducing the level of significance to 0.01 affect the answer to each part of Exercise 9?

12. A restaurant recorded mean daily sales of \$326 during the summer season a year ago. The null hypothesis that mean daily sales during this year's summer season are no greater than last year's is to be tested at the 0.05 significance level. Daily sales during the summer season are normally distributed with a standard deviation of \$10. If the mean this season should prove to be \$330, the probability of rejecting the null hypothesis is to be 0.90.

    (a) State the null hypothesis and the alternative to it.

    (b) State an expression for the critical value such that the probability of a Type I error will be 0.05 when the population mean is \$326.

    (c) State an expression for the critical value for which the probability of a Type II error will be 0.10 when the population mean is \$330.

13. A chemist in a pharmaceutical firm knows that the toxic dosage of an ingredient in a household product is 12 grains. She wants to make sure that the mean dosage in the lot currently being manufactured does not exceed 10 grains. The weights of this ingredient are normally distributed with a standard deviation of 0.2 grain. At the 0.01 level of significance, conduct the correct one-tailed test to determine whether the maximum specification is being exceeded if a sample of 25 observations has a mean dosage of 9.80 grains.

14. For the test described in Exercise 13, find the probabilities of accepting the null hypothesis if the population mean is 9.9032 grains; if it is 9.8532 grains. What type of error is possible in these two circumstances? What are the probabilities of making these errors?

15. For the test described in Exercise 13, find the probability of accepting the null hypothesis if the population mean is 10.0032 grains per dosage. What type of error is possible in this circumstance? What is the probability of making such an error?

Overview

1. What procedure of statistical inference starts with a known $\bar{x}$ and makes a confidence statement about the location of μ?

2. What procedure of statistical inference assumes a value for μ and then checks whether an obtained value of $\bar{x}$ supports the assumption?

3. What type of error do we risk when we (a) reject a null hypothesis? (b) When we reject an alternative hypothesis?

4. What symbol is given the probability with which we risk an error when we (a) reject a null hypothesis? (b) When we reject an alternative hypothesis?

5. What is μ_0? How does it differ from μ, if at all? How does $\bar{x}$ enter the hypothesis testing procedure?

6. In a survey of household incomes for a certain neighborhood, a student selects a random sample of households, determines their individual incomes, and calculates the mean of the sample to be $9152. He then states that the mean income for the entire neighborhood is $9,152 because the sample mean is an unbiased estimator of the population mean. Do you agree? Discuss.

7. A certain procedure is used to measure the hardness of water in parts of calcium carbonate per million parts of water. If the procedure is applied repeatedly on the same water source, the resulting measurements will be normally distributed with a standard deviation of 6 ppm (parts per million). Nine independent readings are made of the hardness of a certain water source. The sample mean is 20.13 ppm. Find the 0.99 confidence interval estimate of the true hardness.

8. A widely used achievement test is given to over a thousand freshmen entering a university. A random sample of 16 scores is selected from this group and the sample mean score is found to be 132. Nationwide, the test is known to have normally distributed scores with a standard deviation of 20 points. Use these national characteristics to find the 0.95 confidence interval estimate of the mean score of the entering freshmen.

9. In Exercise 8, what would be the effect if the sample size were increased to 100? What would the length of the new 0.95 interval estimate be?

Glossary of Equations*

8.1(a) $\mu_L = \bar{x} - z_{\alpha/2} \cdot \frac{\sigma}{\sqrt{n}}$

8.1(b) $\mu_U = \bar{x} + z_{\alpha/2} \cdot \frac{\sigma}{\sqrt{n}}$

When the population standard deviation is known and the population distribution is normal or essentially so, the $(1 - \alpha)$ confidence interval estimate of the mean has lower and upper limits that are the product of the standard normal deviate and the standard error of the mean subtracted from and added to the sample mean. The same limits apply to samples of 30 or more from moderately skewed populations.

*A summary table for Chapters 8 and 9 is given on page 253.

8.3 $E = z_{\alpha/2} \cdot \dfrac{\sigma}{\sqrt{n}}$

With probability $(1 - \alpha)$, the expected maximum error in the interval estimate of the population mean is the product of the standard normal deviate and the standard error of the mean.

8.4 $n = \left(\dfrac{z_{\alpha/2} \cdot \sigma}{E}\right)^2$

To limit the expected maximum error in an interval estimate to a specified amount, the number of observations needed in the sample is the square of the standard normal deviate times the standard deviation divided by the error.

8.5 $z = \dfrac{\bar{x} - \mu_0}{\sigma/\sqrt{n}}$

The standard normal deviate for a sample mean from the hypothesized population mean is the difference between these two means divided by the standard error of the sample mean.

9

Inference for Means – Variance Unknown

This chapter is a continuation of the general topic introduced in the previous chapter—inference about the means of statistical populations. We will continue to discuss estimation and testing hypotheses.

The new feature introduced here is concerned with the variance and, hence, the standard deviation of the statistical populations being considered. In the last chapter we assumed that the value of the population variance is known. In this chapter we assume that this value is not known.

When the population variance is not known, inference procedures for the population mean fall under two headings. One set of procedures is applicable to large samples. The other is applicable to small samples. For most purposes described here, a random sample composed of at least 30 observations is a large sample and one with less than 30 is small.

Large-Sample Techniques

We learned in Chapter 8 that our knowledge of the sampling distribution of the sample mean $\overline{X}$ is the key to forming confidence interval estimates for the population mean μ and to testing hypotheses about it. Specifically, both inference procedures were constructed from three ingredients:

1. The value of the sample mean $\overline{x}$ as determined from a random sample of n observations
2. The value of the standard error of estimate $(\sigma/\sqrt{n})$ as determined from the known population standard deviation σ and the sample size n
3. The known normal shape of the sampling distribution

We learned that there are many statistical populations for which enough history is available to provide known values for the population standard deviation σ. But now we must face the fact that there are many other populations, perhaps the majority, for which we do *not* know the standard deviation. Lacking this knowledge we cannot determine the value of the standard error of estimate, and without the standard error of estimate we cannot use the procedures described in Chapter 8.

All is not lost, however. We began Chapter 8 with a discussion of point estimation. We saw that the sample mean $\bar{x}$ is an effective estimator of the population mean. Why not extend this notion and use the random sample to provide an estimate of the unknown population variance and hence of the standard deviation? In the following section we do so. Then we go on to a section describing estimating and hypothesis testing procedures for the population mean μ based on the estimated variance.

Variance Estimation 9.1

In Chapter 1, the population variance was defined in Equation 1.4 as the mean of the squared deviations from the population mean, that is,

$$\sigma^2 = \frac{\sum_{i=1}^{N} (x_i - \mu)^2}{N}.$$

Recall that in this equation μ is the *population* mean and N is the total number of observations in the *population.* In this chapter, we are assumed not to have access to the entire population and not to know the values of either the population mean μ or variance σ^2. The information we have is contained in a random sample of n observations ($n < N$) from that population.

To find a point estimator of the population variance, we can begin by finding the sample mean $\bar{x}$. Then we can measure the variability of the sample observations with reference to the sample mean. Specifically, the estimator s^2 is

$$s^2 = \frac{\sum_{i=1}^{n} (x_i - \bar{x})^2}{n - 1}, \qquad \textit{9.1}$$

which we will call the **sample variance**.

There are three differences between the right side of Equation 9.1 and the right side of Equation 1.4 as repeated above. First, in the sample estimator s^2, deviations are measured from the sample mean $\bar{x}$ while for the population variance σ^2 the reference is the population mean μ. The change is required because μ is unknown. Second, only the n observations available in the sample are used for s^2 while all N

observations in the population are used for σ^2. Third, the denominator in Equation 9.1 is the number of sample observations less one $(n - 1)$, while for σ^2 the denominator is the number of observations in the population (N).

The first two differences are obviously made necessary by lack of access to the population. To explain the third difference, we can begin by pointing out that a random sample of, say, 45 observations is likely to yield a value of s^2 different from that yielded by any other sample of 45 observations. Sample values of s^2 are just estimates of σ^2 and vary from sample to sample. Just as values of the sample mean $\bar{x}$ form a sampling distribution around the population mean μ, so do the values of the sample variance s^2 form a sampling distribution around the population variance σ^2. Recall that $\bar{x}$ is an unbiased estimator of μ. Similarly, division by $(n - 1)$ rather than by n makes s^2 an *unbiased* estimator of σ^2. The expected value of the sample variance s^2 is the population variance σ^2 when $(n - 1)$ is used in the denominator.

The denominator in Equation 9.1, which is $(n - 1)$, is often called the **degrees of freedom** in the estimator. For example, the value of s^2 from a random sample of 45 observations is said to have 44 degrees of freedom. If we knew the population mean μ, we could use it as a reference for the deviations in the sample $(x - \mu)$ and would have 45 deviations from the *population* mean $x - \mu$ as a basis for estimating the population variance σ^2. Since we don't know the value of the population mean, we measure deviations from the *sample* mean $(x - \bar{x})$. But, from sample to sample, the sample mean $\bar{x}$ varies around the population mean μ. In an estimate based on deviations from the *population* mean $(x - \mu)$, the variation of the sample mean $(\bar{x} - \mu)$ is free to enter the estimate. But in Equation 9.1 this 1 degree of freedom to vary is not present. Hence there are only $(n - 1)$ degrees of freedom. We will have more to say about degrees of freedom in the last half of this chapter and in succeeding chapters.

We are ready now for the next step in preparation for inferences about the unknown population mean μ when the population variance σ^2 is also unknown. Having found the value of the sample variance s^2 for a particular random sample, we go on to estimate the population standard deviation σ from

$$s = +\sqrt{s^2} \qquad \textit{9.2(a)}$$

or

$$s = +\sqrt{\frac{\sum (x - \bar{x})^2}{n - 1}}. \qquad \textit{9.2(b)}$$

We will refer to s as the **sample standard deviation**. Incidentally, although the sample variance s^2 is an unbiased estimate of the population variance σ^2, the sample standard deviation s is a *biased* estimator of the population standard deviation, σ. It is, however, not seriously biased for large samples.

Recall that our original objective in estimating the variance was to overcome our lack of knowledge about the standard error of the mean $(\sigma_{\bar{x}} = \sigma/\sqrt{n})$. Having obtained an estimate for the unknown value of σ by applying Equation 9.2, we can substitute to get the estimated standard error of the mean $(s_{\bar{x}})$, where

$$s_{\bar{X}} = \frac{s}{\sqrt{n}}. \tag{9.3}$$

In the process, we have made use of the number of observations in the sample twice. We used n first to find the number of degrees of freedom in the denominator of Equation 9.2(b). Then we used it again in the denominator of Equation 9.3. In fact, an algebraic equivalent for Equation 9.3 is

■ $$s_{\bar{X}} = +\sqrt{\frac{\sum (x - \bar{x})^2}{n(n-1)}}, \tag{9.4}$$ ■

where n appears twice in the denominator. 9.1

A random sample of 32 entering freshmen at Hoover University has just taken a widely used reading comprehension test. This collection of 32 scores is listed in Table 9.1. From year to year, for the nation as a whole the distribution of scores is essentially normal with a mean of 500 and a standard deviation of 100. For the past several years at Hoover University the means and standard deviations of annual score distributions have differed a great deal and have never been very close to the national values. Based on the sample of 32 scores, the dean of students wants a preliminary estimate of the variance, the standard deviation, and the standard error of the mean for this year's distribution of reading test scores for all entering freshmen at Hoover.

In Table 9.1, the sample mean $\bar{x}$ is shown to be 477. The deviations from this mean are shown in the middle column. The deviations are squared and summed in the final column. The sample variance is, from Equation 9.1,

$$s^2 = \frac{\sum (x - \bar{x})^2}{n - 1} = \frac{194{,}076}{32 - 1}, \text{ or } 6260.52.$$

The sample standard deviation as found from Equation 9.2(a) is

$$s = \sqrt{6260.52}, \text{ or } 79.12.$$

From Equation 9.3, the estimate of the standard error of the mean is

$$s_{\bar{X}} = \frac{s}{\sqrt{n}} = \frac{79.12}{\sqrt{32}}, \text{ or } 13.99.$$

As a check, from Equation 9.4 we get

$$\begin{aligned} s_{\bar{X}} &= \sqrt{\frac{\sum (x - \bar{x})^2}{n(n-1)}} \\ &= \sqrt{\frac{194{,}076}{32(31)}} = \sqrt{195.64}, \text{ or } 13.99. \end{aligned}$$

Table 9.1 *Mean and Squared Deviations for 32 Test Scores*

Reading Comprehension Scores (x)	Deviations from Sample Mean ($x - \bar{x}$)	Squared Deviations ($x - \bar{x})^2$
492	15	225
489	12	144
507	30	900
520	43	1,849
455	−22	484
460	−17	289
318	−159	25,281
427	−50	2,500
470	−7	49
510	33	1,089
548	71	5,041
460	−17	289
490	13	169
464	−13	169
475	−2	4
487	10	100
441	−36	1,296
306	−171	29,241
476	−1	1
358	−119	14,161
507	30	900
496	19	361
651	174	30,276
480	3	9
600	123	15,129
308	−169	28,561
475	−2	4
533	56	3,136
446	−31	961
653	176	30,976
496	19	361
466	−11	121
15,264		194,076

$$\bar{x} = \frac{15{,}264}{32} = 477$$

Incidentally, alternative calculation techniques for the sample variance are available. These are direct counterparts of the techniques discussed in Chapter 3. The exercises at the end of this half chapter present these alternatives.

We have found out how to get an estimate of the standard error of the mean from a random sample. We can now proceed with statistical inference about the population mean μ when the population variance σ^2 is not known.

Estimation of the Population Mean

When we discussed interval estimation of the mean μ in the first part of Chapter 8, we saw in Equation 8.1 that for a confidence coefficient of $(1 - \alpha)$ the upper (μ_U) and lower (μ_L) limits of the interval estimate were

$$\mu_U = \bar{x} + z_{\alpha/2} \cdot \sigma_{\bar{X}}$$

and

$$\mu_L = \bar{x} - z_{\alpha/2} \cdot \sigma_{\bar{X}},$$

where z is the standard normal deviate. We also learned that this interval estimate applies for any sample size from a normally distributed population and for any random sample of at least 30 observations from an unspecified population distribution.

In this chapter we assume that the population standard deviation is unknown, so we must estimate the standard error of the mean. We do so as described above. Then we can substitute our estimate, $s_{\bar{X}} = s/\sqrt{n}$, directly for the correct value, $\sigma_{\bar{X}}$, in Equation 8.1 to get the $(1 - \alpha)$ confidence interval estimate of the population mean.

Statement

For a random sample of at least 30 observations from a population distributed normally or nearly so, the lower (μ_L) and upper (μ_U) limits of the $(1 - \alpha)$ confidence interval estimate of the mean μ are

$$\mu_L = \bar{x} - z_{\alpha/2} \cdot s_{\bar{X}} \quad \text{and} \quad \mu_U = \bar{x} + z_{\alpha/2} \cdot s_{\bar{X}}, \qquad \textbf{9.5}$$

where $\bar{x}$ is the sample mean, $s_{\bar{X}}$ is the sample estimate of the standard error of the mean $(s/\sqrt{n})$, and $z_{\alpha/2}$ is the standard normal deviate.

A minimum of 50 observations is required for populations that are seriously skewed. The distributions of sample means from samples this large or larger will be essentially normal in shape and have nearly stable standard deviations.

For the 32 reading comprehension scores in Table 9.1, we found that the sample mean $\bar{x}$ is 477 and the estimated standard error of the mean $s_{\bar{X}}$ is 13.99. If we want to find the 0.95 confidence interval estimate of the population mean μ, we must use a value of 1.96 for the standard normal deviate $z_{\alpha/2}$. Then we substitute in Equations 9.5:

$$\begin{aligned}\mu_L &= \bar{x} - z_{\alpha/2} \cdot s_{\bar{X}} \\ &= 477 - 1.96(13.99), \text{ or } 449.58,\end{aligned}$$

and

$$\mu_U = \bar{x} + z_{\alpha/2} \cdot s_{\bar{x}}$$
$$= 477 + 1.96(13.99), \text{ or } 504.42.$$

At the 0.95 confidence level, the mean reading comprehension score at Hoover University this year lies in the interval from 449.58 to 504.42. This rather wide interval is a consequence of collecting only 32 scores. Considerably more observations would have reduced the estimate of the standard error of the mean ($s/\sqrt{n}$).

Sample Size

Suppose we wanted to form a new interval estimate on the basis of the mean of a larger sample of scores which will be much narrower than the 449.58 to 504.42 estimate just found. All we need do is adapt Equation 8.4, which is

$$n = \left(\frac{z_{\alpha/2} \cdot \sigma}{E}\right)^2.$$

In this equation we replace the population standard deviation σ with the sample estimate s.

Statement

The number of observations n needed to limit the maximum error in an interval estimate of the mean μ to the quantity E with probability $(1 - \alpha)$ is

$$\blacksquare \quad n = \left(\frac{z_{\alpha/2} \cdot s}{E}\right)^2, \quad \blacksquare \qquad \textbf{9.6}$$

where $z_{\alpha/2}$ is the standard normal deviate and s is the sample standard deviation defined in Equation 9.2.

Equation 9.6 applies only if n turns out to be larger than 30 and the population is not seriously skewed. The original value of s used to enter in the equation should also come from a sample of at least 30 observations.

Now suppose the dean at Hoover University wants an interval estimate of the mean that has a range of only 20 score points which will include the population mean score with 0.99 confidence. Since the new sample mean $\bar{x}$ will be in the center of the 20-point interval estimate, the maximum error E will be 10 score points on either side of $\bar{x}$, that is, $E = 10$. The standard normal deviate $z_{\alpha/2}$ for 0.99 confidence is 2.58. From our original sample of 32 scores in Table 9.1, our estimate of the standard deviation s was 79.12. Then it follows that

$$n = \left(\frac{2.58(79.12)}{10}\right)^2, \text{ or } 417,$$

to the next larger whole number. By selecting an additional 385 test scores and putting them with our original 32 we will have a combined sample of 417. Then we recompute the sample mean $\bar{x}$ and the estimated standard error $s_{\bar{x}}$ for this combined sample and use them to form a 0.99 confidence interval. We should get a 0.99 confidence interval estimate with a range of slightly more or less than 20 points. If we knew the population standard deviation σ, we would be sure of the range. But when we calculate the sample mean $\bar{x}$ and standard deviation s for the new combined sample of 417, we can expect both estimates to differ somewhat from 477 and 79.12, respectively, because of random sampling fluctuation. If s comes out slightly larger than 79.12, this will make $s_{\bar{x}}$ a bit larger than anticipated and extend the interval limits in Equation 9.5 to a range of slightly more than 20 points.

In this example we assume that a sample of 417 observations is not over 10 percent of the freshman class. Until the sample exceeds 10 percent of the population no correction for the fraction that a sample is of its population need be made. The correction for this latter situation will be discussed in a later chapter.

Test of Hypotheses about Population Means

As was the case with interval estimation, the procedures discussed in Chapter 8 apply to hypothesis tests in this chapter so long as the population standard deviation σ is replaced by the sample estimate s. Recall that both two-tailed and one-tailed tests are based on the statistic

$$z = \frac{\bar{x} - \mu_0}{\sigma/\sqrt{n}}$$

from Equation 8.5. The value of this statistic was compared with either one or two critical values.

Statement

When the population standard deviation σ is unknown and the sample contains at least 30 observations,* the statistic

$$z = \frac{\bar{x} - \mu_0}{s/\sqrt{n}}, \qquad 9.7$$

where s is the sample estimate of σ, can be treated as a standard normal deviate regardless of the form of the population distribution.

*A minimum sample of 50 observations is required when the population is seriously skewed.

Placed in the circumstance of not knowing the population standard deviation σ, we find the sample standard deviation s as an estimate of σ. Then we find the value of the test statistic z in Equation 9.7. Finally we compare the sample value of z with critical values from the standard normal table. Except for the use of s, the procedures are identical to those in Chapter 8.

Summary

When the population variance and standard deviation are unknown, interval estimation and hypothesis tests are still possible for the population mean μ. Both types of statistical inference are based on using deviations from the *sample* mean, rather than from the unknown population mean, to form estimates of the population variance σ^2 and standard deviation σ. The estimate of σ^2 is called the sample variance s^2; the estimate of σ is called the sample standard deviation s. The value of s for a particular random sample is used to find an estimate $s_{\bar{X}}$ of the standard error of the mean $\sigma_{\bar{X}}$, which is the key requirement for statistical inference concerning the population mean.

The procedures described in Chapter 8 for making interval estimates and hypothesis tests are the same as those used in this part of this chapter, with one exception. In place of the unknown standard error of the mean $\sigma_{\bar{X}}$, we must substitute the estimate $s_{\bar{X}}$. The resulting procedures apply when the random sample consists of at least 30 observations from statistical populations which are normally or nearly normally distributed. For seriously skewed populations, samples of at least 50 observations are required to normalize the sampling distribution of the sample mean and to stabilize the value of the sample standard deviation.

See Self-Correcting Exercises 9A.

Exercise Set A

1. For samples of a given size n the variance of these samples (values of s^2) form a sampling distribution around what parameter? Is the sample variance an unbiased point estimator

of this parameter? For what other parameter have we studied an unbiased point estimator? What was this estimator?

2. We can illustrate that s^2 as defined is an unbiased estimator of σ^2 as follows. For the population consisting of the numbers 1, 3, 5, and 7, consider all possible samples of size 2 with replacement. (Replacement is necessary to avoid the need for the finite population correction factor referred to in Exercise 4(e) of the Chapter 5 Overview.) The samples, their respective means, and the value of s^2 for each are shown in the following two-way tables:

Samples

x_1 \ x_2	1	3	5	7
1	1,1	1,3	1,5	1,7
3	3,1	3,3	3,5	3,7
5	5,1	5,3	5,5	5,7
7	7,1	7,3	7,5	7,7

$\overline{x}$

x_1 \ x_2	1	3	5	7
1	1	2	3	4
3	2	3	4	5
5	3	4	5	6
7	4	5	6	7

s^2

x_1 \ x_2	1	3	5	7
1	0	2	8	18
3	2	0	2	8
5	8	2	0	2
7	18	8	2	0

Each of the 16 samples (only 10 of which are actually different if order of selection is disregarded) is equally likely to be the sample selected if a sample of size 2 is taken from the original population with replacement. Thus the probability distribution of s^2 is

s^2	$P(s^2)$
0	$4/16$
2	$6/16$
8	$4/16$
18	$2/16$

and so the expected value of s^2 would be

$$(0 \times 4/16) + (2 \times 6/16) + (8 \times 4/16) + (18 \times 2/16) = 80/16 = 5.$$

Verify that this expected value of s^2 is precisely σ^2 of the original population. Thus s^2 is an unbiased estimator of σ^2.

3. Suppose that instead of using the denominator $(n - 1)$ in the formula for s^2 we were to use n. Follow through the steps of Exercise 2 for the same original population to illustrate that $[\Sigma(x - \overline{x})^2]/n$ is not an unbiased estimator of σ^2. Is the expected value or average of your "estimate" larger or smaller than σ^2?

4. For the sampling population in Exercise 2 construct another two-way table showing the respective values of s for the samples and from it obtain a probability distribution for s.

Then compare the expected value of s for this distribution with the actual value for σ to show that s is *not* an unbiased estimator of σ.

5. An alternative to Equation 9.1, a shortcut formula for obtaining s^2, is

$$s^2 = \frac{n\sum x^2 - (\sum x)^2}{n(n-1)}.$$

For the sample which consists of the observations 10,7,7,8,9,7,4,10,11,7, compute s^2 using (a) this alternative formula and (b) Equation 9.1. Your numerical results should be identical.

6. Using the formula given in Exercise 5, determine first s and then $s_{\bar{x}}$ for the sample consisting of 3.4, 4.7, 3.9, and 5.2.

7. Why are Equations 9.5 not applicable to establishing a confidence interval estimate of the mean of the population of either Exercise 2 or Exercise 6? Would knowledge that the populations sampled are normal permit you to use Equations 9.5? What would you have to do to form a confidence interval estimate using Equations 9.5 in these cases?

8. (a) What does each of the following symbols represent:
   (i) s? (iii) $s_{\bar{x}}$?
   (ii) σ? (iv) $\sigma_{\bar{x}}$?
   (b) Which symbols in part (a) represent parameters? Which represent statistics?
   (c) How are s and $s_{\bar{X}}$ related in size? How are σ and $\sigma_{\bar{X}}$ related in size? What does this latter relationship indicate about the relative dispersion of the original population and the sampling distribution of the mean?

9. Regarding Equations 9.5, under what circumstances is $n = 35$:
   (a) A large, that is, "large enough," sample size?
   (b) A small, that is, "not large enough," sample size?

10. How does the procedure for determining a 95 percent confidence interval estimate of μ based on the sample mean of a sample of size 64 differ depending on whether σ is known or unknown?

11. Determine the 95 percent confidence interval estimate of the mean if $\bar{x} = 37$ for a sample of size 100 and
    (a) σ is known and equals 22.
    (b) σ is unknown but $s = 20.5$.

Exercises 12 through 16 refer to the following problem: An interval estimate of the average weight gain by Rhode Island Red hens in 4 weeks on a vitamin-enriched feed is to be made on the basis of a sample mean.

12. Suppose the gains are normally distributed and a sample of size 36 has been found to have a standard deviation of 2 ounces. How large a combined sample must be taken to ensure with 95 percent confidence that:
    (a) The true mean gain is within ½ ounce of the sample mean gain?
    (b) The true mean gain is within 1 ounce of the sample mean gain?

13. Why does applying the same formula to both parts of Exercise 12 result in a useful result for part (a) and an almost useless result for part (b)? The "almost useless" result for part (b) leads you logically to what minimum sample size in this case?

14. If all the information given in Exercise 12 obtains but the origina! distribution of gains

is known to be slightly skewed, how large a sample must be taken to ensure that the true mean gain is within ½ ounce of the sample mean gain?

15. Suppose the gains are thought to be distributed with slight positive skewness and a sample of size 64 has been found to have a standard deviation of 2.2 ounces. What is the minimum sample size that will ensure that the sample and true mean gains differ by at most ½ ounce (a) with 95 percent certainty? (b) With 98 percent certainty? (c) With 99 percent certainty?

16. Suppose the gains are distributed with slight skewness and a sample of size 64 has a standard deviation of 1.9 ounces. What is the minimum sample size to ensure with 98 percent confidence that the sample mean and the true mean (a) differ by less than 0.01 ounce? (b) By at most 0.01 ounce?

Small-Sample Techniques

In the first part of this chapter random samples composed of at least 30 observations each were available (50 in the case of seriously skewed populations). We used a sample estimate of the standard error of the mean $s_{\bar{X}}$, along with the standard normal deviate z, as the basis for statistical inferences about the population mean μ. The samples were large enough so that two things occurred. First, the sampling fluctuations in the standard deviations s and in the standard error of the mean $s_{\bar{X}}$ were reduced to negligible amounts. Second, the shape of the sampling distribution of the sample mean $\bar{X}$ became essentially normal. In effect, these two results permitted us to treat the statistic

$$z = \frac{\bar{X} - \mu}{s_{\bar{X}}}$$

or its equivalent

$$z = \frac{\bar{X} - \mu}{s/\sqrt{n}}$$

as a standard normal deviate for which Appendix A provided confidence coefficients or levels of significance.

When only a small random sample is available, further changes in procedure are necessary. Two important facts must be borne in mind when we are considering making inferences about the population mean μ from small random samples:

1. Only when the statistical population is normally distributed or nearly so can inferences about the population mean μ be made safely.

2. Even for normally distributed populations, the test statistic which serves as the basis for inferences concerning μ is *not normally distributed.*

The first statement rules out inferences based on procedures discussed in this chapter for the mean μ of a seriously skewed population when a random sample of less than 50 observations is available. We must collect at least 50 observations and use the methods discussed in the first part of this chapter. The second statement implies that, although we cannot use the z statistic, inferences about the mean μ are possible for a random sample of less than 30 observations from a population that is normally distributed or approximately so. This includes any population which has a single mode and tails off reasonably symmetrically in both directions from the mode. Even though inferences about μ are possible in these circumstances, the basic statistic is *not* distributed normally. Consequently, we must work with a new family of distributions.

9.2 The Family of *t* Distributions

To illustrate the new type of distribution needed now, we return to the cement sack filling machine we considered in the last half of Chapter 8. We know that the content weights of sacks filled by this machine at a given setting are normally distributed. Suppose we also know that the mean fill μ being delivered at the given setting is 101 pounds. We want to study sampling distributions involving the sample mean $\bar{x}$ under three sets of conditions. First, we will assume that we know that the population standard deviation σ is 0.90 pound and we want to know how the statistic

$$\frac{\bar{x} - \mu}{\sigma/\sqrt{n}}$$

is distributed for random samples each containing the same number of observations ($n = 15$, for example). Second, we will assume that we do *not* know that the population standard deviation σ is 0.90 pound. We don't know what its numerical value is. We want to study how the statistic

$$\frac{\bar{x} - \mu}{s/\sqrt{n}}$$

is distributed for large random samples each of a given size ($n = 40$, for example). Third, we will assume that σ is again unknown and we want to study the sampling distribution of

$$\frac{\bar{x} - \mu}{s/\sqrt{n}}$$

for small random samples each of a given size ($n = 10$, for instance).

In the first case (σ known), we can simulate the sampling distribution of the statistic

$$\frac{\bar{x} - \mu}{\sigma/\sqrt{n}}$$

by randomly selecting, say, 1500 fill weights and dividing them up into 100 samples of 15 observations each. For each sample we can find the mean $\bar{x}$. Since we already know that μ is 101 pounds and σ is 0.90 pound and n is 15 observations, we can then find the value of the quantity

$$\frac{\bar{x} - \mu}{\sigma/\sqrt{n}}$$

for each sample. Then we can prepare a frequency polygon for these 100 values of the quantity. We know from Chapters 7 and 8 that the frequency polygon will be essentially normally distributed with a mean of zero and a standard deviation of 1. Hence we can interpret the quantities of the form

$$\frac{\bar{x} - \mu}{\sigma/\sqrt{n}}$$

as z, standard normal deviates, and Appendix A can be used to make probability calculations.

In the second case (normal population, σ not known, large sample), we have learned that the statistic

$$\frac{\bar{x} - \mu}{s/\sqrt{n}}$$

can also be interpreted as z, a standard normal deviate. The large sample size keeps the estimated standard deviation s essentially constant from sample to sample and nearly equal to the population standard deviation σ. Hence Appendix A is applicable here as well as in the first case.

In the third case (normal population, σ not known, small sample), the random sample is so small that the fluctuations in the values of the sample standard deviation s can no longer be ignored. As we move from one sample of, say, 10 observations to others of the same size, we note substantial changes in s. In the statistic

$$\frac{\bar{x} - \mu}{s/\sqrt{n}},$$

the bottom part, $s/\sqrt{n}$, will vary substantially from sample to sample rather than remaining constant or essentially so, as it did in the previous two cases. As a result the entire fraction will have a sampling distribution characterized by greater variability than was true in either of the previous cases. The standard deviation of the sampling distribution will no longer be equal to 1 as it is for the standard normal deviate. The standard deviation will instead be greater than 1. In fact, the smaller the sample size n on which the sampling distribution is based, the larger the standard deviation of that distribution. Consequently, there is a different sampling distribution for each value of n. We will refer to these collectively as the **family of t distributions**. We can now summarize what we have said for the third case:

Statement

For statistical populations which are essentially normal, statistical inferences about the population mean μ from an independent random sample of less than 30 observations are based on the statistic

$$t = \frac{\bar{x} - \mu}{s/\sqrt{n}}. \qquad \textbf{9.8}$$

This statistic has a different sampling distribution for each different sample size n.

To clarify the nature of t distributions further, we have illustrated a few in Figure 9.1. In the figure note that each of the three t distributions has a mean of zero ($\mu = 0$) and that each tails off symmetrically in both directions from the single mode. The standard normal distribution is also shown to have a mean of zero and a single mode and to tail off symmetrically in both directions from the mode. But the standard normal distribution has a standard deviation of 1 ($\sigma = 1$) while the standard deviations of t distributions are greater than 1. In Figure 9.1, the smaller the sample size, the larger the standard deviation and, hence, the larger the proportion of area in the tails. On the other hand, as the sample size n grows larger, t distributions shrink in tail area and increase in central area. In fact: As sample size n grows large without limit, the sampling distribution of t approaches that of the standard normal deviate z.

We have described different t distributions by referring to the sample size n on which each is based. But when we go to use tables for these distributions, we find they usually are presented for various *degrees of freedom*. In the first part of this

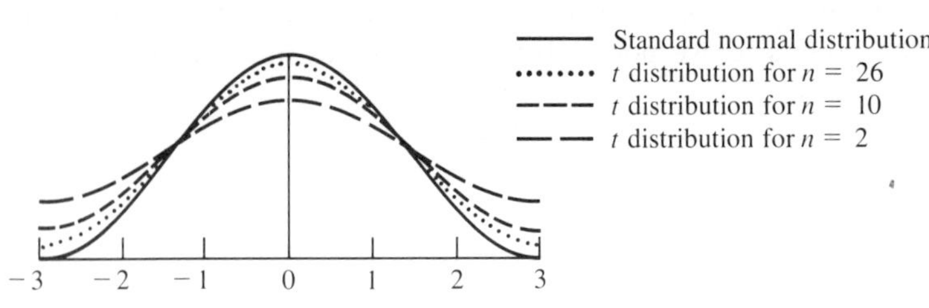

Figure 9.1 *Three t distributions and the standard normal distribution*

chapter we saw that using the sample mean $\bar{x}$ as the reference for deviations reduces the number of degrees of freedom in the sample to $(n - 1)$. For inferences about μ based on the t statistic defined in Equation 9.8, the degrees of freedom for t are $(n - 1)$, where n is the number of observations in the sample. In Figure 9.1 the three t distributions are for 25, 9, and 1 degrees of freedom, respectively.

Appendix B presents values of t for selected probability areas and degrees of freedom. For example, consider the t distribution with 5 degrees of freedom (df). The entry "-2.571" in the body of the table that is three entries to the right of "5" is interpreted by referring to the column heading "$t_{0.025}$." The probability is 0.025 of selecting a value of t equal to or less than -2.571 from a t distribution with 5 degrees of freedom. Also for 5 degrees of freedom, under "$t_{0.975}$" we see that the probability is 0.025 $(0.025 = 1 - 0.975)$ of selecting a value of t equal to or greater than 2.571. Hence the probability is 0.95 that t will lie within the interval $-2.571 < t < 2.571$ for 5 degrees of freedom. On the other hand, for 10 degrees of freedom the same probability (0.95) can be seen to apply to the shorter interval $-2.228 < t < 2.228$. As degrees of freedom increase, we can see by referring to the entries under $t_{0.025}$ and $t_{0.975}$ that the t distributions have less and less dispersion. In fact, in the bottom row, for $(n - 1)$ large without limit, the probability is 0.95 for the interval $-1.96 < t < 1.96$ and this is identical to the 0.95 interval for the standard normal deviate, that is, $-1.96 < z < 1.96$. This is the reason why it is conventional to treat the t statistic as a standard normal deviate when there are 29 or more degrees of freedom in the sample.

In Appendix B, each row presents information about 10 probability areas for an entire t distribution. Each row describes areas for an entirely different distribution than the one described on any other row. By comparison, Appendix A gives far more detailed information about probability areas for only one distribution, the standard normal distribution. A detailed table similar to Appendix A could be prepared for every one of the 34 t distributions partially described by a row in Appendix B. The space requirements would, however, be excessive when you consider that the information in Appendix B covers virtually all the cases you meet in practice.

We have seen that the sample mean $\bar{x}$ can be incorporated in a statistic (t) with a known sampling distribution when the population standard deviation σ is unknown and the sample size is less than 30 observations. The population from which the observations came is assumed to be normally distributed or nearly so. We will now see how to use the t distribution for making statistical inferences about the unknown value of a population mean μ. 9.2

Tests of Hypotheses about Population Means 9.3

A long-standing training program for telephone installers has been conducted by instructors lecturing directly to small classes. A new instructional system based on programmed learning materials and videotaped TV lectures has been developed. The old and new instructional systems are being studied to find out which is better.

Under the old system, each student's combined score on a series of written and performance tests has been recorded for a number of years. The mean score under the old system of instruction has remained virtually stable at a value of 445. The standard deviation is known and the distribution shape is essentially normal.

The training director wants to perform a one-tailed hypothesis test in which the objective is to establish that the new system is better than the old. The hypothesis statement is

$$H_0: \ \mu \leq 445;$$
$$H_A: \ \mu > 445.$$

The level of significance is set at 0.01.

The first group of 16 students has just completed the new course. These students were given the same series of tests as was used under the old system. Their total scores appear in Table 9.2. Their mean score ($\bar{x} = 459$) also appears in the table along with the deviations and squared deviations from the mean.

Table 9.2 Performance Scores for 16 Installers

Scores (x)	Deviations $(x - \bar{x})$	Squared Deviations $(x - \bar{x})^2$
468	9	81
455	−4	16
457	−2	4
493	34	1156
491	32	1024
456	−3	9
464	5	25
447	−12	144
427	−32	1024
448	−11	121
468	9	81
492	33	1089
439	−20	400
444	−15	225
452	−7	49
443	−16	256
7344		5704

$$\bar{x} = \frac{7344}{16} = 459$$

Although the standard deviation σ is known for the old population of scores, we cannot assume that the same value applies to the score distribution for the new training program. On the other hand, the same series of examinations can be expected to produce an essentially normal distribution for scores based on the new program, just as it did for the old one. The normal assumption for the population, the small sample size ($n = 16$), and the unknown population variance satisfy the requirements for a t distribution. Hence the statistic is

$$t = \frac{\bar{x} - \mu_0}{s/\sqrt{n}}$$

with $n - 1 = 15$ degrees of freedom. The sample variance is

$$s^2 = \frac{\sum (x - \bar{x})^2}{n - 1} = \frac{5704}{15}, \text{ or } 380.27,$$

so that

$$s = +19.50$$

and

$$s_{\bar{x}} = \frac{s}{\sqrt{n}} = \frac{19.50}{\sqrt{16}}, \text{ or } 4.875.$$

The sample value of the test statistic is therefore

$$t = \frac{459 - 445}{4.875}, \text{ or } 2.872.$$

From Appendix B, the critical value for t with 15 degrees of freedom at the 0.01 level of significance is

$$t_{0.99,15} = +2.602.$$

The t distribution with 15 degrees of freedom appears in Figure 9.2. The shaded portion to the left of $+2.602$ is the acceptance region for the null hypothesis.

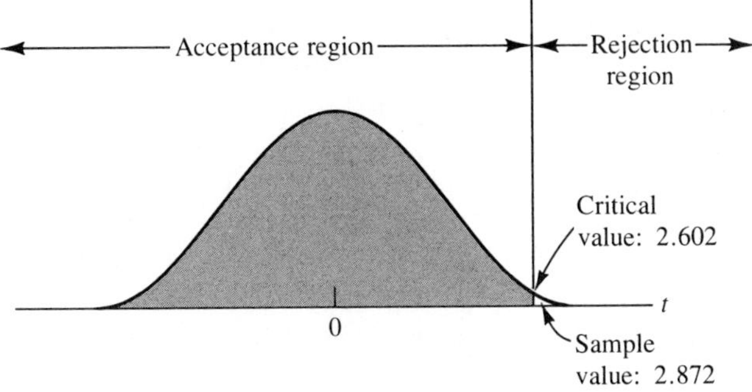

Figure 9.2 *A t distribution*

Since the sample value, 2.872, exceeds the critical value we must accept H_A. The evidence from this sample is that the population mean score for the new training program is greater than it was for the old program. The new program seems to be more effective.

If a 0.95 confidence interval estimate of the population mean score had been desired instead of a hypothesis test, the same concepts described in connection with Equations 9.5 apply. All we need do for the small-sample case being discussed is to replace $z_{\alpha/2}$ in those equations with $t_{\alpha/2}$. For 15 degrees of freedom,

$$t_{\alpha/2} = t_{0.025}, \text{ or } 2.131,$$

neglecting algebraic sign. The limits are

$$\mu_L = \bar{x} - t_{\alpha/2}s_{\bar{X}} = 459 - 2.131(4.875), \text{ or } 448.61,$$

and

$$\mu_U = \bar{x} + t_{\alpha/2}s_{\bar{X}} = 459 + 2.131(4.875), \text{ or } 469.39.$$

Hence the 0.95 interval estimate of the population mean is

9.3 $$448.61 \le \mu \le 469.39.$$

Summary

When the population variance is unknown and you have a small sample, the test statistic for inferences about the population mean μ follows a t distribution provided that the population is normally distributed. As in the large-sample case described in the first part of this chapter, the test statistic is the deviation of the sample mean $(\bar{x} - \mu)$ divided by the estimated standard error of the mean $(s_{\bar{X}} = s/\sqrt{n})$. When the sample is large, the test statistic is interpreted as a standard normal deviate. When the sample is small, the test statistic is interpreted as t provided that the population is normally distributed or nearly so.

A t distribution is a bell-shaped curve similar to the standard normal distribution. There is a different t distribution for every different sample size. The smaller the sample, the larger the standard deviation of the related t distribution. Most tables for t distributions are indexed by referring to degrees of freedom, rather than to the number of observations n in the sample. For the purposes discussed in this chapter, the degrees of freedom (df) are one less than the sample size (df $= n - 1$).

See Self-Correcting Exercises 9B.

Exercise Set B

1. What line of approach should you take in constructing a confidence interval estimate of the mean of a population with severe positive skewness and unknown standard deviation —based on the mean of a sample of size 30?

2. What statistic should you use to create a confidence interval estimate for the mean of a population which is seriously skewed and has:
   (a) A known standard deviation based on the mean of a sample of size 16?
   (b) A known standard deviation based on the mean of a sample size of 36?
   (c) An unknown standard deviation based on the mean of a sample of size 36?
   (d) An unknown standard deviation based on a sample of size 64?

3. What statistic should you use to create a confidence interval estimate for the mean of a population which is normal or nearly so and has:
   (a) Known σ based on a sample of size 16?
   (b) Known σ based on a sample of size 100?
   (c) Unknown σ based on a sample of size 25?
   (d) Unknown σ based on a sample of size 50?

4. Determine the numerical value of:
   (a) $t_{0.025}$ for 3 df; 15 df; 27 df.
   (b) $t_{0.05}$ for 7 df; 30 df.
   (c) $t_{0.995}$ for 8 df; 24 df.
   (d) $t_{0.50}$ for 3 df; 10 df.

5. What is the probability that for a sample of 12 from a normal population a t value is:
   (a) Less than 1.363?
   (b) Greater than -3.106?
   (c) No more than zero?

6. A sample of size 4 from a normal population has a mean of 6 and a standard deviation of 2.5. Construct:
   (a) A 90 percent confidence interval estimate of μ.
   (b) A 98 percent confidence interval estimate of μ.
   (c) A 99 percent confidence interval estimate of μ.

7. A sample of size 25 from a normal population has a mean of 6 and a standard deviation of 2.5. Construct:
   (a) A 90 percent confidence interval estimate of μ.
   (b) A 98 percent confidence interval estimate of μ.
   (c) A 99 percent confidence interval estimate of μ.

8. A sample of size 121 from a normal population has a standard deviation of 12. We can have 95 percent confidence that the true mean and the sample mean differ by at most what numerical value
   (a) Using t?
   (b) Using z?

9. Referring to Exercise 8, construct a 95 percent confidence interval estimate of μ if the sample mean is found to be 96.45 (a) by using the t statistic and (b) by using the z score.

10. Which test statistic, the t or the z, is more commonly used in cases such as Exercise 9? Why?

11. The mean daily wage of a random sample of 25 unskilled workers in a city is $17.50. The sum of squared deviations of the 25 observations from the sample mean is 240.
    (a) What is the value of the sample variance?
    (b) What is the value of the sample standard deviation?
12. Refer to Exercise 11:
    (a) For what parameter is the answer to part (a) an estimate? Is it unbiased?
    (b) For what parameter is the answer to part (b) an estimate? Is it unbiased?
13. Use the t table in Appendix B to answer the following questions:
    (a) What are the values of t for a t distribution with 17 degrees of freedom such that 5 percent of the area is in each tail?
    (b) In a t distribution with 21 degrees of freedom, what is the value of t that will be exceeded with the probability 0.05?
14. How many degrees of freedom are there in a t distribution in which the probability of t being greater than $+2.718$ is 0.01?

Overview

1. The distribution of lifetimes for a type of transistor has moderate positive skewness. A random sample of 100 from a large shipment of these transistors has a mean lifetime of 1190 hours and a standard deviation of 90 hours. What is the 0.95 confidence interval estimate of the mean lifetime for the shipment?
2. In a factory, a new work-station design is being tested for productivity. Observations are made on output during each of 16 hours. The sample mean is 236 units of output per hour and the sample standard deviation is 20 units per hour. Assume that the population of hourly outputs is normally distributed and find the 0.99 confidence interval estimate of the population mean hourly output.
3. For the situation described in Exercise 2, construct a 0.99 confidence interval estimate of μ assuming that the sample mean of 236 and sample standard deviation of 20 came from a sample of size 36.
4. The net weight printed on boxes of one brand of butter is 1 pound. A random sample of these boxes has been selected from several supermarkets. The sample mean net weight is 15.95 ounces and the sample standard deviation is 0.12 ounce. What is the 0.99 confidence interval estimate of the mean net weight:
    (a) If the sample is of size 25?
    (b) If the sample is of size 50?
    (c) If the sample is of size 100?
5. The shelves in the mail room of a mail order firm have been designed for a mean package length of 16 inches. The manager wants to test the hypothesis that the mean length is actually 16 inches. She selects a random sample of six packages and finds their lengths to be 13, 14, 14, 14, 17, and 18 inches to the nearest inch. Assume that package lengths are normally distributed.
    (a) Find the sample mean and standard deviation.

(b) Find the limits of the acceptance region for a two-tailed test of the hypothesis that the population mean is 16 inches. Use a significance level of 0.02.

(c) Should the null hypothesis be rejected? Why?

6. Nickel-cadmium battery packs for a certain brand of cassette recorder are very expensive to produce. It has been suggested that a single random sample of only five such packs be selected from a large production lot in order to obtain a confidence interval estimate of the mean lifetime for the lot. The sample standard deviation must be used in place of the unknown population value.

   (a) What shape can the population distribution be excepted to have?

   (b) What is the correct course of action with respect to the recommended procedure?

7. In past years, the mean score on a national reading examination has remained virtually constant at 80. The national distribution of scores always has marked negative skewness, and the standard deviation varies materially from year to year. This year a random sample of 100 scores has been selected for a two-tailed test of the hypothesis that the population mean score is 80. The sample mean is 83.5 and the sample standard deviation is 12.

   (a) What are the correct equations for finding the limits of the acceptance region? Explain your choice.

   (b) Perform the test described for a 0.05 level of significance.

8. In a large department store, the mean balance due on active credit accounts was $214 for this month a year ago. The standard deviation of balances due has remained essentially constant at $42 for several years. The credit manager wants to perform a two-tailed test, at the 0.01 level, of the hypothesis that the mean balance due has not changed. He has selected a random sample of 36 accounts and has found the sample mean to be $199.

   (a) In terms of $\bar{x}$ what are the correct equations for the limits of the acceptance region? Explain your choice.

   (b) Complete the test and state your conclusion.

9. Refer to Exercise 8:

   (a) If the population mean balance due on active credit accounts is now $234, which type of error is possible? What is the probability of making such an error with the test described in that exercise?

   (b) If the population mean balance due is $206, which type of error is possible? What is the probability of making such an error?

10. For the situation in Exercise 8, the acceptance region was described in terms of values of the sample mean.

    (a) What would the values of the acceptance region limits be if they were stated in terms of the standard normal variate z instead of the sample mean?

    (b) What would the procedure be if z, rather than $\bar{x}$, were used as the test statistic?

11. A firm makes a popular model of stereo FM radio that has a mean defect-free life of 2.3 years. A random sample of 9 of the most popular competitive model has a mean defect-free life of 1.8 years with a standard deviation of 0.7 year. The distribution from which the observations came is normal. At the 0.05 level of significance test the hypothesis that will allow the firm to claim a longer mean life, if the test is successful.

12. A coin-operated machine that pours milk by the cup is set to deliver 8 ounces per cup. A random sample of 10 cups has a mean weight of 8.60 ounces and a standard deviation of 0.5 ounce. This sample is to be used in a two-tailed hypothesis test that the setting is correct. Should this hypothesis be rejected at the 0.05 level of significance?

13. A plant engineer has stated that the time to assemble a new product should be 12 minutes on the average. It is known that assembly time distributions for very similar products have marked skewness and standard deviations which vary little from 0.6 minute. A random sample of assembly times has been collected to make a two-tailed test of the engineer's statement at the 0.01 level of significance. The mean of the sample is 10.3 minutes. What conclusion is justified by the test if (a) the sample size is 16? (b) If the sample size is 36? (c) If the sample size is 100?
14. Last year, the state tourist bureau estimated that visitors spent an average of $138 in the state. This year a random sample of 100 visitors has spent an average of $134 and the sample standard deviation is $20. The population is known to be positively skewed. Perform a two-tailed test of the hypothesis that this year's population mean is $138. The level of significance is 0.02.

Glossary of Equations

9.1 $$s^2 = \frac{\sum_{i=1}^{n} (x_i - \bar{x})^2}{n - 1}$$

The sample variance is the sum of squared deviations from the sample mean divided by one less than the number of observations.

9.4 $$s_{\bar{x}} = +\sqrt{\frac{\sum (x - \bar{x})^2}{n(n - 1)}}$$

The estimated standard error of the sample mean is the square root of the sample variance divided by the number of observations in the sample.

9.6 $$n = \left(\frac{z_{\alpha/2} \cdot s}{E}\right)^2$$

With probability $(1 - \alpha)$, the number of sample observations needed to limit the expected maximum error in an interval estimate of the population mean to a specified amount is the square of the standard normal deviate times the estimated standard deviation divided by the size of the error.

9.8 $$t = \frac{\bar{x} - \mu}{s/\sqrt{n}}$$

When sampling is from a normal population with unknown variance, the sample mean and variance can be incorporated in a t statistic.

Summary Table

Population Means: Interval Estimates and Hypothesis Tests

Population	Sample Size	σ Known	σ Unknown
Essentially normal	Large ($n > 30$)	Interval: $\bar{x} \pm z_{\alpha/2} \cdot \frac{\sigma}{\sqrt{n}}$ Test: $z = \frac{\bar{x} - \mu_0}{\sigma/\sqrt{n}}$	Interval: $\bar{x} \pm z_{\alpha/2} \cdot \frac{s}{\sqrt{n}}$ Test: $z = \frac{\bar{x} - \mu_0}{s/\sqrt{n}}$
	Small ($n < 30$)		Interval: $\bar{x} \pm t_{\alpha/2} \cdot \frac{s}{\sqrt{n}}$ Test: $t = \frac{\bar{x} - \mu_0}{s/\sqrt{n}}$
Moderately skewed	$n > 30$	Interval: $\bar{x} \pm z_{\alpha/2} \cdot \frac{\sigma}{\sqrt{n}}$ Test: $z = \frac{\bar{x} - \mu_0}{\sigma/\sqrt{n}}$	Interval: $\bar{x} \pm z_{\alpha/2} \cdot \frac{s}{\sqrt{n}}$ Test: $z = \frac{\bar{x} - \mu_0}{s/\sqrt{n}}$
Seriously skewed	$n > 50$*	Interval: $\bar{x} \pm z_{\alpha/2} \cdot \frac{\sigma}{\sqrt{n}}$ Test: $z = \frac{\bar{x} - \mu_0}{\sigma/\sqrt{n}}$	Interval: $\bar{x} \pm z_{\alpha/2} \cdot \frac{s}{\sqrt{n}}$ Test: $z = \frac{\bar{x} - \mu_0}{s/\sqrt{n}}$

**A minimum sample of 30 usually is sufficient when σ is known.*

10

Statistical Inference for Classification Data

In this chapter we turn our attention to statistical inference for nominal scale data. In Chapter 6 we encountered the binomial distribution, which dealt with outcomes of repeated samples from binary populations. Subsequent chapters dealt with continuous variables and introduced testing of hypotheses and construction of confidence intervals. These major tools of inference were examined in the context of continuous variables. Now we want to see how they are applied to attribute data. We turn first to testing a hypothesis about a binary population and find that we need only apply the basic procedures of hypothesis testing to the binomial sampling distribution. Then we find that for certain sizes of sample the normal distribution may be used as an approximation to the binomial probability distribution. Hypothesis tests can then be carried out in terms of standard scores.

In the second half of the chapter a new sampling distribution, called chi square, is introduced. This sampling distribution enables us to extend our study of inference for nominal (or classification) data beyond the binary case. It also permits us to test a hypothesis of independence between the nominal variables in a two-way classification table.

Whenever our interest centers on qualities that are distinguished by classification, attribute data arise. When there are just two classes, as for example users and non-users of a product, we have a binary population. If persons are classed as current users, former users, and those who never used a product, we have three rather than two attributes. The sampling distribution that underlies statistical inference in the first case is the binomial distribution, which we have already encountered. Sample outcomes in the second case would be called multinomial—more than two categorical outcomes. This is one of the extensions taken up in the second half of the chapter.

Another question about attribute data arises when we ask, for example, whether the distribution of males by product use class is the same as for females. This question asks whether the attribute *product use* is independent of the attribute *sex*. Testing for independence also requires an extension of the sampling distributions to which we have been exposed.

Hypothesis Testing for Proportions

Suppose that, in the past, 25 percent of families served by a long-distance moving company have been dissatisfied in some way with the moving service. Reasons for dissatisfaction could include actual costs greatly exceeding estimates given, damage to belongings, failure to observe schedule, delays in settling claims, discourtesy of employees, and so forth. The moves originate from a large number of offices, or agents, throughout the country. These agents arrange the move, hire the packers, and process claims on moves they originate. In an effort to find what practices and procedures contribute to customer satisfaction, the parent company has decided to sample 16 recent moves from each of many offices. Their idea is to find offices whose services are resulting in above and below average customer satisfaction. Management specialists will then study the operations of these offices with the objective of recommending to all offices those practices and procedures that appear to reduce customer dissatisfaction. The immediate problem is to find the above and below average offices.

Assuming that the number of moves originating in an office is large in relation to the sample size of 16 and that each move has the same probability of inclusion in the sample, the number dissatisfied out of 16 is approximately binomially distributed. The company would be concerned to exclude from the above or below average groups any offices whose true percentage of dissatisfied customers could reasonably be inferred to be 25 percent. In terms of the hypothesis testing procedure used in preceding chapters, we have:

1. State the null hypothesis, H_0, and the alternative, H_A:

$$H_0: \quad \pi = 0.25;$$

$$H_A: \quad \pi < 0.25 \text{ or } \pi > 0.25.$$

2. Select a level for α, the level of significance. Alpha is the probability of accepting H_A when H_0 is true. In the moving company example, a low α would seem appropriate. We will work with $\alpha = 0.02$.
3. Select the test statistic. The assumptions mentioned earlier lead to designation of the binomial as the appropriate distribution.
4. Find the critical value(s) of the test statistic. The binomial distribution of interest is the binomial with $\pi = 0.25$ and $n = 16$. We want a lower critical level (r_L) such that $P(r \leq r_L \mid n = 16, \pi = 0.25) \leq 0.010$ and an upper critical level (r_U) such that $P(r \geq r_U \mid n = 16, \pi = 0.25) \leq 0.010$. Consulting the binomial table we find

$$P(r \leq 0 \mid n = 16, \pi = 0.25) = 0.0100$$

$$P(r \leq 1 \mid n = 16, \pi = 0.25) = 0.0100 + 0.0535 = 0.0635.$$

The lower rejection level must be zero—that is, $r_L = 0$—in order for $P(r \leq r_L)$ not to exceed 0.010 when $\pi = 0.25$. At the upper level we see that

$$P(r \geq 9 \mid n = 16, \pi = 0.25) = 0.0058 + 0.0014 + 0.0002 = 0.0074$$

$$P(r \geq 8 \mid n = 16, \pi = 0.25) = 0.0197 + 0.0058 + 0.0014 + 0.0002 = 0.0271.$$

Clearly, r_U must be 9 for $P(r \geq r_U)$ not to exceed 0.010 when $\pi = 0.25$. The rule for decision, then, is this:

1. If $1 \leq r \leq 8$, accept H_0 and exclude the office from further consideration.
2. If $r = 0$, accept H_A and include the office in the below average customer dissatisfaction group.
3. If $r \geq 9$, accept H_A and include the office in the above average customer dissatisfaction group.

The actual probability of accepting H_A when H_0 is true will be $0.0100 + 0.0074 = 0.0174$. Because the binomial is a discrete distribution, we will not end up with the maximum alpha permitted but with an alpha error somewhat below that.

The one-tailed hypothesis test using the binomial distribution is carried out much as we carried out each side of the two-tailed test above. The difference, as before, is that all the allowable alpha error is placed in one tail of the binomial distribution.

A problem can occur in the two-tailed binomial test that does not happen with a continuous test statistic. Suppose alpha had been set at 0.05 in the moving service example. If we set r_U so that

$$P(r \leq r_U \mid n = 16, \pi = 0.25) \leq 0.025$$

and r_L so that

$$P(r \geq r_L \mid n = 16, \pi = 0.25) \leq 0.025,$$

the decision rule would have to be $r_L = 0$ and $r_U = 9$, just as for $\alpha = 0.02$. However, a rule with $r_L = 0$ and $r_U = 8$ would limit alpha to $0.0100 + 0.0271 = 0.0371$, which is still less than 0.05. Where, as in most cases, a different action is called for when H_0 is rejected in favor of the alternative that $\pi < \pi_0$ than when H_0 is rejected in favor of the alternative that $\pi > \pi_0$, we believe it makes sense to require that the $\alpha/2$ limitation be met on each side.

10.1 Normal Approximation to the Binomial

When the sample size is adequately large, binomial probabilities can be closely approximated by normal probabilities. Figure 10.1 shows binomial distributions of

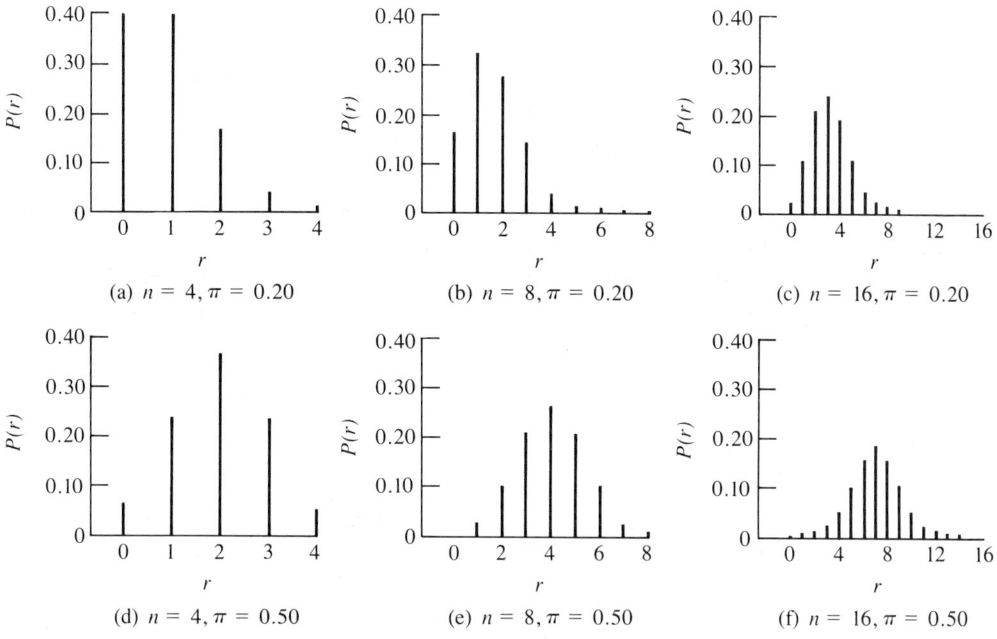

Figure 10.1 *Selected binomial distributions*

r for $n = 4$, 8, and 16 and $\pi = 0.20$ and 0.50. The graphs for $\pi = 0.20$ show skewness decreasing with increasing n. Also present is a definite suggestion of the normal distribution shape as n increases–most evident in the $\pi = 0.50$ graphs. Of course, binomial distributions are *discrete*, while the normal distribution is a continuous function. But as n is increased, the ordinates for possible values of r in the binomial are more numerous–giving a suggestion of continuity. In fact, as n increases, binomial distributions may be closely approximated by the normal distribution as long as π and $(1 - \pi)$ have positive values. Here is a rule of thumb used by many statisticians:

Rule The normal distribution provides satisfactory approximations to binomial probabilities when the values of $n\pi$ and $n(1 - \pi)$ both equal or exceed 5.

The rule of thumb means that if $\pi = 0.10$ we need to have a sample size of at least 50, while for $\pi = 0.50$ a sample size of 10 would be needed before the normal approximation would be satisfactory.

Table 10.1 presents calculations of normal probabilities to approximate the binomial distribution of r for $n = 16$ and $\pi = 0.50$. The normal distribution employed is one with a mean and standard deviation that match the mean and standard deviation of the binomial being approximated. These are

$$\mu_r = n\pi = 16(0.50) = 8.0$$

and

$$\sigma_r = \sqrt{n\pi(1-\pi)} = \sqrt{16(0.50)(0.50)} = 2.0.$$

The binomial is a *discrete* distribution with probabilities only for integer values of r, while the normal distribution is a *continuous* function. Probability in the continuous case is measured by proportionate area within the limits of desired intervals. Thus the integer value of $r = 3$, for example, becomes under the normal approximation the interval 2.5 to 3.5, the integer value $r = 4$ becomes 3.5 to 4.5, and so on.

Table 10.1 *Normal Approximation to the Binomial Distribution for $n = 16$ and $\pi = 0.50$*

Integer	Interval		z Limits		Cumulative Normal Probability		Normal Probability	Binomial
(r)	$r - 0.5$	$r + 0.5$	Lower	Upper	Lower	Upper	in Interval	Probability
0	−0.5	0.5	−4.25	−3.75	0.0000*	0.0001*	0.0001	0.0000
1	0.5	1.5	−3.75	−3.25	0.0001*	0.0006*	0.0005	0.0002
2	1.5	2.5	−3.25	−2.75	0.0006*	0.0030	0.0024	0.0018
3	2.5	3.5	−2.75	−2.25	0.0030	0.0122	0.0092	0.0085
4	3.5	4.5	−2.25	−1.75	0.0122	0.0401	0.0279	0.0278
5	4.5	5.5	−1.75	−1.25	0.0401	0.1056	0.0655	0.0667
6	5.5	6.5	−1.25	−0.75	0.1056	0.2266	0.1210	0.1222
7	6.5	7.5	−0.75	−0.25	0.2266	0.4013	0.1747	0.1746
8	7.5	8.5	−0.25	0.25	0.4013	0.5987	0.1974	0.1964
9	8.5	9.5	0.25	0.75	0.5987	0.7734	0.1747	0.1746
10	9.5	10.5	0.75	1.25	0.7734	0.8944	0.1210	0.1222
11	10.5	11.5	1.25	1.75	0.8944	0.9599	0.0655	0.0667
12	11.5	12.5	1.75	2.25	0.9599	0.9878	0.0279	0.0278
13	12.5	13.5	2.25	2.75	0.9878	0.9970	0.0092	0.0085
14	13.5	14.5	2.75	3.25	0.9970	0.9994*	0.0024	0.0018
15	14.5	15.5	3.25	3.75	0.9994*	0.9999*	0.0005	0.0002
16	15.5	16.5	3.75	4.25	0.9999*	1.0000*	0.0001	0.0000
							1.0000	1.0000

These values were obtained from a more extensive table than Appendix A.

In the second two columns of Table 10.1 the intervals corresponding to the values of r are stated. Then to carry out the normal approximation for the entire binomial distribution we need only find the normal probabilities in these intervals under a normal distribution with a mean of $n\pi = 8.0$ and a standard deviation of $\sqrt{n\pi(1-\pi)} = 2.0$. This is done in the subsequent columns by calculating the standard score (z) for the lower and upper limits of each interval and then subtracting the cumulative normal probability for the lower limit from the cumulative normal probability for the upper limit of the interval.

As an example, the normal approximation to the binomial probability for $r = 11$ is found as follows:

Interval for r: $r - 0.5$ to $r + 0.5$

$$z_L = \frac{(r - 0.5) - n\pi}{\sqrt{n\pi(1 - \pi)}} = \frac{(11 - 0.5) - 16(0.50)}{\sqrt{16(0.50)(0.50)}} = \frac{10.5 - 8.0}{2.0} = 1.25$$

$$z_U = \frac{(r + 0.5) - n\pi}{\sqrt{n\pi(1 - \pi)}} = \frac{(11 + 0.5) - 16(0.50)}{\sqrt{16(0.50)(0.50)}} = \frac{11.5 - 8.0}{2.0} = 1.75$$

$$P(r = 11) \approx P(Z \leq 1.75) - P(Z \leq 1.25) = 0.9599 - 0.8944 = 0.0655.$$

The normal approximation of 0.0655 is quite close to the binomial probability of 0.0667, which is shown in the final column. The entire set of binomial probabilities, of course, comes from Appendix E. **10.1**

Binomial Tests Using the Normal Approximation **10.2**

For an example that can be related to the normal approximations in Table 10.1, suppose that National Football League teams win 11 of 16 games against American Football League competition over a certain period. We want to test the hypothesis, at $\alpha = 0.10$, that the leagues are evenly matched. "Evenly matched" implies that the underlying probability of a National League victory is 0.50. The steps in the hypothesis test are:

1. H_0: $\pi = 0.50$; H_A: $\pi < 0.50$ or $\pi > 0.50$.
2. $\alpha = 0.10$.
3. Test statistic: normal approximation to the binomial is appropriate because $n\pi = 16(0.50) = 8$ and $n(1 - \pi) = 16(1 - 0.50) = 8$, which are both greater than the rule of thumb minimum of 5.
4. Accept H_0 if $-1.64 \leq z \leq 1.64$.

   $$z_{\alpha/2} = z_{0.05} = -1.64 \qquad \text{and} \qquad z_{1-\alpha/2} = z_{0.95} = 1.64.$$

   Accept H_A if $z < -1.64$ or $z > 1.64$.
5. Find sample value of z and follow decision rule from step 4. Here we want to know if $P(r \geq 11)$ is less than $\alpha/2$, or $0.10/2 = 0.05$. Eleven wins exceeds the mean of the binomial that prevails when H_0 is true, and the upper-tail rejection region for H_0 begins at $z = 1.64$. In order to reject H_0, the entire "slice" of probability for $r = 11$ in Figure 10.2 must lie in the rejection region. Thus we will be concerned with the z value for the *lower* limit of the interval representing $r = 11$. This is the standard score

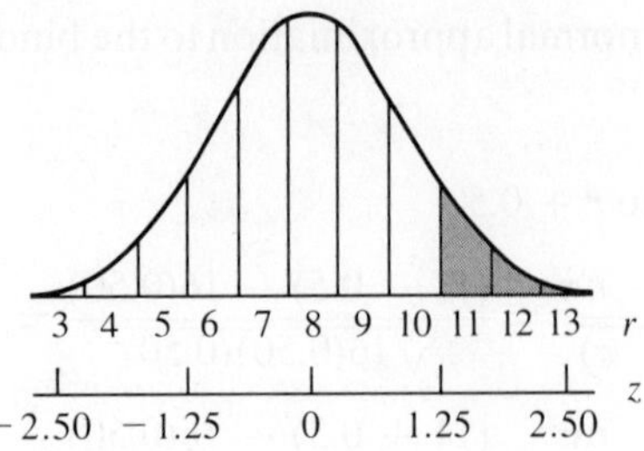

Figure 10.2 *Normal approximation to binomial for n = 16 and π = 0.50*

$$z = \frac{(r - 0.5) - n\pi}{\sqrt{n\pi(1-\pi)}} = \frac{11 - 0.5 - 16(0.50)}{\sqrt{16(0.50)(0.50)}} = \frac{10.5 - 8.0}{2.0} = 1.25.$$

Accept H_0 that $\pi = 0.50$ because $1.25 \leq 1.64$.

Had we been concerned with a lower-tail rejection region for H_0, we would have calculated

$$z = \frac{(r + 0.5) - n\pi}{\sqrt{n\pi(1-\pi)}}$$

from the observed data to check with the critical z level. For $P(r \leq r_{\text{obs}})$ to be less than $\alpha/2$, the probability in the interval $r - 0.5$ to $r + 0.5$ must lie wholly in the rejection region. For this reason we check against the z value for the *upper* limit of the interval when concerned with a *lower*-tail rejection region.

The procedures for binomial tests using the normal approximation can now be summed up:

Statement

In a two-tailed test for a single proportion using the normal approximation to the binomial, the observed result will lie in the lower-tail rejection region if

■ $$z = \frac{(r + 0.5) - n\pi}{\sqrt{n\pi(1-\pi)}} < z_{\alpha/2}$$ ■ ***10.1***

and in the upper-tail rejection region if

■ $$z = \frac{(r - 0.5) - n\pi}{\sqrt{n\pi(1-\pi)}} > z_{1-\alpha/2}.$$ ■ ***10.2***

When the rejection region for a hypothesis test of a single proportion is entirely in the lower tail, reject the null hypothesis if

$$z = \frac{(r + 0.5) - n\pi}{\sqrt{n\pi(1-\pi)}} < z_{\alpha}.$$

When the rejection region is entirely in the upper tail, reject the null hypothesis if

$$z = \frac{(r - 0.5) - n\pi}{\sqrt{n\pi(1 - \pi)}} > z_{1-\alpha}.$$

To illustrate a one-tailed test, suppose a local dairy has enjoyed 40 percent preference for the taste of its premium ice cream in the past. Recently, an advertising campaign emphasizing the superior taste has been conducted. A random sample of 1350 out of 350,000 consumers in the market area produces 594, or 44.0 percent, who prefer the taste of the local product. Test the null hypothesis, using $\alpha = 0.10$, that the brand-preference share has not increased.

1. Null hypothesis: $\pi \leq 0.40$; alternative hypothesis: $\pi > 0.40$.
2. $\alpha = 0.10$.
3. Normal approximation to binomial is appropriate.
4. Rejection region: upper tail; reject null hypothesis if $z > z_{0.90} = 1.28$.

$$z = \frac{(r - 0.5) - n\pi}{\sqrt{n\pi(1 - \pi)}} = \frac{(594 - 0.5) - 1350(0.40)}{\sqrt{1350(0.40)(0.60)}} = \frac{53.5}{18.0} = 2.97.$$

5. Reject null hypothesis because $2.97 > 1.28$. **10.2**

Confidence Interval for a Proportion

Suppose the dairy in our example wanted to establish the 0.95 confidence interval for the true proportion of consumers preferring the taste of its premium ice cream. The confidence interval for a proportion is a special case of the confidence interval for a mean when the population distribution is not normal and the population standard deviation is not known. In Chapter 9 such a confidence interval was constructed from

$$\bar{x} \pm z_{\alpha/2}\, s/\sqrt{n}$$

and was regarded as an approximate method to be used for continuous variables only when the sample size is large. As we saw in Chapter 6, the sample proportion of successes, r/n, is the sample mean, and is given the symbol p for sample proportion. The sample standard deviation is, then, approximately $\sqrt{p(1 - p)}$.* The procedure of Chapter 9, Equation 9.5, restated for a proportion, is thus:

*Remember that the standard deviation of a binary population is $\sqrt{\pi(1 - \pi)}$. The sample proportion p is an estimate of π. The square root of the average squared deviation of the binary observations in the sample is $\sqrt{p(1 - p)}$. The sample standard deviation, as previously defined, would be $\sqrt{p(1 - p)}.\ \sqrt{n/(n - 1)}$.

Statement

A $(1 - \alpha)$ symmetrical confidence interval for the unknown proportion of successes (π) in a binary population can be constructed from a large sample of n independent observations from the population by

$$p \pm z_{\alpha/2}\sqrt{\frac{p(1-p)}{n}}, \qquad (10.3)$$

where p is the sample proportion of successes, r/n.

To apply the procedure of Equation 10.3, the following minimum sample sizes are recommended:

Lesser of p or $(1 - p)$	0.01	0.02	0.05	0.10	0.20	0.30	0.40	0.50
Minimum sample size n	10,000	5000	2000	900	300	150	60	30

The large minimum sample sizes as p departs from 0.50 are required to maintain a reasonable degree of accuracy of $\sqrt{p(1-p)}$ as an estimate of the binary population standard deviation of $\sqrt{\pi(1-\pi)}$. This accuracy is adversely affected by skewness in binary populations.

In the dairy example, 594 out of 1350 consumers were found to prefer the marketer's brand. For the 0.95 confidence interval for the population proportion preferring the marketer's brand, we have

$$p = \frac{r}{n} = \frac{594}{1350} = 0.44$$

$$\begin{aligned} p \pm z_{0.025}\sqrt{\frac{p(1-p)}{n}} &= 0.44 \pm 1.96\sqrt{\frac{0.44(0.56)}{1350}} \\ &= 0.44 \pm 1.96(0.0135) \\ &= 0.44 \pm 0.0265 \\ &= 0.4135 \text{ to } 0.4665. \end{aligned}$$

Determining Sample Size

In Chapter 8, a method was developed for finding the sample size required to limit the maximum error in an interval estimate of a population mean to a specified quantity (E) with probability $(1 - \alpha)$. Equation 8.4 for the necessary sample size was

$$n = \left(\frac{z_{\alpha/2}\sigma}{E}\right)^2,$$

which can also be expressed as

$$n = \frac{z_{\alpha/2}^2 \sigma^2}{E^2}.$$

Since the variance of a binary population is $\pi(1 - \pi)$, the comparable expression in dealing with proportions is

$$n = \frac{z_{\alpha/2}^2 \pi(1 - \pi)}{E^2}.$$

Note that π, the true population proportion, is the population mean we are called on to estimate. If we knew π for substitution in the formula above, we would not be considering how large a sample we needed to estimate π. However, it is the product, $\pi(1 - \pi)$, that effectively determines n once $z_{\alpha/2}$and E are specified. This product never exceeds $0.50(1 - 0.50) = 0.25$, so this figure can be used to obtain a "fail-safe" estimate of required sample size. Suppose the marketing manager in our dairy example wanted an estimate of brand-preference proportion to be within 0.02 of the true proportion with 0.95 confidence. We could calculate

$$n = \frac{z_{\alpha/2}^2(0.25)}{E^2} = \frac{(1.96)^2(0.25)}{(0.02)^2} = 2401.$$

Now, no matter what proportion actually occurs in the sample of 2401 consumers, the product $1.96\sqrt{p(1 - p)/n}$ used in establishing the 95 percent confidence interval will not exceed 0.02. For example, if the sample should yield $p = 0.30$,

$$1.96\sqrt{\frac{0.30(0.70)}{2401}} = 0.0183.$$

For $p = 0.40$,

$$1.96\sqrt{\frac{0.40(0.60)}{2401}} = 0.0196.$$

And for $p = 0.50$,

$$1.96\sqrt{\frac{0.50(0.50)}{2401}} = 0.0200.$$

When we are dealing with proportions that are known to be quite small (or large), using 0.25 in the sample size formula may be overly conservative. Suppose an estimate is desired of the proportion of licensed drivers in a state who were guilty of traffic violations during a year. Accuracy desired is within 0.01 with 99 percent confidence. Suppose those designing the sample of motor vehicle department records are quite sure that the porportion does not lie below 0.05 or exceed 0.15, then an appropriate sample size could be calculated from

$$n = \frac{(2.58)^2(0.15)(0.85)}{(0.01)^2} = 8487.$$

As long as the proportion of violations in the sample of 8487 driver records does not exceed 0.15, then $2.58\sqrt{p(1 - p)/n}$ will not exceed 0.01. The sample size resulting from the procedure above should be checked against the minimum sample size table given in connection with the normal confidence interval procedure of Equation 10.3. For a valid application, the calculated sample size must exceed the minimum requirement of the table for any sample proportion anticipated. For example, the sample size of 8487 would not be adequate for a normal confidence interval procedure if the sample proportion turned out to be much below 0.02. In the example we assumed that the motor vehicle department knew enough about violation rates to rule out this possibility. Similarly, the sample size of 2401 in the dairy example would not be sufficient for a normal confidence interval procedure if the sample proportion turned out to be much below 0.05 or above 0.95—values that are virtually impossible in the context of that example. Sample size determination for proportions can then be summarized:

Statement

The required number of observations n to limit the maximum error in a normal interval estimate of the population proportion π to a specified quantity (E) with probability $(1 - \alpha)$ is

$$n = \frac{z_{\alpha/2}^2(0.25)}{E^2}. \qquad \textbf{10.4}$$

In cases where Equation 10.4 is overly safe, you should substitute for 0.25 a maximum value of $\pi(1 - \pi)$ for the situation at hand. The resulting n should be checked in the minimum sample size table to make sure that a *normal* interval estimate is appropriate for proportions regarded as possible.

Summary

A test of a hypothesis concerning a proportion can be carried out by using probabilities from the binomial table. The null hypothesis is accepted if the appropriate tail probability, which always includes the observed r, exceeds $\alpha/2$ in a two-tailed test or α in a one-tailed test.

The normal probability distribution with $\mu = n\pi$ and $\sigma = \sqrt{n\pi(1 - \pi)}$ may be used to approximate binomial probabilities as long as $n\pi$ and $n(1 - \pi)$ both exceed 5. Because the binomial is discrete and the normal is a continuous distribution, the normal approximation to the probability of any integer r is the area under the normal distribution from $r - 0.5$ to $r + 0.5$. Therefore, in calculating a sample value of z to check against a critical z value, we evaluate z at $r - 0.5$ in connection with an upper-tail critical level and evaluate z at $r + 0.5$ in connection with a lower-tail critical level.

Under certain conditions of minimum sample size, you can establish confidence intervals for a proportion by an approximation based on the normal distribution. You can also determine the sample size required to ensure that the error embodied in a subsequent confidence interval statement does not exceed a specified quantity.

See Self-Correcting Exercises 10A.

Exercise Set A

1. From Appendix E, determine lower and upper critical values for a two-tailed hypothesis test on a binomial distribution at a level of significance of 5 percent if:

   (a) $\pi = 0.40$ and $n = 14$ (b) $\pi = 0.15$ and $n = 15$

2. What is the actual probability of rejecting a true null hypothesis in each part of Exercise 1? By what symbol is this quantity denoted? Is it equal to the maximum value permitted in either case?

3. From Appendix E, determine r_U as appropriate and state the acceptance range for a binomial distribution upper-tail hypothesis test at a level of significance of 2 percent if:

   (a) $\pi = 0.50$ and $n = 12$ (b) $\pi = 0.85$ and $n = 16$

4. What is the actual probability of rejecting a true null hypothesis in each part of Exercise 3? Is it equal to the maximum value permitted in either case?

5. Prior to the discussion of hypothesis testing with a binomial distribution, lower and upper critical values were always located symmetrically with respect to the hypothesized mean. But in the text example with $\pi = 0.25$ and $n = 16$, the mean is 4 while r_L and r_U at the 0.02 level of significance are 1 and 9, respectively, and clearly not symmetrically located with respect to 4. Why is this so? For what value(s) of π are r_L and r_U located symmetrically with respect to the mean?

6. Using the rule of thumb for determining the appropriateness of applying the normal curve approximation of the binomial distribution, the normal curve approximation is applicable:

   (a) Within what range of values for π if the sample size is (i) 8; (ii) 50; (iii) 100?

   (b) For what minimum sample size if (i) $\pi = 0.6$; (ii) $\pi = 0.05$; (iii) $1 - \pi = 0.03$?

7. When using the normal curve approximation to the binomial we consider, for example, the probability of obtaining 7.5 to 8.5 successes in some specified number of trials, rather

than that of obtaining precisely eight successes. Why is it necessary to assume that a precise value may be represented by a range of values? To use the normal curve approximation to determine the probability of fewer than four successes in 22 independent trials, what range of values would you consider? For at most four successes?

8. (a) For a binomial distribution with $n = 30$ and $\pi = 0.6$, find μ_r and σ_r.
   (b) Using the normal curve approximation to estimate $P(15 \leq r \leq 18 \mid n = 30, \pi = 0.6)$, we evaluate (i) $P(\_\_\_\_ \leq X \leq \_\_\_\_)$ or equivalently (ii) $P(\_\_\_\_ \leq Z \leq \_\_\_\_)$.

9. Applying the normal curve approximation to the binomial distribution as an estimate of the following probabilities, we would give the area under the normal curve over what interval on the horizontal (x) axis?
   (a) $P(r = 27 \mid n = 50, \pi = 0.7)$
   (b) $P(1 < r < 6 \mid n = 12, \pi = 0.5)$
   (c) $P(1 \leq r \leq 6 \mid n = 12, \pi = 0.5)$
   (d) $P(r \geq 70 \mid n = 100, \pi = 0.75)$

10. State your x-value interval endpoint(s) in each part of Exercise 9 as z scores.

11. Converting from the discrete r values to the continuous x values and then to z values seems a rather involved procedure. Why, with all this "involvement," is approximation to the binomial distribution considered to be a laborsaving device when appropriate?

12. Using the normal approximation to the binomial, find each probability referred to in Exercise 9.

13. Compare your results for parts (b) and (c) of Exercise 9 with those you would obtain from Appendix E. Which values are exact (to four decimal places) and which are only approximate? How much do the values differ for part (b)? For part (c)?

14. Before the adoption of a new type of packaging, Muncho Cruncho candy bars had 20 percent of the candy bar market in a certain area. A sampling of current brand preference there is taken; of 805 candy bar purchases, 158 are Muncho Cruncho bars. Test at a level of significance of 0.10 the null hypothesis that Muncho Cruncho now has at *least* 20 percent of the market.

15. Several years after the introduction of a new chemical process at an industrial plant, the plant doctor notices what she considers to be an unusually high incidence of strokes among workers involved with that process. She conducts a hypothesis test at the 10 percent level of significance by comparing the total number of stroke cases among workers under age 40 who have been involved with the process (21 cases among 1946 workers) with the plantwide stroke rate (0.008) for workers of the same age group.
    (a) What are H_0 and H_A?
    (b) Why is such a high level of significance selected? What might the consequences be of erring in choosing H_0? Of erring by choosing H_A? Would a lower level of significance be more appropriate? Justify your conclusion.
    (c) Should the doctor reject H_0? Justify your conclusion.

Chi-Square Tests Involving Frequencies

In this section we present a new test statistic which is especially useful in tests of hypotheses about classification data. The test statistic is called chi square (pronounced "keye," to rhyme with *eye*). It enables us to go beyond binomial tests to consider hypotheses about proportions of observations in a population falling into more than two classes. We will introduce a problem involving three classes first, and will see how the chi-square distribution permits tests of the kind mentioned. Then we will see how the chi-square distribution is used in yet another situation—a test for independence in cross-classification tables.

Chi-Square Tests for One-Way Tables

Suppose 150 randomly selected schoolchildren are asked to indicate a preference among three flavors of gum. Their preferences are tabulated here:

Flavor	Number Preferring
A	68
B	43
C	39
	150

The sponsor of the study is interested in whether the flavors enjoy equal preference among all schoolchildren and is willing to run a 0.05 risk of rejecting this hypothesis when it is true. We then have

$$H_0\colon \quad \pi_A = \pi_B = \pi_C = 1/3;\ \alpha = 0.05.$$

Once a hypothesis has been specified, we can derive a set of *expected* frequencies under the hypothesis. We reason that *if* H_0 *were true,* repeated samples of 150 observations would produce $1/3\,(150) = 50$ preferences for each flavor *on the average.* We have, then, a table of actual frequencies and a corresponding table of expected frequencies under the null hypothesis:

	Preference	
Flavor	Actual (f_o)	Expected (f_e)
A	68	50
B	43	50
C	39	50

We can now think of our hypothesis test problem as one of evaluating the probability of observing an outcome at least as extreme (that is, different from the expected result under the null hypothesis) as our sample result. If the probability of such a result is small—that is, less than a predesignated α level—we reject the null hypothesis. Otherwise the null hypothesis is accepted.

Definition For comparing observed frequencies (f_o) with frequencies expected under a null hypothesis (f_e), the following statistic is an index of discrepancy (or disagreement) between the observed and expected frequencies.

$$\text{Index} = \sum\left[\frac{(f_o - f_e)^2}{f_e}\right]$$

From the definition of the index of disagreement, you can see that for a given set of f_es, the index will be zero if the observed frequencies coincide with the expected frequencies. The index will take on larger values as the observed frequencies differ more from the expected frequencies. If the null hypothesis leading to the expected frequencies were true, repeated samples would still lead to different sets of observed frequencies and different values for the index of disagreement. Indeed, the probability distribution of these observed indexes is known—provided we know how many independent opportunities exist for differences between f_o and f_e. This number, called the *degrees of freedom,* depends in part on the number of classes for which f_o and f_e are compared. In the present situation (one-way tables) the degrees of freedom (m) is one less than the number of classes.

Statement The index of discrepancy between observed frequencies and frequencies expected under a null hypothesis is distributed as chi square (χ^2) with degrees of freedom (m) equal to the number of independent opportunities for difference.

$$\blacksquare \quad \chi^2_m = \sum\left[\frac{(f_o - f_e)^2}{f_e}\right] \quad \blacksquare \qquad \textbf{10.5}$$

There is a different chi-square distribution for every value of m, the number of independent opportunities for difference. The number of such independent opportunities is called the **degrees of freedom.** Appendix C shows different chi-square distributions for various numbers of degrees of freedom. The column headings are areas under the distributions, or probabilities, cumulated from the left. The body of the table contains values of chi square that cut off the areas given in the column headings.

The graph at the top of Appendix C shows how to read the table. There the chi-square distribution is shown for 5 degrees of freedom. The probability is 0.90 that a value of chi square (χ^2) selected at random from this distribution will be less than 9.24. Only 10 percent of the time will values of χ^2 selected randomly from this distribution exceed 9.24. The value 9.24 appears in the table under $\chi^2_{0.90}$ in the row for 5 degrees of freedom.

Figure 10.3 shows the density functions for two chi-square distributions. The first is for 5 degrees of freedom ($\chi^2_{m=5}$) and the second is for 10 degrees of freedom ($\chi^2_{m=10}$). The expected value of χ^2 is equal to the degrees of freedom, m. The mean of the chi-square distribution for $m = 5$ is 5.0, and the mean of the chi-square distribution for $m = 10$ is 10.0.

Returning to the preferences of schoolchildren for chewing gum flavors, the calculation of the observed chi square (χ^2_{obs}) is

Flavor	f_o	f_e	$f_o - f_e$	$(f_o - f_e)^2/f_e$
A	68	50	18	6.48
B	43	50	-7	0.98
C	39	50	-11	2.42
				$9.88 = \chi^2_{\text{obs}}$

There are three rows in the table, so the degrees of freedom are $3 - 1 = 2$. Consulting this row of the chi-square table, we find that $\chi^2_{0.99} = 9.21$. The discrepancy between the sets of actual and expected frequencies is sufficiently large that the probability of a larger discrepancy, if the null hypothesis of equal preference were true, is less than 0.01.

Since the chi-square statistic relates here to a measure of discrepancy (or disagreement) between *actual* frequencies and *expected* frequencies under the null hypothesis, the chi-square test is inherently one-tailed. We reject the null hypothesis when the discrepancy is large, that is, so large that the probability of a larger discrepancy *when the null hypothesis is true* is less than α, the allowable Type I error risk.

With $\alpha = 0.05$, the level specified earlier, the critical value of χ^2 with $m = 2$ is $\chi^2_{0.95} = 5.99$. Our observed chi square exceeds this level, so we reject the hypothesis

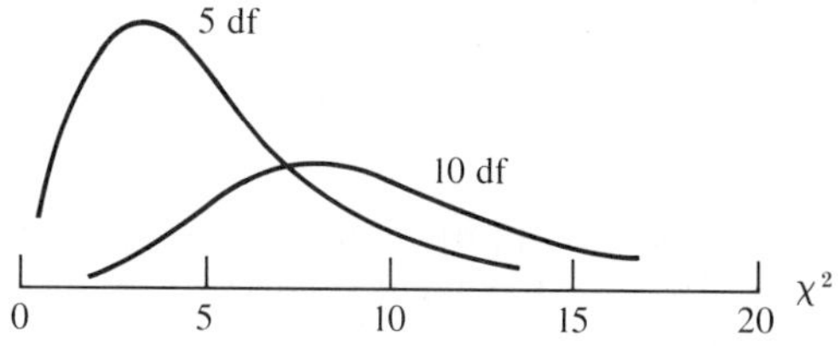

Figure 10.3 *Two chi-square distributions*

that $\pi_A = \pi_B = \pi_C = 1/3$. The sample data are sufficient to conclude that preference for the flavors is not equal among all the schoolchildren.

We can now summarize the chi-square test for a one-way table.

Rules

1. Null hypothesis: specify values for π_i $(i = 1,2,\ldots,r)$, where r is the number of classes in the one-way table. Note that only $r - 1$ of these can be independently specified, since $\sum_i \pi_i = 1.0$.
2. Specify α, the level of significance for the test.
3. The test statistic is χ^2 as in Equation 10.5, as long as each f_e in the table equals or exceeds 5. This requirement is necessary because a continuous distribution (χ^2) is being used to approximate the distribution of indexes of discrepancy. The f_o involved in these indexes can take on integer values only. Random sampling is assumed.
4. Find $\chi^2_{1-\alpha}$ for the degrees of freedom, $m = r - 1$. The rejection region for the null hypothesis is $\chi^2_{\text{obs}} \geq \chi^2_{1-\alpha}$.
5. From π_i determine the expected frequencies from n times π_i. Calculate χ^2_{obs} and accept the null hypothesis if $\chi^2_{\text{obs}} < \chi^2_{1-\alpha}$. If $\chi^2_{\text{obs}} \geq \chi^2_{1-\alpha}$ reject the null hypothesis.

Chi-Square Test of Independence

A frequently encountered situation is illustrated by the following example. A test was run on approval of a proposed editorial policy among a random sample of 300 subscribers to a magazine. The magazine staff questioned whether approval of the proposed policy would be the same among female as among male subscribers. The sample results are shown in the following table:

Attitude	Female	Male	Total
Approve	122	88	210
Disapprove	38	52	90
Total	160	140	300

The sample data suggest that approval of the proposed policy is higher among women than among men, $122/160 = 76$ percent versus $88/140 = 63$ percent. The data are only a sample, however, and we should check to see if the suggestion is warranted. We do this by testing the null hypothesis that approval is the same among women as among men.

The best estimate of the proportion of all subscribers approving the policy is $^{210}/_{300} = 0.70$. If approval were the same among women as men, 0.70 is our best estimate of what the proportion approving in each group would be.

As with our earlier examples of chi-square tests, the chi-square statistic is calculated from a set of differences between actual frequencies and expected frequencies under the null hypothesis. The expected frequencies are calculated from the marginal totals in the frequency table. They are estimates consistent with a null hypothesis of independence between the classes of events in the two-way table. In our example that means independence between approval and sex of subscriber—that approval does not vary by sex.

Recall from Chapter 4 that when independence prevails, joint probabilities can be obtained from products of marginal probabilities. This is expressed in the multiplication law for independent events, $P(A \cap B) = P(A) \cdot P(B)$. In working our expected frequencies under a null hypothesis of independence, we follow this law except that the probabilities are estimates from the sample data. Using $\hat{P}$ for estimated probability,

$$f_e = \hat{P}(A \cap B) \cdot n$$
$$= [\hat{P}(A) \cdot \hat{P}(B)] \cdot n.$$

In our example,

	Observed (f_o)			Expected (f_e)	
	Female	Male	Total	Female	Male
Approve	122	88	210	$^{210}/_{300}(^{160}/_{300})\ 300 = 112$	$^{210}/_{300}(^{140}/_{300})\ 300 = 98$
Disapprove	38	52	90	$^{90}/_{300}(^{160}/_{300})\ 300 = 48$	$^{90}/_{300}(^{140}/_{300})\ 300 = 42$
Total	160	140	300		

Note, as a practical matter, that once any one of the expected frequencies in a 2 by 2 table is found, the remaining ones can be determined by subtractions from the marginal totals of the original table. This suggests what is indeed the case, that there is only one degree of freedom, or independent opportunity for difference between f_o and f_e in a 2 by 2 table.

The expected and observed frequencies, along with the calculation of chi square for our example are

f_o	f_e	$f_o - f_e$	$(f_o - f_e)^2/f_e$
122	112	10	0.89
38	48	−10	2.08
88	98	−10	1.02
52	42	10	2.38
			$6.37 = \chi^2_{obs}$

With one degree of freedom, $\chi^2_{0.95}$, a commonly selected critical level, is 3.84. With a 5 percent risk of concluding that approval depends on sex when independence really prevails, we could conclude that the proportions approving the proposal differ by sex.

We can now state the chi-square test of independence between the nominal variables of a two-way frequency table:

Statement

When a two-way frequency table has r rows and c columns, the test statistic under a null hypothesis of independence between row and column variables is

$$\chi^2 = \sum \left[\frac{(f_o - f_e)^2}{f_e} \right]$$

with $(r - 1)(c - 1)$ degrees of freedom. The sample must be random and each expected frequency must equal or exceed 5.

Chi-square tests of independence from two-way tables may involve tables of larger dimension than 2×2. As an example Table 10.2 shows a tabulation of frequency of newspaper purchase by educational level of head for 411 families randomly selected from all families in a community.

Table 10.2 *Frequency of Newspaper Purchase by Educational Level*

Newspaper Purchase	Primary	High School	Post High School	Total
Daily	42	47	96	185
Once or twice a week	24	32	35	91
Occasionally	17	11	12	40
Never	59	21	15	95
Total	142	111	158	411

We will test the null hypothesis that frequency of newspaper purchase is independent of educational level among all families in the community. If this hypothesis were true, then the frequencies expected in each cell of the table would be n times the products of the appropriate marginal probabilities. Using the marginal frequencies from the sample to estimate the marginal probabilities, the expected frequencies under independence are:

	Primary	High School	Post High School	Marginal Probability
Daily	63.9	50.0	71.1	185/411
Once or twice a week	31.4	24.6	35.0	91/411
Occasionally	13.8	10.8	15.4	40/411
Never	32.8	25.7	36.5	95/411
Marginal probability	142/411	111/411	158/411	

An example calculation of an expected frequency is

$$\begin{aligned} f_e\,(\text{Daily} \cap \text{Primary}) &= \hat{P}(\text{Daily}) \times \hat{P}(\text{Primary}) \times 411 \\ &= {}^{185}\!/_{411} \times {}^{142}\!/_{411} \times 411 \\ &= 63.9. \end{aligned}$$

Our index of discrepancy is

$$\chi^2_{\text{obs}} = \sum \left[\frac{(f_o - f_e)^2}{f_e} \right] = 56.33.$$

The degrees of freedom for the test are $(r - 1)(c - 1) = (4 - 1)(3 - 1) = 6$ for our table. For 6 degrees of freedom $\chi^2_{0.995} = 18.55$. The chi-square value of 56.33 is one that would hardly ever occur if frequency of newspaper purchase were independent of educational level of heads of family. We conclude that newspaper purchase frequency differs by educational level among all families in the community.

Figure 10.4 shows the percentage distributions of purchase frequency within each educational class in a way designed to bring out the pattern in the sample data.

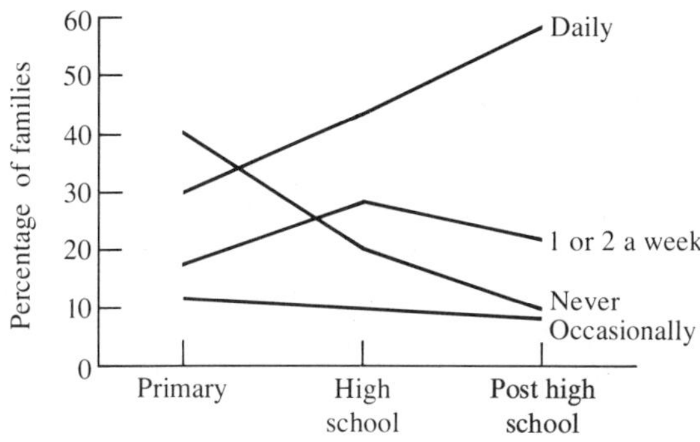

Figure 10.4 *Percentage distributions of newspaper purchase frequencies by educational level*

We see that the percentage of families purchasing daily increases with increasing educational level and that the percentage of families never purchasing a newspaper decreases markedly as educational level increases. The null hypothesis of independence that we rejected for the population would be represented on this type of graph as a series of horizontal lines for each class of newspaper purchase frequency.

Summary

The chi-square distribution can be used to test the agreement between observed frequencies for nominal data and frequencies expected under a specified null hypothesis. An index of discrepancy involving sums of ratios of squared differences (between observed and expected frequencies) to expected frequencies is distributed as a chi-square statistic whose probability distribution is known and has been tabled. The chi-square test extends our capabilities beyond binary populations in that it allows us to test hypotheses about the population distribution of relative frequencies in any number of classes of a nominal variable. The number of degrees of freedom for these chi-square tests is one less than the number of classes of the variable.

The chi-square statistic also permits tests of independence between the nominal variables of a two-way frequency table. In a test of independence, the null hypothesis is that the population relative frequencies among the classes of the one variable are the same for all classes of the other variable. In population terms this is a hypothesis that the relative frequency distribution by rows is the same for each column in the table (and vice versa). The degrees of freedom for chi-square tests of independence are the product of one less than the number of rows and one less than the number of columns in the two-way table.

See Self-Correcting Exercises 10B.

Exercise Set B

1. For a chi-square distribution with 12 degrees of freedom, determine the probability that a value of chi square selected at random from this distribution is:
   (a) Less than 3.57.
   (b) Less than or equal to 3.57.
   (c) Not in excess of 6.30.
   (d) In excess of 3.07.
   (e) In excess of or equal to 26.22.
   (f) Equal to 21.03.
2. In a chi-square distribution with 7 degrees of freedom, what value will be exceeded with probability 0.05? With probability 0.95?
3. Determine the value which is exceeded by 90 percent of the scores in a chi-square distribution having the number of degrees of freedom equal to (a) 2; (b) 20; (c) 120.

4. Determine the value of the mean of a chi-square distribution having the number of degrees of freedom equal to (a) 5; (b) 12; (c) 30.

5. Why, for any fixed level of significance, is it reasonable to expect the critical value of chi square to increase as the number of degrees of freedom increases?

6. Why, for any fixed number of degrees of freedom, is it reasonable to expect the value of chi square to decrease as the level of significance increases?

7. Why is the chi-square test necessarily a one-tailed test?

8. In a bicentennial year study of impressions of the United States with which foreign tourists ended their visits, the following responses were obtained. Results were tabulated according to whether they came from members of group tours or from persons who had traveled independently.

Response	Group	Independent
Generally favorably impressed	78	34
Favorable impressions balanced by unfavorable	36	22
Generally unfavorably impressed	41	19
	155	75

Test at a level of significance of 0.05 the null hypothesis that foreign tourists' impressions of the United States are not influenced by their mode of touring.

9. In an attitude study of recently divorced persons, responses to a question concerning likelihood of another marriage were obtained and tabulated as follows:

Response	Duration of Recent Marriage Less Than 3 Years	3 Years or More
Unlikely to marry again	47	96
Likely to marry again	52	131

Test at a level of significance of 0.01 the null hypothesis that attitudes toward the likelihood of remarriage are independent of duration of the most recent marriage.

10. A manufacturer of a meat seasoning is satisfied with the spiciness of his seasoning when consumers are divided evenly among those who want more spice, those who want less spice, and those who consider the spiciness "about right." Recently, taste tests were conducted in three cities with the following results:

Reaction	City *A*	City *B*	City *C*
Want more spice	17	27	25
About right	21	17	22
Want less spice	22	16	13

Test the hypothesis that the reactions of customers are independent of city of residence.

11. Test the data for each city in Exercise 10 against the theoretical population desired by the manufacturer.

Overview

1. During the past year, a company enjoyed a brand-preference proportion in a certain market of 0.30. A sampling of current brand preference is to be taken. If α is set at 0.10, express the rejection region as a level of z
   (a) for a test of the null hypothesis that the firm's current brand preference in the market is 0.30, and
   (b) for a test of the null hypothesis that its current brand preference is not greater than 0.30.

2. A random sample of 805 consumers in the situation of Exercise 1 yields 265 who prefer the company's brand. Carry out the test in (b) of that exercise.

3. Use, *if appropriate,* both the normal curve approximation to the binomial and Appendix E to determine:
   (a) $P(7 \leq r \leq 12 \mid n = 15, \pi = 0.4)$
   (b) $P(r = 2 \mid n = 12, \pi = 0.5)$
   (c) $P(3 < r < 7 \mid n = 10, \pi = 0.6)$
   (d) $P(r \geq 10 \mid n = 36, \pi = 0.5)$
   Compare your results where both methods are applicable.

4. Find the normal approximations to the following binomial probabilities and compare the results with the true binomial probabilities from Appendix E:
   (a) $P(r \geq 12 \mid n = 16, \pi = 0.5)$
   (b) $P(r < 7 \mid n = 14, \pi = 0.6)$
   (c) $P(r = 6 \mid n = 16, \pi = 0.5)$

5. Consider the outcomes of samples of size 100 taken from a process which is producing 10 percent defective products. Use the normal approximation to the binomial probability distribution to find: (a) $P(r < 4)$; (b) $P(r = 8)$; (c) $P(r > 14)$.

6. In a random sample of 100 students, 64 agree with recent criticisms of selections of speakers for an "ecological awareness" program. Test the hypothesis, at $\alpha = 0.05$, that student attitude is evenly divided on the issue.

7. In an acceptance sampling procedure a random sample of 144 parts was taken from each large lot of parts and the lot quality declared unacceptable if eight or more defectives were found. What is the probability under this scheme of:
   (a) Declaring a lot with 4 percent defectives unacceptable?
   (b) Declaring a lot with 10 percent defectives acceptable?

8. By the rule of thumb given in the text, how large a sample size is required in order for the normal curve approximation to the binomial to be applicable if:
   (a) $\pi = 0.1$?
   (b) $\pi = 0.01$?
   (c) $\pi = 0.9$?
   (d) $\pi = 0.99$?

9. What is the smallest value of π for which a sample size of 20 is adequate? The largest value?

10. In a random sample of 50 residents of Plymouth Union, Vermont, 27 preferred a certain brand of maple-flavored syrup to pure maple syrup. Determine the sample mean and standard deviation.

11. Referring to Exercise 10, determine a 95 percent confidence interval estimate of the true

proportion of Plymouth Union residents who prefer this brand of flavored syrup to the local product. Is the sample sufficiently large to warrant the application of Equation 10.3.?

12. Make a "fail-safe" estimate of the sample size required in the syrup preference test of Exercise 10 to ensure at a level of confidence of 98 percent that the true and sample proportions differ by no more than 0.05. Does your answer in any way reflect the numerical values given in the statement of Exercise 10? On what two values only does your "fail-safe" sample size depend?

13. Use your result of Exercise 12 to determine the maximum error in a 98 percent confidence interval estimate of π if for a sample of this size the sample proportion is (a) 0.6; (b) 0.2.

14. An equipment manufacturer changed the service features connected with his guarantee. A random sample of guarantee holders under the old and under the new system produced the following figures:

Response	Old	New
Satisfied with service	145	155
Dissatisfied	55	45

Can the manufacturer conclude that there is any association between satisfaction and the change in service features? Explain.

15. A questionnaire was distributed by Nolo Contendere College to a random sample of its alumni 2 years after their graduation. The following responses were obtained, indicating the graduates' levels of satisfaction with their educational experiences at Nolo Contendere:

Response	Business	Science and Engineering	Liberal Arts	Fine Arts
Highly satisfactory	49	47	64	22
Quite satisfactory	57	32	67	25
Somewhat unsatisfactory	24	18	42	19
Very unsatisfactory	18	13	22	12

Test at a level of significance of 0.05 the null hypothesis that their level of satisfaction is independent of their area of study at Nolo Contendere.

16. A random sample of families in four neighborhoods produced the following data on new arrivals. Test the hypothesis that the proportion of families who took up residence in the past year is the same in all four neighborhoods.

Lived in Neighborhood	Neighborhood 1	2	3	4
Less than 1 year	24	2	19	15
1 year or more	126	78	71	65

Glossary of Equations

10.1 $z_L = \dfrac{(r + 0.5) - n\pi}{\sqrt{n\pi(1 - \pi)}}$

The z score for a binomial variable r, used in finding the lower-tail normal approximation.

10.2 $z_U = \dfrac{(r - 0.5) - n\pi}{\sqrt{n\pi(1 - \pi)}}$

The z score for a binomial variable r, used in finding the upper-tail normal approximation.

10.3 $p \pm z_{\alpha/2}\sqrt{\dfrac{p(1 - p)}{n}}$

A large sample normal approximation to the $(1 - \alpha)$ confidence interval for a fixed but unknown proportion of successes in a binary population based on observing r successes in n independent draws from the population ($p = r/n$, the sample proportion).

10.4 $n = \dfrac{z^2_{\alpha/2}(0.25)}{E^2}$

Formula for the required sample size to limit the maximum error in estimating a population proportion to within $\pm E$ with probability $(1 - \alpha)$.

10.5 $\chi^2_m = \sum\left[\dfrac{(f_o - f_e)^2}{f_e}\right]$

The chi-square statistic for testing a hypothesis that a set of actual frequencies (f_o) was generated by independent draws from a population leading to a set of expected frequencies (f_e).

11

Regression and Correlation

In this chapter, we are concerned with techniques based on **bivariate observations**. A single number constitutes a univariate observation, and a pair of numbers from the same unit of observation constitutes a bivariate observation. The weight of an automobile is an example of a univariate observation. The weight and gasoline consumption of that automobile are an example of a bivariate observation, but the weight of one automobile and the gasoline consumption of another are not. To qualify as a bivariate observation the numbers in the pair must represent measurements of two properties of the same unit of observation.

This chapter investigates association between the two variables of interest in a collection of bivariate observations. It involves detecting and describing association when the variables are measured on either interval or ratio scales. We want to know, for example, how gasoline consumption varies with the weights of automobiles.

Linear Regression

Whenever we are trying to detect association between two variables, both of which are measured on interval or ratio scales, we may be able to find a functional relationship between the two variables that is expressed in the form of an algebraic equation. A widely used technique for finding a straight-line relationship from a set of bivariate observations is called **linear regression**.

For an example of linear regression, suppose that an analyst for an insurance company seeks a relationship between the number of customer calls, or visits, an agent makes in a week and the number of customers to whom the agent makes a sale. The analyst reasons that the number of sales should depend to some degree on the number of calls made. This leads her to designate sales as the dependent variable (Y) and calls as the independent variable (X).

Table 11.1 Observations of Calls and Sales

Agent	Number of Calls (x)	Number of Sales (y)
1	26	11
2	13	7
3	21	8
4	37	20
5	17	9
6	20	12
7	17	4
8	28	16
9	28	11
10	6	2
11	23	11
12	25	7
13	38	18
14	33	14
15	12	2
16	30	15
17	30	10
18	10	4
19	18	7
20	21	10

In this example the 20 observations in Table 11.1 constitute a random sample from the complete population of interest to the analyst. She is interested in the underlying relationship for the company as a whole. Note also that each of the 20 observations is *bivariate* since it consists of a pair of numbers describing two characteristics of a specific agent.

The Scattergram

It is usually worthwhile before beginning a linear regression analysis to plot a **scattergram**. This will give an indication of the functional relationship, if any, between the variables. It is customary to plot values of the dependent variable (y) on the vertical axis and of the independent variable (x) on the horizontal axis. A point is placed at the intersection of each of the observed values. The scattergram for our insurance sales example appears in Figure 11.1. Note that a given point represents the simultaneous observation of two characteristics of a certain agent. Thus the point $(10, 4)$, where x is 10 and y is 4, shows the numbers of calls and sales for the eighteenth agent on the list.

In the scattergram in Figure 11.1, the points appear to fall reasonably close to an imaginary straight line running from lower left to upper right. Their failure

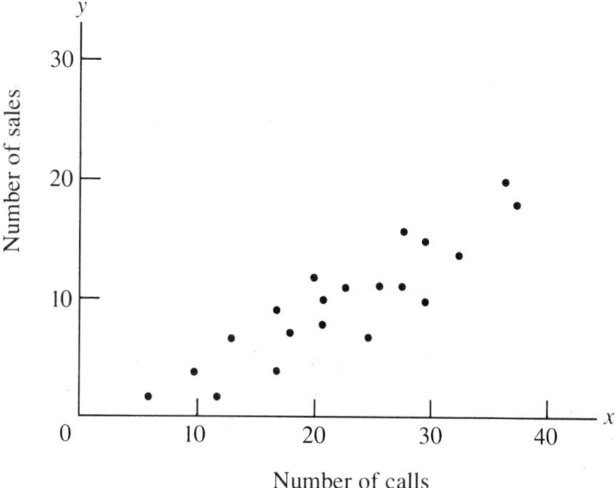

Figure 11.1 *Scattergram for insurance data*

to fall even closer may be the net effect of many other factors that we could reasonably expect to exert some influence on sales.

Linear Equations

11.1

One way to find a straight line to describe the relationship between calls and sales in our sample is simply to draw one that we judge to be a good fit by its visual appearance on the scattergram. This technique does not constitute part of linear regression but is a useful point of departure. One such line is shown in Figure 11.2.

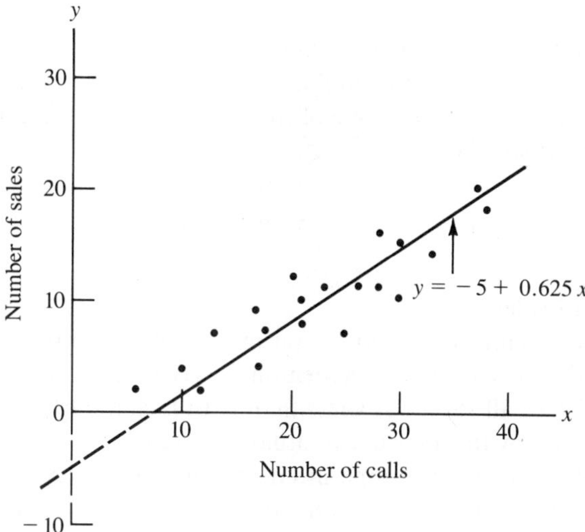

Figure 11.2 *Line fitted by eye*

This line cuts the extended y axis at *minus* 5 and cuts the x axis at 8. The coordinates of these points are $(0,-5)$ and $(8,0)$, respectively.

Although drawing a line on a scattergram may be visually satisfactory, we must be able to express a straight line in the form of an equation for later efficiency. The **slope-intercept** form of the equation of a straight line is convenient for this purpose. In general, this form is

$$y = a + bx,$$

where a is the intercept and b is the slope. On the graph, the intercept (the y intercept in the strict sense) is the value of y at which the line cuts the y axis. In Figure 11.2, y is -5 when x is 0. Hence $a = -5$ for our specific line.

The slope is the number of units of change in y relative to a unit increase in x. For example, the line goes through the points (40,20) and (8,0). The value of y changes by a positive 20 points $(20 - 0 = 20)$ while x is increasing by 32 points $(40 - 8 = 32)$. The ratio of the change in y to the increase in x yields both the value and the algebraic sign of the slope; that is,

$$b = \frac{20 - 0}{40 - 8} = \frac{+20}{+32} = +0.625.$$

We now know the values of both the y intercept and the slope for the straight line in Figure 11.2. The value of a is -5 and the value of b is 0.625. The resulting equation of the line is

$$y = -5 + 0.625x. \qquad \textit{11.1}$$

This equation summarizes the data in Table 11.1 and can be used to approximate the number of sales resulting from any number of calls within the range of the data. For instance, if an agent is known to be one of those who made 30 calls during the week, the approximation would be

$$y = -5 + 0.625(30) = 13.75,$$

which rounds to 14 sales.

The line seems to indicate that if a salesman made no calls during the week, he could expect his sales to be *negative* 5, a meaningless concept. Similarly, the line shows negative values of y for all values of x less than 8. In fact, the true relationship would never drop below zero by the logic of the situation. We have extended the line so that we can illustrate the convenient slope-intercept form of its equation. This illogical result shows the danger in using a functional relationship outside the range of observations on which the relationship is based.

The Least-Squares Line

There are major shortcomings in fitting a straight line to data by eye as we did above. Probably no two people will produce exactly the same line under these circumstances. Furthermore, there is no agreed-upon standard with which to compare the lines drawn by eye. We need a procedure that will produce only one line for a given set of data. We also want that line to be, in some sense, the *best-fitting* line for the given set of points. The **least-squares** technique produces such a line and constitutes part of standard linear regression procedures.

In Figure 11.3, we repeat the scattergram for the insurance example and show the line for which the least-squares regression equation is

$$y_c = -1.69 + 0.51x. \qquad \textit{11.2}$$

Also shown in Figure 11.3 are the deviations of the observations from the line in the y direction (vertical). These deviations are the errors that result when we use the least-squares equation to approximate the number of sales (y_c) that will result from a given number of calls (x). For example, the agent who made 20 calls actually made 12 sales. In this observation, the value of y is 12 for an x value of 20. If we substitute an x value of 20 in Equation 11.2, we find that y_c is only 8.51. Hence the deviation ($y - y_c$) is 12 − 8.51, or 3.49. The approximation based on 20 calls falls 3.49 short of the actual sales for this observation.

Together the deviations for the 20 observations shown in Figure 11.3 represent the *dispersion* of the data about the line shown. Any other line would have a similar set of deviations. We can build a measure of dispersion on these deviations. Then it would seem reasonable to define the best-fitting line as that line from which the

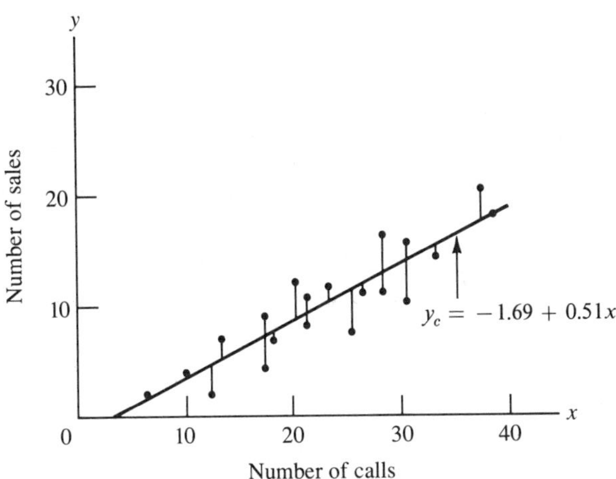

Figure 11.3 *Least-squares line and deviations*

dispersion of the entire collection of data is a minimum. The least-squares line is just such a line.

Definition Of all lines that can represent the functional relationship for a given collection of bivariate observations, the **least-squares line** is the one for which the sum of the squared deviations of the data from the line is a minimum. In this sense, the least-squares line is the line of best fit.

In symbols, suppose we let the equation of the least-squares straight line be

$$y_c = a + bx, \qquad \textbf{\textit{11.3}}$$

where a is the intercept and b is the slope. Then for a given set of observations,

$$\sum (y - y_c)^2 \qquad \textbf{\textit{11.4}}$$

is the sum of the squared y deviations. In order for Equation 11.3 to be the *least-squares* line for a given set of data points, the sum of squared y deviations as defined in Expression 11.4 must be smaller than is the comparable sum for any other line. The values of a and b which minimize the squared deviations are found from

$$b = \frac{n\left(\sum xy\right) - \left(\sum x\right)\left(\sum y\right)}{n\left(\sum x^2\right) - \left(\sum x\right)^2} \qquad \textbf{\textit{11.5 (a)}}$$

and

$$a = \bar{y} - b\bar{x}. \qquad \textbf{\textit{11.5 (b)}}$$

For our insurance example, the 20 pairs of numbers that constitute our observations appear under x (calls) and y (sales) in Table 11.2. The table shows how to calculate some of the values to be substituted in Equations 11.5. In the column headed x^2, the squared value of x for every observation is listed. These squares can be obtained from Appendix H. In the column headed xy, the product of x and y for each observation is shown. Then the columns are totaled to give us

$$\sum x = 453, \qquad \sum x^2 = 11{,}733,$$

$$\sum y = 198, \qquad \sum xy = 5238.$$

Finally, since there are 20 observations, $n = 20$. The given set of observations has

Table 11.2 *Calculations Required for Least-Squares Procedure*

Observation	x	y	x^2	xy
1	26	11	676	286
2	13	7	169	91
3	21	8	441	168
4	37	20	1369	740
5	17	9	289	153
6	20	12	400	240
7	17	4	289	68
8	28	16	784	448
9	28	11	784	308
10	6	2	36	12
11	23	11	529	253
12	25	7	625	175
13	38	18	1444	684
14	33	14	1089	462
15	12	2	144	24
16	30	15	900	450
17	30	10	900	300
18	10	4	100	40
19	18	7	324	126
20	21	10	441	210
Total	453	198	11,733	5238

been used to find every value we need to substitute in Equations 11.5 in order to find the values of a and b that define the least-squares line.

We next substitute the sums from Table 11.2 for our 20 observations into Equations 11.5:

$$b = \frac{20(5238) - 453(198)}{20(11{,}733) - (453)^2} = 0.51156$$

and

$$a = \frac{198}{20} - 0.51156\left(\frac{453}{20}\right) = -1.68683.$$

To two-place accuracy, the least-squares line for the insurance example is, then,

$$y_c = -1.69 + 0.51x,$$

which is the same as Equation 11.2. Because -1.69 and 0.51 satisfy Equations 11.5 for our data, we know that no other straight line can fit the data as well. We know that these two values for the intercept and slope produce the smallest value of Expression
11.4 possible for any straight line fitted to the 20 observations. Hence the line is 11.1
called either the least-squares line or the least-squares regression line.

Other Relationships

The relationship between Y and X in our insurance example is called a **positive linear** relationship because a straight line appears to be the proper shape of the function and because that line has a positive slope (the value of b is positive). Not all linear relationships are positive, however. Studies have shown that as family income increases, the consumption of bread tends to decrease. A scattergram for these two variables might be like the one in Figure 11.4(a). Here we have a **negative linear** relationship and the value of b will be negative.

Figure 11.4(b) shows an example of a *perfect* positive linear relationship. It is perfect because no data point deviates from the straight regression line. Such a relationship would be very nearly realized by simultaneously reading the temperatures of several different containers of water on well-calibrated Fahrenheit and Celsius thermometers.

Equations 11.5 are proper for finding the least-squares regression lines for any scattergram to which a linear function applies. These formulas would apply to all three of the scattergrams we have discussed so far.

The two scattergrams in Figure 11.5 are examples of curvilinear relationships. The one in part (a) might represent sales as increasing amounts of advertising are applied, for instance. The underlying relationships appear to be represented by curved lines rather than straight ones. In neither case do all the points fall exactly on the curved line. Consequently, neither one is a perfect curvilinear relationship.

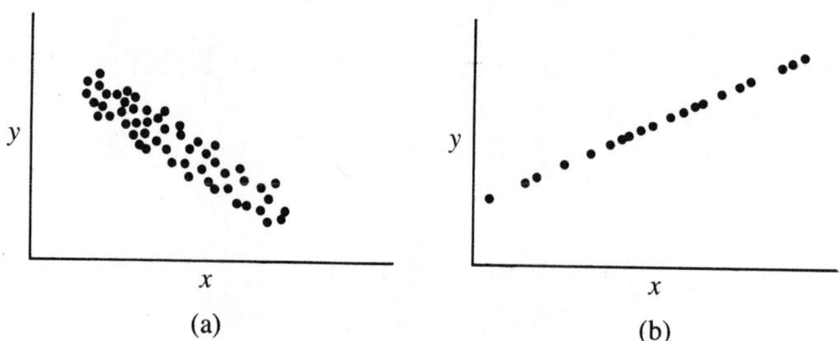

Figure 11.4 *Linear relationships*

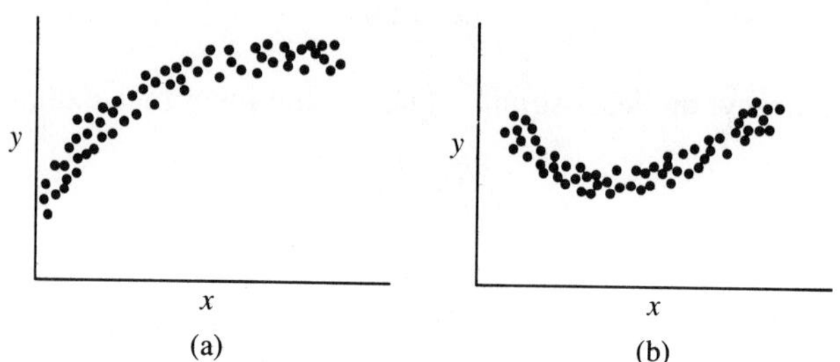

Figure 11.5 *Curvilinear relationships*

The Standard Error of Estimate

11.2

In describing how to find the least-squares regression line illustrated in Figure 11.3, we pointed out that this line minimizes the sum of squared deviations from the line as defined in Expression 11.4. We can use this same expression as the basis for a measure of dispersion of the data about the regression line.

Definition

The **standard error of estimate** ($s_{y \cdot x}$) is the standard deviation of the predictive errors measured with reference to the least-squares regression line. In symbols,

$$s_{y \cdot x} = +\sqrt{\frac{\sum (y - y_c)^2}{n - 2}}. \qquad \textit{11.6}$$

In Figure 11.3, we can consider the least-squares regression line to be a *moving* sample mean, the value of which depends on the number of calls (x) an agent makes. The standard error of estimate is the special standard deviation with respect to this moving mean that applies in least-squares regression. Note the division by ($n - 2$), rather than by n, in Equation 11.6. This is necessary to make $s^2_{y \cdot x}$ an unbiased estimate of its population counterpart.

The purpose of Table 11.3 is to find the sum of squared deviations from the regression line [$\sum(y - y_c)^2$] for our insurance example. The two columns on the left repeat the 20 bivariate observations. Each value in the third column is found by substituting the value of x for that observation into the least-squares regression formula, Equation 11.2. For example, in the first observation the agent made 26 calls ($x = 26$). The regression line approximation is 11.614 sales [$y_c = -1.68683 + 0.51156(26)$]. The fourth column shows the deviations of actual sales from regression line approximations, and the fifth column is the square of these deviations.

We can substitute 94.44 from Table 11.3 into Equation 11.6 to find the standard error of estimate:

$$s^2_{y \cdot x} = \frac{94.44}{20 - 2} = +5.2467$$

or

$$s_{y \cdot x} = 2.29 \text{ sales.}$$

The approach taken in Table 11.3 to find the standard error of estimate works best when values of ($y - y_c$) are small integers. Otherwise, as in the insurance example, we must carry along several decimal places to avoid serious rounding errors. In the latter situation, an alternative calculation technique is to find the sum of squared deviations in the numerator of Equation 11.6 from

Table 11.3 Calculation of Sums Required for the Standard Error of Estimate

x	y	y_c	$(y - y_c)$	$(y - y_c)^2$
26	11	11.614	−0.614	0.377
13	7	4.963	2.037	4.149
21	8	9.056	−1.056	1.115
37	20	17.241	2.759	7.612
17	9	7.010	1.990	3.960
20	12	8.544	3.456	11.944
17	4	7.010	−3.010	9.060
28	16	12.637	3.363	11.310
28	11	12.637	−1.637	2.680
6	2	1.382	0.618	0.382
23	11	10.079	0.921	0.848
25	7	11.102	−4.102	16.826
38	18	17.752	0.248	0.062
33	14	15.195	−1.195	1.428
12	2	4.452	−2.452	6.012
30	15	13.660	1.340	1.796
30	10	13.660	−3.660	13.396
10	4	3.429	0.571	0.326
18	7	7.521	−0.521	0.271
21	10	9.056	0.944	0.891
453	198	198.000	0	94.444

$$\blacksquare \quad \sum (y - y_c)^2 = \sum (y^2) - a\left(\sum y\right) - b\left(\sum xy\right), \quad \blacksquare \qquad \textbf{\textit{11.7}}$$

or, for our example,

$$\sum (y - y_c)^2 = 2440 - (-1.68683)(198) - (0.51156)(5238), \text{ or } 94.44.$$

Calculation of the sums is illustrated in Table 11.4.

Table 11.4 Calculation of Sums Required in Equation 11.7

x	y	y^2	xy
26	11	121	286
13	7	49	96
21	8	64	168
37	20	400	740
17	9	81	153
20	12	144	240
17	4	16	68

Table 11.4 (*continued*)

x	y	y^2	xy
28	16	256	448
28	11	121	308
6	2	4	12
23	11	121	253
25	7	49	175
38	18	324	684
33	14	196	462
12	2	4	24
30	15	225	450
30	10	100	300
10	4	16	40
18	7	49	126
21	10	100	210
453	198	2440	5238

Use of Equation 11.7 avoids having to find the individual values of y_c for each
observation and keeps rounding errors to a minimum. **11.2**

Summary

A bivariate observation is the values of two variables that apply to a single unit of observation. A person's age and weight are an example. Linear regression is a technique for finding and describing a relationship between two variables represented in a collection of bivariate observations. Each variable is measured on an interval or a ratio scale.

When a straight line constitutes a satisfactory representation of the relationship between two variables, the equation for such a line can be found so that the dispersion about that line is a minimum. This is called the least-squares regression line. A measure of the dispersion of the data about the least-squares regression line is called the standard error of estimate.

See Self-Correcting Exercises 11A.

Exercise Set A

1. Give the equations of the following lines in the form $y = a + bx$:
   (a) The line has slope -4 and intersects the y axis at 2.
   (b) The line passes through the origin and has slope $3/5$.
   (c) The line passes through the points (0,7) and (7,0).
   (d) The line passes through the points (0,3) and $(7,-3)$.
2. What do you know about the size of the standard error of estimate for the least-squares regression line as compared with the standard error of estimate for any other straight line "fitted" to the bivariate data?
3. Construct a scattergram on graph paper for the bivariate data given below. Does your scattergram indicate whether a linear regression is to be assumed, or would a curvilinear regression seem to be preferred?

x	3	3	3	5	4	5	6	8
y	8.5	7.5	8.0	12.0	9.5	10.0	13.0	16.5

4. For the data pairs in Exercise 3, determine the least-squares regression line.
5. Refer to the equation you obtained in Exercise 4:
   (a) Predict a y value for an x value of 7.
   (b) Predict a y value for an x value of 6. How does your prediction compare with the observed value of y associated with the observed value 6 of x? To what might this difference be attributed?
   (c) Why would it be risky to predict a y value for an x value of 12 or an x value of 1 based on the least-squares line obtained for the data in Exercise 3?
6. For the regression line you obtained in Exercise 4, determine the standard error of estimate. What is known about the standard error of estimate of any other line "fitted" to these data?
7. A moving firm wishes to find a relation between estimated weight of an upcoming job and the time it will take to load the goods on the truck. Data kept on 10 jobs resulted in the following figures, where x is weight in tons and y is hours:

$$\sum x = 20, \quad \sum y = 30, \quad \sum xy = 70;$$
$$\sum x^2 = 50, \quad \sum y^2 = 85.$$

   Determine the least-squares linear relationship.
8. Suppose that for all the observed data pairs you obtained a least-squares regression line such that each pair of values could be represented by a point lying precisely on that line. What would the sum of the squared deviations be? What would the standard error of estimate be?
9. What does a large standard error of estimate reveal about the fit of the least-squares regression line to the scattergram? What does a small standard error of estimate suggest?

10. Which of the following pairs of measures do *not* represent bivariate observations? Justify your conclusion for each one that does not.
    (a) The age and circumference of a tree.
    (b) An individual's head size and average daily sugar consumption.
    (c) The age of a wife and that of her husband.
    (d) An individual's IQ and educational level.
    (e) An individual's educational level and annual income.
    (f) The acreage of a national park and the number of annual visitors to the park.
    (g) The average daily rainfall in Joliet, Illinois, and the average daily rainfall in Aberdeen, Scotland.
    (h) The population of Joliet and the average daily high temperature there.
    (i) An individual's age and religious preference.

Hypothesis Tests

11.3

In regression analysis, the independent variable (X) must be known without error. On the other hand, the dependent variable (Y) is a random variable. So that we can make inferences about population parameters, three assumptions concerning values of the dependent variable are necessary:

1. For any given value of the independent variable X there is a population of values of the dependent variable Y. At the designated x value, the mean of the dependent variable is obtained when that value of x is substituted in the *population* regression equation

$$y_C = \alpha + \beta x.*$$

   Call this mean $\mu_{y \cdot x}$.

2. At the designated value of x, the population of y values is normally distributed with a standard deviation that is called the *population standard error of estimate,* for which we will use the symbol $\sigma_{y \cdot x}$.

3. Alternatively, we can say that for *any* value of x the population of y values is normally distributed with a mean of

$$\mu_{y \cdot x} = \alpha + \beta x$$

   and a standard deviation of $\sigma_{y \cdot x}$.

*Don't confuse α and β with probabilities of Type I and Type II errors in hypothesis testing. Here α and β are the intercept and slope of the population regression equation.

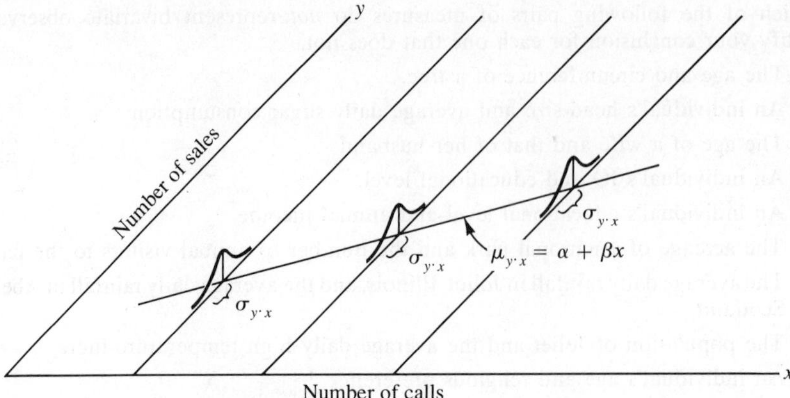

Figure 11.6 *Linear regression model*

This set of assumptions is illustrated in Figure 11.6. Note there that the mean value ($\mu_{y \cdot x}$) of the dependent variable moves in accordance with the population regression equation as x changes. Note also that for any given value of x the standard deviation of the y values ($\sigma_{y \cdot x}$) is the same and is the population standard error of estimate. This figure shows the form of regression population from which our insurance sales sample is assumed to come.

To understand what happens under sampling conditions, consider Figure 11.7. The sample least-squares regression line for the data in Table 11.1 is shown along with sample regression lines from two other random samples, each composed of 20 observations from the same population. The entire regression line changes as we go from one sample to the next. Both the slope (b) and the intercept (a) fluctuate under sampling conditions. Furthermore, no sample regression line is likely to be the same as the population regression line. From the model based on the conditions enumerated

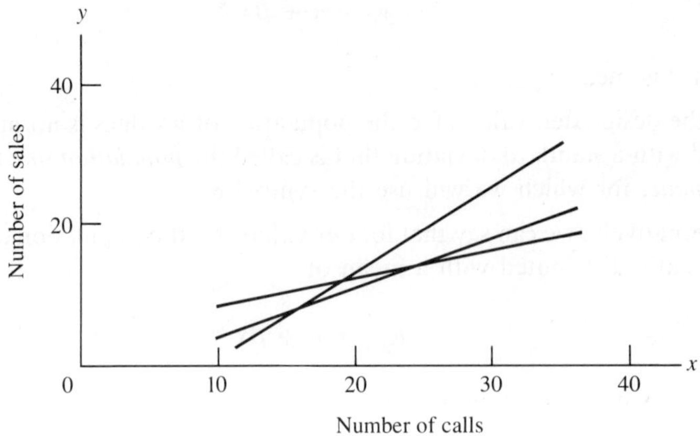

Figure 11.7 *Regression lines from three samples*

above, mathematical statisticians have found the sampling distributions for a and b. These are useful in making inferences about the unknown values of α and β for the population regression line.

Perhaps the most often required test statistic based on the model discussed in this section is the one that applies to the slope of the sample regression line.

Statement

Consider the regression situation in which the conditional populations of y values are centered on the population regression line, are normally distributed about that line, and have standard deviations equal to the standard error of estimate ($\sigma_{y \cdot x}$). Then the slopes (b) of regression lines from independent random samples can be transformed to the statistic

$$t = \frac{b - \beta}{s_b}, \qquad \textit{11.8(a)}$$

where the standard error of b is

$$s_b = \frac{s_{y \cdot x}}{\sqrt{\sum (x - \bar{x})^2}} \qquad \textit{11.8(b)}$$

and where β is the slope of the population regression line. Here $s_{y \cdot x}$ is defined by Equation 11.6 and $\sum(x - \bar{x})^2$ is the sum of the squared deviations of the x values from their mean. The statistic in Equation 11.8(a) forms a t distribution with $(n - 2)$ degrees of freedom.

The statistic in Equation 11.8(a) can be used to test a hypothesis about any presupposed value of the population slope (β). The most frequent test, however, is performed for the hypothesis that β is zero. If this hypothesis cannot be rejected, then we would be forced to conclude that there is no evidence of a linear regression relationship in the population.

For our insurance example, we have already found that b is 0.512 and $s_{y \cdot x}$ is 2.291. We also must find $\sum(x - \bar{x})^2$. We could do so by finding and summing the individual squared deviations from $\bar{x}$. It is usually easier, however, to use the relationship

$$\sum (x - \bar{x})^2 = \sum x^2 - \frac{\left(\sum x\right)^2}{n}. \qquad \textit{11.9}$$

For our example,

$$\sum x^2 - \frac{\left(\sum x\right)^2}{n} = 11{,}733 - \frac{453^2}{20}, \quad \text{or} \quad 1472.55.$$

Then

$$s_b = \frac{2.291}{\sqrt{1472.55}}, \quad \text{or} \quad 0.0597,$$

and Equation 11.8(a) becomes

$$t = \frac{0.512 - 0}{0.0597}, \quad \text{or} \quad 8.58.$$

The slope of the sample regression line (0.512) is 8.58 standard errors greater than a slope of zero. We can reject the hypothesis of no linear regression in the population at a level of significance well beyond 0.005 because, in Appendix B, $t_{.995} = 2.878$ for $n - 2 = 18$ degrees of freedom. Instead we accept the one-tailed alternative that
11.3 β is greater than zero and a linear relationship exists in the population.

Prediction

Our two major concerns so far in this chapter have been to find characteristics of the sample linear regression equation and to test hypotheses about the population regression relationship. Another major concern in regression analysis is the problem of making point and interval predictions based on the sample linear regression equation.

11.4 Point Prediction

For the insurance data in Table 11.1 we found that the sample linear regression equation was $y_c = -1.69 + 0.51x$. Suppose that we want to predict what the number of sales will be when the number of calls is 35. In terms of our regression analysis, we want to know the value of y when x takes on the value 35. We can predict the value of y by substituting 35 for x in the sample regression equation:

$$y_c = -1.69 + 0.51(35), \text{ or } 16.2.$$

Our *point prediction* is that 16.2 sales will result from 35 sales calls or visits.

Although it can be shown that Equation 11.2 produces the most likely single-number prediction for a stated value of x, from previous experience we know that we can place no confidence in y taking on this exact value. For this reason we prefer prediction intervals for which we can state a known level of confidence.

Interval Prediction

When sampling, we have learned to expect successive sample estimates to fluctuate around the parameter being estimated. But when we examine the linear regression model described in conjunction with Figure 11.6, another source of fluctuation becomes evident in the population itself. The values of y are dispersed about the population regression equation with a standard deviation equal to the standard error of estimate ($\sigma_{y \cdot x}$). Hence even if we knew the population regression equation,

$$\mu_{y \cdot x} = \alpha + \beta x,$$

we could not predict a specific value of y without error. We would only know the mean value of y at the designated value of x. This would, of course, be the value of $\mu_{y \cdot x}$ found by substituting the given value of x in the population regression equation. Consequently, when we base a prediction of the value of y on a sample regression equation, two sources of prediction error must be taken into account. As indicated in Figure 11.7, sample regression lines fluctuate about the population regression line. Moreover, individual y values are scattered above and below the sample regression line.

Suppose we have a random sample as in Table 11.1. Then the $(1 - \alpha)$ prediction interval for an individual value of the dependent variable y associated with a specific value of the independent variable x' is

$$■ \quad y_c \pm t_{\alpha/2} \sqrt{s_{y \cdot x}^2 + \frac{s_{y \cdot x}^2}{n} + s_b^2(x' - \bar{x})^2}. \quad ■ \qquad \mathbf{11.10}$$

In this expression, y_c is the point estimate found by substituting the given value of the independent variable x' into the sample least-squares regression equation and $t_{\alpha/2}$ has $(n - 2)$ degrees of freedom. The symbol $s_{y \cdot x}$ is the sample standard error of estimate and s_b is the standard error of b, which was defined in Equation 11.8(b).

The three terms under the radical in Equation 11.10 can be viewed as three sources of prediction error. The first term $s_{y \cdot x}^2$ arises from the dispersion of the y values around the population regression line. The second term represents the possible error in the mean location of the sample regression line. The third term is the possible error in the slope of the sample regression line.

Suppose we let x' be 35 calls and suppose that we want a 0.95 prediction interval estimate of the number of resulting sales. In the previous section we found that y_c is 16.2 when 35 is substituted in the sample regression equation. Furthermore, $t_{.975}$ for 18 degrees of freedom is 2.101, and we have already found that $s_{y \cdot x}^2$ is 5.247 and s_b^2 is 0.003563. Therefore the 0.95 prediction interval for y, given that x' equals 35 calls, is

$$16.2 \pm 2.101\sqrt{5.247 + \frac{5.247}{20} + 0.003563(35 - 22.65)^2}$$

or

$$16.2 \pm 5.2.$$

Hence the interval runs from 11.0 to 21.4 sales. A salesman who plans to make 35 calls can expect 11 to 21 sales to result from these calls with 0.95 confidence.

Consider the effect of the term under the radical that is farthest to the right in Expression 11.10. If we elected to predict y for a value of x' even farther from $\bar{x}$ than 35 is from 22.65, this term would be larger. If x' were selected closer to the value of $\bar{x}$, it would be smaller. Since all other values in the equation remain unchanged for a given sample, the effect of this term is to widen the prediction interval as the deviation of the selected value of the independent variable x' from the sample mean $\bar{x}$ increases. This extra width compensates for the greater variability of the extremes of fluctuating sample regression lines as compared with their central
11.4 regions (see Figure 11.7).

11.5 Correlation

When we are dealing with random samples, a basic distinction must be made between correlation analysis and regression analysis. In regression analysis the dependent variable Y is a random variable, but values of the independent variable are assumed to be known without error. In our insurance example, for instance, the number of calls x each agent made could have been assigned in advance by his supervisor. On the other hand, in correlation analysis both the dependent and the independent variables are assumed to be random variables. For instance, a unit of observation such as a student or a consumer is selected at random and the values of x and y that happen to be associated with that unit are recorded as the observation. In regression analysis we study the random variation of the dependent variable for fixed values of the independent variable. In correlation analysis we study the joint variation of two random variables.

As an example of a correlation analysis, suppose a study is being made of the relationship between daily family income X and weekly food expenditures Y. A random sample of 20 families in a certain market area is selected, and the data shown in Table 11.5 are recorded. Note that it is the families that are selected at random.

Consequently, both X and Y are free to vary jointly at random. The scattergram for these 20 data points appears in Figure 11.8.

Table 11.5 *Economic Data for a Sample of Families*

Family	Income (x)	Food Expense (y)	Family	Income (x)	Food Expense (y)
1	49	47	11	47	54
2	46	55	12	49	57
3	48	57	13	54	58
4	38	35	14	48	65
5	63	73	15	56	66
6	58	64	16	51	53
7	40	38	17	55	69
8	55	73	18	28	18
9	66	87	19	63	76
10	54	68	20	35	49

The **sample correlation coefficient** r is a dimensionless measure of the degree of association between two random variables that are linearly related. The value of this coefficient ranges from -1 through zero to $+1$. Suppose that the scattergram in Figure 11.4(b) represents the observations in a correlation analysis. For these data the value of r is $+1$ because the slope of a least-squares line fitted to them is positive and all points fall exactly on this line. The relationship is perfect and positive. If the relationship were perfect but the slope of the least-squares line were negative, then the value of r would be -1. A value of r equal to zero indicates no relationship between the variables. By inspection, we can see that the value of r for the data in Figure 11.8 lies between 0 and $+1$, probably considerably closer to $+1$.

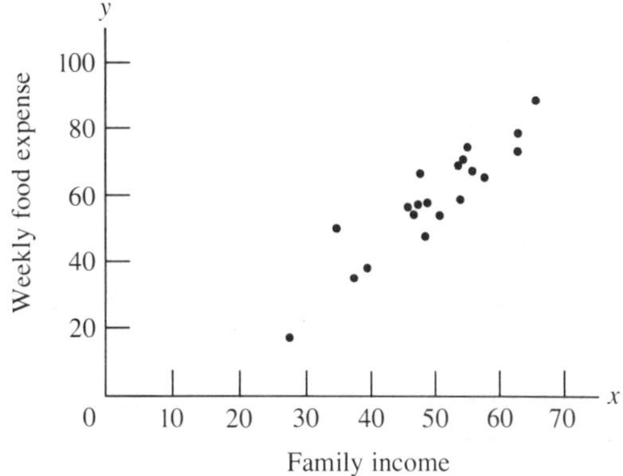

Figure 11.8 *Scattergram for consumer expense data*

The square of the sample correlation coefficient, r^2, is called the **coefficient of determination.** The coefficient of determination is the proportion of the total variation in one variable that is accounted for by covariation with the other variable. This coefficient ranges only from 0 to +1; it can never be negative.

An equation for the sample correlation coefficient which usually minimizes calculation difficulties is

$$r = \frac{n\left(\sum xy\right) - \left(\sum x\right)\left(\sum y\right)}{\sqrt{\left[n\sum x^2 - \left(\sum x\right)^2\right]\left[n\sum y^2 - \left(\sum y\right)^2\right]}}. \qquad \textit{11.11}$$

For our example,

$$n = 20, \qquad \sum xy = 60{,}943,$$
$$\sum x = 1003, \qquad \sum y = 1162,$$
$$\sum x^2 = 52{,}065, \qquad \sum y^2 = 72{,}280,$$

and

$$r = \frac{20(60{,}943) - (1003)(1162)}{\sqrt{[20(52{,}065) - (1003)^2][20(72{,}280) - (1162)^2]}} = 0.92.$$

The coefficient of determination is

$$r^2 = (0.92)^2 = 0.8464.$$

This means that 84.6 percent of the variance in the 20 observations of food expense is attributable to the observed differences in annual income or, alternatively, that 84.6 percent of the variance in income is attributable to covariation with the food expense data.

To make hypothesis tests about the value of the population correlation coefficient (ρ), we must note some characteristics of the sampling distribution of the sample correlation coefficient (r). When there is no correlation in the population ($\rho = 0$), the variable

$$t = \frac{r - 0}{s_r}, \qquad \textit{11.12(a)}$$

where

$$s_r = \sqrt{\frac{1 - r^2}{n - 2}}, \qquad \textit{11.12(b)}$$

conforms essentially to a t distribution with $(n - 2)$ degrees of freedom under a fairly broad range of conditions. For our example, suppose we want to test the null hypothesis that the population coefficient of correlation is no greater than zero against the one-tailed alternative that it is positive. Then

$$s_r = \sqrt{\frac{1 - (0.92)^2}{20 - 2}}, \quad \text{or} \quad 0.09238,$$

and

$$t = \frac{0.92}{0.09238}, \quad \text{or} \quad +9.96.$$

This is well beyond any tabled value of t for 18 degrees of freedom in Appendix B. Consequently we can accept the hypothesis of positive correlation in the population at the 0.005 level of significance. 11.5

Summary

When linear regression relationships are based on sample data, we usually want to test hypotheses about the values of the slope and intercept for the population from which the sample came. The test of the hypothesis that the slope of the population regression line is equal to zero (that there is no linear relationship) is often of most interest.

Predictions of values of the dependent variable that will result from specified values of the independent variable are frequently sought in regression analysis. Though point predictions based on sample regression equations are possible, interval predictions are more informative. A confidence level can be found for the latter but not for the former.

When both variables in a sample of bivariate observations are random variables, the coefficient of correlation is used to measure the degree of linear association between them. The coefficient of correlation takes the sign of the slope of the least-squares line fitted to the data and varies in value from zero for no relationship to ± 1 for a perfect relationship where all data fall on the line. The coefficient of determination is the proportion of variation in one variable that is attributable to co-variation with the other variable. Tests of hypotheses are given for the coefficient of correlation.

See Self-Correcting Exercises 11B.

Exercise Set B

1. Which of the following represent statistics? Which represent parameters?

(a) α (d) b (g) y_c
(b) r (e) a (h) r^2
(c) ρ (f) $\mu_{y \cdot x}$ (i) s_r

2. Indicate with a sketch a least-squares line representing a perfect negative correlation of two random variables. Is the slope of such a line necessarily equal to -1? Is the slope necessarily negative?

3. Why can the coefficient of determination never (a) be negative? (b) Never exceed 1?

4. What is indicated by the value of a coefficient of determination being (a) almost 1? (b) Almost zero? (c) Zero?

5. Sample data are recorded of each of the following types. For each, in response to what sort of problem might correlation analysis be appropriate?
   (a) For students at a large university, weekly class-hour load and grade point average.
   (b) For eighth-grade students in the Tulsa school system, IQ and days absent from school during the past year.
   (c) For army inductees, height and boot size.
   (d) For 3-week-old laboratory rats, total lifetime protein intake (by weight) and brain weight.

6. In Exercise 5, what might the question of concern be in each case to require regression analysis?

7. Suppose 98 percent prediction intervals for y, based on a sample for which $\bar{x} = 7$ and $\bar{y} = 21.2$, are given for $x = 9.2$, $x = 3.6$, and $x = 12$.
   (a) Which prediction interval is the narrowest?
   (b) Which is the widest?
   (c) Justify your conclusions to parts (a) and (b).

8. Given the following pairs of random variables:

x	8	7	4	12	9	3	8	7	10
y	2.5	3	3.5	1	2	3.5	2	2.5	3

   (a) To what sort of analysis should these data be subjected to determine whether a linear relationship exists between them?
   (b) What is the number of degrees of freedom for this sample?
   (c) Determine, at a level of significance of 0.02, whether there is a linear relationship between the two variables.

9. For the data in Exercise 8, determine the sample least-squares regression equation.

10. For the pairs of variables in Exercise 8, test, at a level of significance of 0.05, the hypothesis that the slope of the population regression line is -0.35. Make a two-tailed test.

11. Use your results from Exercise 9 and Expression 11.10 to determine a 0.95 prediction interval for the value of y associated with the given value of the independent variable (x') equal to (a) 7; (b) 9; (c) 2. Why do the intervals differ in width?

12. Why would it not be sound to apply the methods of Exercise 11 to $x' = 14$?

13. Find the coefficient of determination for the data pairs in Exercise 8. What proportion of the variation in y is *not* accounted for by covariation with x?

14. The following table shows the number of minutes the caution light is on during a complete cycle of traffic lights at an intersection and the number of accidents during a month at that intersection:

Intersection	Yellow Light (minutes/cycle) (x)	Accidents per Month (y)
1	1.0	11
2	1.0	12
3	1.2	12
4	1.2	13
5	1.2	15
6	1.4	12
7	1.4	14
8	1.8	15
9	1.8	16
10	1.8	16

(a) Find the equation of the least-squares sample regression line.

(b) Find the number of accidents to be expected from a setting of 1.6 minutes.

15. Find the standard error of estimate and the coefficient of determination for the sample in Exercise 14. Discuss the meaning of the coefficient of determination.

16. At the 0.05 level of significance, test the hypothesis that there is no linear relationship of accident rate to light timing in Exercise 14.

Overview

1. The following terms and symbols are used with respect to which type of analysis: regression or correlation?

(a) Independent and dependent variables (e) s_r

(b) Coefficient of determination (f) s_b

(c) Covariation (g) ρ

(d) Bivariate analysis (h) β

2. By using a forecast of daily high temperatures, the owner of an ice cream stand wants to estimate how many gallons of ice cream she will sell. For a 2-week period, she has kept track of the amount of ice cream sold each day and the temperature that was forecast on the previous day. She has arrived at the following relationship:

$$y = -25 + 0.5x,$$

where

y = gallons per day of ice cream sold,

x = the day's forecast temperature.

During the 2-week period, sales ranged from 13 to 38 gallons per day and high temperatures ranged from 85° to 98°.

(a) If tomorrow's forecast is for a temperature of 96°, how much ice cream will be sold according to the estimating equation?

(b) If tomorrow's forecast is for a temperature of 108°, how much ice cream will be sold?

3. A major appliance company was trying to develop a relationship between the total income of a household and the number of appliances in that household. The company surveyed 50 households and arrived at the following set of data (y = number of appliances per household; x = income per household in thousands of dollars):

$$\sum x = 490, \quad \sum y = 210, \quad \sum x^2 = 12{,}000, \quad \sum xy = 462.$$

(a) Find the equation of the least-squares line.

(b) Find the estimated number of appliances owned in a household that has an income of $8000.

4. A stock analyst, trying to find a way to predict small changes in the Dow-Jones index, noticed the following facts: when automobile production (x) did not change, the Dow-Jones index (y) rose 2 percent; when automobile production rose 1 percent, the Dow-Jones index rose 6 percent; when automobile production dropped 1 percent, the Dow-Jones index dropped 2 percent. From this information alone, determine the change in the Dow-Jones index per unit change in automobile production.

5. A supervisor collected the following data on the number of days employees had been on a new job assembling a component and the average time spent per assembly. The employees were randomly selected. Find the equation of the least-squares regression line for these observations.

Employee	Days on Job (x)	Minutes per Assembly (y)
1	1	8
2	3	5
3	2	6
4	4	4
5	1	10
6	5	2
7	3	5
8	3	4

6. What is the standard error of estimate for the sample in Exercise 5?

7. What is the coefficient of determination for the sample in Exercise 5?

8. Each of twenty-five students in a chemistry lab section is required to perform a series of four experiments in which a fixed quantity of reagents in constant proportion are exposed to varying amounts of a catalyst, and the time required for the reaction to terminate is measured. With x (in milligrams) denoting the quantity of catalyst used and y (in seconds) denoting the reaction time, the following values are obtained for the 100 experiments:

$$\sum x = 1250 \qquad \sum y = 392$$

$$\sum x^2 = 19{,}640 \qquad \sum y^2 = 3104$$

$$\sum xy = 7128$$

(a) Determine the sample correlation coefficient. What sort of correlation is present, if any? Is the linear relationship significant?

(b) Determine the coefficient of determination. What proportion of the variation in reaction time cannot be accounted for by covariation with the amount of catalyst used?

9. For the data in Exercise 8, letting x be the independent variable and y the dependent, find the equation of the least-squares sample regression line, and determine y_c for $x' = 11.3$ mg., assuming such an estimation is valid. What sort of estimate is preferable to this point estimate, and what further information must we have to make this more useful sort of estimate? How can this information be obtained?

10. Assuming that the chemistry students have provided us with their raw data, and we have utilized it and our results of Exercise 9 to determine that $s_{y \cdot x} = 0.7$, construct a 95% confidence interval for y for (a) $x' = 11.3$; (b) $x' = 12.2$.

11. Why is your confidence interval estimate narrower in Exercise 10(b) than in 10(a)? How is the length of the interval related to the location of x' with respect to the mean $\bar{x}$?

Glossary of Equations

11.3 $y_c = a + bx$

The sample least-squares regression line has a y intercept of a and a slope of b.

11.5(a)

$$b = \frac{n\left(\sum xy\right) - \left(\sum x\right)\left(\sum y\right)}{n\left(\sum x^2\right) - \left(\sum x\right)^2}$$

11.5(b) $a = \bar{y} - b\bar{x}$

These usually are the most convenient equations to find the slope and intercept of the sample least-squares regression line.

11.6

$$s_{y \cdot x} = +\sqrt{\frac{\sum (y - y_c)^2}{n - 2}}$$

The sample standard error of estimate is the square root of the sum of squared deviations of the values of the dependent variable from the sample least-squares regression line divided by the number of observations in the sample less two.

11.7

$$\sum (y - y_c)^2 = \sum (y^2) - a\left(\sum y\right) - b\left(\sum xy\right)$$

The sum of squared deviations of the values of the dependent variable from the regression line can be found from the slope and intercept and certain sums from the sample observations.

11.8(a) $t = \dfrac{b - \beta}{s_b}$

Under suitable assumptions the difference between slopes of sample regression lines and the population slope divided by the standard error of the sample slope forms a t distribution with $(n - 2)$ degrees of freedom.

11.8(b) $s_b = \dfrac{s_{y \cdot x}}{\sqrt{\sum (x - \bar{x})^2}}$

The standard error of the slope of a sample regression line is the sample standard error of estimate divided by the square root of the sum of squared deviations of observations of the independent variable from the sample mean.

11.10

$$y_c \pm t_{\alpha/2}\sqrt{s_{y \cdot x}^2 + \frac{s_{y \cdot x}^2}{n} + s_b^2(x' - \bar{x})^2}$$

The $(1 - \alpha)$ prediction interval for a single value of the dependent variable associated with a given value of the independent variable (x') is centered on the point prediction (y_c) from the sample regression line.

11.11

$$r = \frac{n\left(\sum xy\right) - \left(\sum x\right)\left(\sum y\right)}{\sqrt{\left[n \sum x^2 - \left(\sum x\right)^2\right]\left[n \sum y^2 - \left(\sum y\right)^2\right]}}$$

The sample correlation coefficient (r) can usually be found most conveniently with this equation.

11.12(a) $t = \dfrac{r - 0}{s_r}$

The statistic for testing the hypothesis that two variables are not correlated is the ratio of the sample correlation coefficient to its standard error as estimated from the sample. The ratio is distributed as t.

11.12(b) $s_r = \sqrt{\dfrac{1 - r^2}{n - 2}}$

The estimated standard error of the sample correlation coefficient is the square root of the unexplained proportion of sample variance for the y variable divided by the number of degrees of freedom.

Part Four
Additional Topics

12

Differences Between Means

Our previous discussions of statistical inference for population means have been limited to questions about a single population mean. Now we will consider inferences concerning possible differences in population means.

The first part of the chapter describes how to make statistical inferences about the difference in two population means from sample results. In one case, two simple random samples covering only the variable under study constitute the observations. In the other case, each observation from one population is paired with one from the other by matching on some related variable or variables. For example, pairs of identical twins can be used to compare the effect of a drug on blood pressure. One twin from each pair is randomly selected to receive the drug while the other is given a placebo. Each set of twins is automatically matched on many genetic and environmental variables which could affect reaction to the drug.

The last part of the chapter describes how to test hypotheses about the differences in more than two population means.

Difference in Two Population Means

We first take up the fairly common problem of testing whether or not two statistical populations have the same mean. Selection of the correct test procedure depends on whether or not the standard deviations of the two populations are known.

Population Standard Deviations Known

An automatic drilling machine is set to drill holes 1.50 inches deep in engine blocks coming down an assembly line. At any single setting, the machine is known to produce a normal distribution of hole depths with a standard deviation of 0.02 inch. A similar machine farther down the line drills a wider hole centered on the first hole, but only 1 inch deep. The second machine produces a normal distribution of hole depths with a standard deviation of 0.01 inch. It is important that the narrower hole be 0.50 inch deeper than the wider hole.

The depths of 20 narrow holes are measured immediately after the first machine is finished. Their mean depth is 1.51 inches. A gauge measures the depth of 25 of the wide holes drilled by the second machine and finds the sample mean to be 0.99 inch. Is it reasonable to assume that the mean depth of the inner hole is 0.50 inch? We can summarize this information in a table:

	Wide Hole	Narrow Hole
Standard deviation (inches)	$\sigma_1 = 0.01$	$\sigma_2 = 0.02$
Sample size	$n_1 = 25$	$n_2 = 20$
Mean depth (inches)	$\bar{x}_1 = 0.99$	$\bar{x}_2 = 1.51$

We can frame the problem as a hypothesis test. Suppose we call the population mean depths for the wide and narrow holes μ_1 and μ_2, respectively. Then the requirement that the narrow hole be 0.50 inch deeper than the wide one, on the average, can be reflected in the null hypothesis:

$$H_0: \quad \mu_2 - \mu_1 = 0.50 \text{ inch.}$$

Presuming that detecting an error in either direction is important, we form the two-tailed alternative:

$$H_A: \quad \mu_2 - \mu_1 < 0.50, \quad \text{or} \quad \mu_2 - \mu_1 > 0.50.$$

For independent random samples of n_1 and n_2 observations from normally distributed populations with means μ_1 and μ_2 and variances σ_1^2 and σ_2^2, the differences in sample means, $\bar{x}_2 - \bar{x}_1$, will be normally distributed. Furthermore, the distribution of differences in sample means will have as its mean

$$\mu_{\bar{x}_2 - \bar{x}_1} = \mu_2 - \mu_1$$

and as its standard deviation

$$\sigma_{\bar{x}_2 - \bar{x}_1} = \sqrt{\frac{\sigma_2^2}{n_2} + \frac{\sigma_1^2}{n_1}}. \qquad \textbf{12.1}$$

The standard deviation of this population of differences in sample means is called the **standard error of the difference between means**. It is then true that the statistic z will be distributed as the standard normal deviate, where

$$z = \frac{(\bar{x}_2 - \bar{x}_1) - (\mu_2 - \mu_1)}{\sigma_{\bar{x}_2 - \bar{x}_1}}.$$

For our example,

$$z = \frac{(1.51 - 0.99) - (0.50)}{\sqrt{\frac{0.0004}{20} + \frac{0.0001}{25}}} = \frac{0.02}{0.0049}, \quad \text{or} \quad 4.08 \text{ standard errors.}$$

The difference in sample means (0.52) exceeds the specification of 0.50 inch by more than four standard errors of the difference between sample means of $n_1 = 25$ and $n_2 = 20$ observations. If the level of significance α is 0.01, then, from Appendix A, $z_{\alpha/2} = 2.58$ and our sample result is significant at well beyond this level. We accept H_A, the alternative to the null hypothesis. In practice, since the difference between 1.51 and 0.99 is greater than 0.50, we would conclude that $\mu_2 - \mu_1 > 0.50$ and that the small holes are deeper than 0.50 inch on the average. The formal summary can now be stated:

Statement

Consider two normal distributions with means μ_1 and μ_2 and standard deviations σ_1 and σ_2. Consider also a sample mean $\bar{x}_1$ based on n_1 observations from the first population and a sample mean of $\bar{x}_2$ based on n_2 observations from the second. The difference in sample means can be incorporated in the statistic

$$\blacksquare \quad z = \frac{(\bar{x}_1 - \bar{x}_2) - (\mu_1 - \mu_2)}{\sigma_{\bar{x}_1 - \bar{x}_2}}, \quad \blacksquare \qquad \textbf{12.2}$$

where z is a value of the standard normal deviate and $\sigma_{\bar{x}_1 - \bar{x}_2}$ is as defined in Equation 12.1.

Several examples in Chapter 8 showed how a standard normal deviate can be used in both one-tailed hypothesis tests and confidence interval estimates. Since the procedures with the z statistic from Equation 12.2 are identical to those used earlier, we will not discuss them further.

The two populations from which the sample means in Equation 12.2 came were assumed to be normally distributed. Even when the population distributions are unknown, Equation 12.2 can be used provided that n_1 and n_2 are each greater than 30 and are nearly equal. Note that the values of the population standard deviations must be known for these lower limits on sample sizes to apply.

Population Standard Deviations Unknown

Large Samples When the populations sampled are normally distributed, we can use sample estimates of the population variances as substitutes for σ_1^2 and σ_2^2 in Equation 12.1 provided that each of the two samples has 30 or more observations in it.

Suppose we have a sample of n_1 ($n_1 \geq 30$) observations from a normal population and suppose this sample has a mean of $\bar{x}_1$ and a variance of s_1^2. A second normal population has yielded an independent random sample of n_2 ($n_2 \geq 30$) observations with a mean of $\bar{x}_2$ and a variance of s_2^2. We want to test the null hypothesis that the difference between the two population means is equal to a certain value. We find $\bar{x}_1 - \bar{x}_2$, the difference in sample means. We also find the standard error of the difference as estimated from the sample variances; that is,

$$s_{\bar{x}_1 - \bar{x}_2} = \sqrt{\frac{s_1^2}{n_1} + \frac{s_2^2}{n_2}}. \qquad \textit{12.3}$$

Then the statistic

$$z = \frac{(\bar{x}_1 - \bar{x}_2) - (\mu_1 - \mu_2)}{s_{\bar{x}_1 - \bar{x}_2}} \qquad \textit{12.4}$$

belongs to an essentially normal distribution if each of the two samples is large. From this point on, hypothesis testing procedures and interval estimation procedures are identical to those discussed for standard normal deviates in Chapters 8 and 9.

Small Samples The only procedure presented here assumes that the two populations in question are normally distributed with equal variances.

A health club director has developed two reducing diets. Two weeks ago a random sample of 5 people started on diet *A* and 11 people started on diet *B*. Today, when the individual weight losses were checked, the following table was prepared:

	Diet *A*	Diet *B*
Sample size	$n_1 = 5$	$n_2 = 11$
Mean weight loss (pounds)	$\bar{x}_1 = 4.2$	$\bar{x}_2 = 5.6$
Sample standard deviation (pounds)	$s_1 = 0.7$	$s_2 = 0.9$

The health director has no preconceptions about which diet should produce the greater weight loss. Consequently, a two-tailed test of the null hypothesis that $\mu_1 - \mu_2 = 0$, where μ_1 and μ_2 are the means of the two populations of weight losses, is appropriate. The alternative hypothesis, then, is $\mu_1 > \mu_2$ or $\mu_2 > \mu_1$. The 0.05 level of significance is selected.

We begin with assumptions that the two populations of weight losses are normally distributed and have the same variances ($\sigma_1^2 = \sigma_2^2$). There is no highly sensitive test for these two assumptions with samples as small as these. On the other hand, the two distributions are not likely to be either heavily skewed or radically different in their variances. Under the circumstances, the test we are about to describe is safe because it is not highly sensitive to moderate departures from the two assumptions we have made.

The first step is to calculate a pooled estimate of the common population variance (σ^2) from the two sample estimates (s_1^2 and s_2^2). We find this estimate ($\bar{s}^2$) by taking a weighted mean of the sample variances in which the degrees of freedom serve as weights. The pooled estimate is

$$\bar{s}^2 = \frac{(n_1 - 1)s_1^2 + (n_2 - 1)s_2^2}{n_1 + n_2 - 2}. \qquad 12.5$$

For our example,

$$\bar{s}^2 = \frac{4(0.49) + 10(0.81)}{14}$$

or

$$\bar{s}^2 = 0.7186,$$

from which

$$\bar{s} = 0.848 \text{ pound.}$$

The second step is to find the standard error of the difference between sample means. The procedure is to use Equation 12.3 with $\bar{s}^2$ substituted for s_1^2 and s_2^2. After simplifying, the result is

$$\bar{s}_{\bar{x}_1 - \bar{x}_2} = \bar{s}\sqrt{\frac{1}{n_1} + \frac{1}{n_2}}. \qquad 12.6$$

A bar is placed over s on the left side to distinguish it from the left side of Equation 12.3. For our example,

$$\bar{s}_{\bar{x}_1 - \bar{x}_2} = 0.848\sqrt{1/5 + 1/11} = 0.848\sqrt{0.2909}, \quad \text{or} \quad 0.457.$$

If the populations are normally distributed with equal variances, then the statistic

$$t = \frac{(\bar{x}_1 - \bar{x}_2) - (\mu_1 - \mu_2)}{\bar{s}_{\bar{x}_1 - \bar{x}_2}}$$

is a t statistic with $(n_1 + n_2 - 2)$ degrees of freedom. In our example, we want to test the hypothesis that $\mu_1 - \mu_2 = 0$, so

$$t = \frac{(4.2 - 5.6) - 0}{0.457}, \quad \text{or} \quad -3.06,$$

with $5 + 11 - 2 = 14$ degrees of freedom. Since the level of significance α is 0.05 for the two-tailed test, the critical values from Appendix B are -2.145 and $+2.145$. On the basis of this experiment we reject the null hypothesis of equal population means. Diet B apparently produces a greater weight loss than diet A. We can summarize the test procedure just described in the following statement:

Statement

Assume that two normally distributed populations have equal variances ($\sigma_1^2 = \sigma_2^2$). A random sample of n_1 observations is selected from one population and a random sample of n_2 is selected from the other. The difference in sample means $(\bar{x}_1 - \bar{x}_2)$ can be incorporated in the statistic

$$t = \frac{(\bar{x}_1 - \bar{x}_2) - (\mu_1 - \mu_2)}{\bar{s}_{\bar{x}_1 - \bar{x}_2}}, \qquad 12.7$$

which is a t statistic with $(n_1 + n_2 - 2)$ degrees of freedom. The standard error $\bar{s}_{\bar{x}_1 - \bar{x}_2}$ is defined in Equations 12.5 and 12.6.

Matched Pairs

An oil company executive believes that his firm's new "regular" gasoline will yield more miles per gallon than the older type. To test his belief he selects one pair of automobiles from each of five weight classes. He then selects one member of each pair at random and assigns it to the old type of gasoline. The other member of the pair uses the new type. As we will see, using matched pairs eliminates much variability attributable to weight differences and thereby improves sensitivity of the test to differences in gasoline mileage.

Each automobile makes several runs of a designated course at 60 miles per hour, after which the miles per gallon of gasoline consumed on all the runs are calculated. The data are shown in Table 12.1 along with the differences (d) in mileage between the old and new types. The mean mileages for each type as well as the difference in means (1.8 miles per gallon) are shown.

Table 12.1 *Mileages from Two Types of Gasoline*

	Automobile Weight					
Gasoline Type	Very Heavy	Heavy	Medium	Light	Very Light	Mean
New	13.3	13.4	15.7	17.6	19.0	15.8
Old	12.5	12.5	13.7	14.9	16.4	14.0
New minus old (d)	0.8	0.9	2.0	2.7	2.6	1.8

Regardless of what the data show, the executive reasons that if the new gasoline is no better than the old, the mean of the population of differences in gasoline mileage (μ_d) should be zero or negative. Since he is interested in establishing that, at the minimum, the new gasoline produces more miles per gallon than the old, he forms the hypothesis statements

$$\text{Null hypothesis } (H_0)\text{:} \quad \mu_d \leq 0;$$
$$\text{Alternative hypothesis } (H_A)\text{:} \quad \mu_d > 0.$$

He would, of course, prefer to accept H_A because a large positive mean difference would indicate that the new gasoline produces greater mileage.

We now want to think about how sample values of paired differences (d) and means of paired differences ($\bar{d}$) are distributed. Consider the population of all values of paired differences that could result from the test. This population of d values will have some mean μ_d and some standard deviation σ_d, both of which are unknown. Next consider samples selected at random from the population of d values and consider the mean of each sample ($\bar{d}$). (Note that $\bar{d}$ is 1.8 for our example.) If we assume that the original population of d values is normally distributed, then the distribution of values of $\bar{d}$ for samples of n differences will also be normally distributed. The sampling distribution will have a mean of μ_d and a standard error of $\sigma_{\bar{d}}$, or $\sigma_d/\sqrt{n}$.

We do not know the value of σ_d, the standard deviation of the population of differences. But our situation with respect to this population is identical to that discussed in conjunction with the t distribution and Equation 9.8, page 244. We can estimate the population standard deviation σ_d from the sample by finding

$$s_d = \sqrt{\frac{\sum (d - \bar{d})^2}{n - 1}}. \qquad \textbf{12.8}$$

Then we can estimate the standard error of the sampling distribution of mean differences ($\sigma_{\bar{d}}$) by finding

$$s_{\bar{d}} = \frac{s_d}{\sqrt{n}}. \qquad \textbf{12.9}$$

Finally, the statistic

$$t = \frac{\bar{d} - \mu_d}{s_{\bar{d}}}$$

belongs to a t distribution with $(n - 1)$ degrees of freedom.

The calculations necessary to find the value of $s_{\bar{d}}$ for our gasoline mileage example appear in Table 12.2. In this table note that n is 5, the number of *differences,* and not 10, the number of mileage observations. The value of the test statistic is

$$t = \frac{\bar{d} - \mu_d}{s_{\bar{d}}} = \frac{1.8 - 0}{0.406}, \quad \text{or} \quad 4.43.$$

There are 4 degrees of freedom associated with the value of t. For a 0.01 level of significance and an upper-tail test, we see from Appendix B that $t_{.99}$ is 3.747. Since 4.43 is greater than 3.747, the executive can reject the null hypothesis and accept the alternative that at 60 miles per hour the new gasoline delivers more miles per gallon than does the old. The test procedure is summarized in the following statement.

Statement

Consider a set of n matched pairs. One treatment (A) is applied to one member of each pair. Another treatment (B) is applied to the other member. Assignment of members to treatments is made at random. Within each pair, the response to B is subtracted from A to obtain a sample of n differences (d). If these values are from a normal population with a mean of μ_d, then the statistic

$$t = \frac{\bar{d} - \mu_d}{s_{\bar{d}}} \qquad \textit{12.10}$$

belongs to a t distribution with $(n - 1)$ degrees of freedom. In this equation $\bar{d}$ is the mean of the n sample differences and $s_{\bar{d}}$ is as defined in Equations 12.8 and 12.9.

Instead of using the matched-pairs test for the gasoline mileage problem, the oil company executive could have used the approach we described in the section just prior to this one. He could have paid no attention to automobile weight. Rather, he could have selected 10 automobiles at random, assigned 5 of them the new gasoline at random, and let the remaining 5 use the old gasoline. Had he done this, it is highly unlikely that the new gasoline would have demonstrated its mileage superiority.

Under the circumstances described, the t test discussed in conjunction with Equation 12.7 would be far less sensitive than the one we performed using Equation 12.10. All the mileage variability attributable to differences in vehicle *weight* would

Table 12.2 *Standard Error of the Mean Difference* ($s_{\bar{d}}$)

d	$d - \bar{d}$	$(d - \bar{d})^2$
0.8	−1.0	1.00
0.9	−0.9	0.81
2.0	0.2	0.04
2.7	0.9	0.81
2.6	0.8	0.64
9.0	0.0	3.30

$\bar{d} = 1.8 \qquad n = 5$

$$s_d^2 = \frac{\sum (d - \bar{d})^2}{n - 1}$$

$$= \frac{3.30}{4}, \quad \text{or} \quad 0.825.$$

$$s_{\bar{d}}^2 = \frac{s_d^2}{n} = \frac{0.825}{5}, \quad \text{or} \quad 0.165.$$

$$s_{\bar{d}} = \sqrt{0.165}, \quad \text{or} \quad 0.406 \text{ mile per gallon.}$$

remain in the denominator of Equation 12.7, whereas matching on weight removes this variability from the denominator of Equation 12.10. Although the symbols differ, the numerators of the two equations produce identical values for identical data. When there really is a difference attributable to the effect of an extraneous variable such as automobile weight, we can expect numerators under the two approaches to be the same but the denominator in Equation 12.10 to be smaller. As a result we can expect larger t values from the matched-pairs approach. This of course presumes that variation in the factor used to match the pairs (weight in this case) would cause variability in the response (mileage) if it were left uncontrolled.

Even with the matched-pairs approach, we could have used a technique that might be expected to be even more sensitive to differences in mileage from the two gasolines. Instead of using a pair of different automobiles in each weight class, we could have used the same automobile with each type of gasoline. Then differences other than weight that cause variations in mileage (such as age and condition of engines and running gear) would have been controlled also. Whether the old or the new fuel is to be used first in such a test should be decided by a random process to minimize the possibility of a systematic carry-over from one treatment to the next. For example, suppose that under random selection five autos that started the test with the old gasoline and then switched to the new achieved better mileages with the new fuel. It might later be found that the old fuel left a residue that was largely responsible for the better performance with the new fuel.

Summary

Occasionally we need to test hypotheses about, or make interval estimates of, the difference in the means of two populations. In the first case discussed, the populations were normally distributed with known standard deviations. The standard normal deviate z was shown to apply. This deviate also is a satisfactory approximation when the populations are not normal–provided that each sample contains at least 30 observations. The sample sizes should be nearly equal.

In the second case discussed, the populations were normally distributed with unknown standard deviations. As was the case with the single sample in Chapters 8 and 9, sample estimates of the standard deviations are used. When both samples contain at least 30 observations, the standard normal deviate is a satisfactory approximation. When both samples are small, the only case discussed assumes that both normally distributed populations have the same standard deviation. Under this assumption, a t distribution serves as the basis for statistical inference.

To test for a difference between the means of two populations, a matched-pairs test can provide much greater sensitivity to a real difference than the t test which would be performed when matching is disregarded. However, the increased sensitivity will be present only if the pairs are matched by controlling a source of variability that would otherwise obscure differences in means.

See Self-Correcting Exercises 12A.

Exercise Set A

1. Suppose you have the following data on samples from two populations:

	Type A	Type B
Standard deviation (cm)	$\sigma_1 = 0.5$	$\sigma_2 = 0.4$
Sample size	$n_1 = 16$	$n_2 = 12$
Mean length (cm)	$\bar{x}_1 = 4.2$	$\bar{x}_2 = 5.0$

(a) For Equation 12.2 to be useful in hypothesis testing with these data, what else must you know about the two original populations being sampled?

(b) Find the standard error of the difference between means.

(c) Test at a level of significance of 0.05 the null hypothesis that the samples come from populations having the same mean. State and justify your conclusion.

2. For the data given in Exercise 1, test at a level of significance of 0.05 the null hypothesis that the mean of the type *B* population exceeds the mean of the type *A* population by 0.3 centimeters.

3. Tests sponsored by a consumer advocacy group to compare two brands of tires produced the following results:

	Brand *A*	Brand *B*
Standard deviation (thousands of miles)	$s_1 = 6.2$	$s_2 = 5.1$
Sample size	$n_1 = 32$	$n_2 = 36$
Mean life (thousands of miles)	$\bar{x}_1 = 28.2$	$\bar{x}_2 = 28.9$

(a) Why can't you determine the value of $\sigma_{\bar{x}_2-\bar{x}_1}$? What estimate must you use instead? What are the minimum values for n_1 and n_2 for this estimate to be applicable? What else must you assume about the populations sampled?

(b) Determine the estimated standard error of the difference.

(c) Test at a level of significance of 0.01 the null hypothesis that the two brands of tire have the same mean life. State and justify your conclusion.

4. An Environmental Protection Agency (EPA) test to determine whether or not a certain industrial plant on the bank of the Pawtuxet River seriously contributes to the pollution of that river consists of taking randomly timed samples of water 50 yards upstream and 50 yards downstream of the plant's "clean waste" discharge pipes. Due to the expense involved, only a few samples are taken; and due to container breakage in transit, 12 samples of upstream water reach the lab but only 10 samples of downstream water do. The results of the test for a certain pollutant are as follows:

	Upstream	Downstream
Standard deviation (ppm)	$s_1 = 1$	$s_2 = 2.2$
Sample size	$n_1 = 12$	$n_2 = 10$
Mean pollutant concentration (ppm)	$\bar{x}_1 = 8$	$\bar{x}_2 = 10.2$

Suppose that the EPA would consider a concentration rise of 2 parts per million (ppm) of this pollutant to be serious pollution by this plant.

(a) What two populations are being compared here?

(b) State the null and alternative hypotheses which the EPA would be interested in testing here.

(c) Is a one-tailed hypothesis test appropriate?

(d) What assumptions must you make about the original populations in order to use Equation 12.5?

(e) Making the necessary assumptions as determined in part (d) find the pooled estimate of the common population variance.

(f) Convert the difference $\bar{x}_2 - \bar{x}_1$ to the appropriate test statistic.

(g) How many degrees of freedom are present?

(h) At a level of significance of 0.02, what is the critical value of the test statistic for this number of degrees of freedom?

(i) State whether or not the EPA should, at a 2 percent level of significance, consider the plant as a serious contributor of this pollutant to the Pawtuxet.

5. Two types of automobile tires have been tested by exposing them to the same driving conditions for an equal number of months. Four tires of the first type lasted a mean of 85.25 weeks with a variance of 25. Five tires of the second type lasted a mean of 88.00 weeks with a variance of 30.25. Assume normal populations with equal variances. The first type is to be used unless the second type is shown to be definitely superior. At the 0.05 level of significance, conduct a one-tailed hypothesis test that will show whether or not the second type is superior.

6. A physician is studying the effectiveness of a reducing regimen. "Before and after" figures have been obtained from a random sample of adult males from 40 to 50 years of age. The data are (in pounds):

253 and 248	210 and 211
203 and 195	185 and 180
194 and 192	194 and 196
239 and 236	225 and 214
187 and 175	176 and 169

At the 0.01 level of significance, test the null hypothesis that adult males weigh at least as much after the regimen as they did before. Assume normal populations.

Analysis of Variance

In the first part of this chapter, we discussed how to test whether the observed difference between two sample means indicates a difference between the two population means or is merely a sampling fluctuation. Now we will consider a general class of techniques for simultaneously testing for differences among several population means. The general class of techniques is called **analysis of variance** (ANOVA). We will discuss just one of the many types: one-way analysis of variance. In one-way analysis there is only one factor—that is, one basis for distinguishing among populations. In our first example below, the factor is poster advertising. Within any factor there are several **levels**. In the first example four posters, or levels, are used.

General Description

In this section we will examine the logic on which analysis of variance is based. We begin by describing an example. Then we illustrate the notation for analysis of variance and the explanation which follows by referring to the example.

Suppose a marketing executive for a chain of small convenience stores is considering ways of advertising a certain product. She has four different posters, prepared for display near the product, that she wants to test for differences in effectiveness. On the first day of the test, a different poster is displayed near the product in each of four markets and the numbers of product sold that day are recorded. On the second day, each poster is moved to a market not involved in the first day of the experiment and the same type of data are collected. On the third and final day, the procedure is repeated at a third set of four markets.

There are good reasons for using a different set of markets every day. If the same market were used for the same poster during all three days of the experiment, we could not be certain whether it was the posters or the markets that were the source of any differences in sales. If posters were rotated among the same set of markets on successive days, the possibility of an effect from one poster carrying over to its successor could not be ruled out.

The sales results from the experiment appear in Table 12.3. The mean number of packages sold for poster D, 50, is higher than the mean for any other poster. The mean number of packages sold for posters B and C are very close and well below the means for the other two posters. The mean for poster A is fairly close to that for poster D. The purpose of the analysis of variance to be described is to test whether these differences in sample means reflect differences in the means of the statistical populations from which they came or are merely sampling fluctuations.

Table 12.3 *Number of Packages Sold*

	Poster			
Day	*A*	*B*	*C*	*D*
1	55	32	38	47
2	43	42	35	48
3	43	40	38	55
Level totals	141	114	111	150
Grand total =	516			
Level means	47	38	37	50
Grand mean =	43			
Number of observations per level	3	3	3	3
Total number of observations =				12

Notation and Terminology To discuss analysis of variance effectively, we need a way to designate observations and the several types of means we must refer to. We also need a standard set of terms. In Table 12.4, the data of Table 12.3 have been transformed to the standard notation we will use.

Table 12.4 Notation

Replication	Level 1	2	3	4	
1	x_{11}	x_{12}	x_{13}	x_{14}	
2	x_{21}	x_{22}	x_{23}	x_{24}	x_{ij}
3	x_{31}	x_{32}	x_{33}	x_{34}	
Level totals	C_1	C_2	C_3	C_4	
Grand total $= T$					
Level means	$\bar{x}_1$	$\bar{x}_2$	$\bar{x}_3$	$\bar{x}_4$	
Grand mean $= \bar{x}$					
Number of observations per level	n_1	n_2	n_3	n_4	
Total number of observations $=$					n

In one-way analysis of variance there can be only one factor, but it may be investigated at any number of levels greater than 1. We will use the symbol k to designate the number of levels. In our example, the use of posters constitutes the factor and there are four levels of this factor, that is, four types of poster. Hence k is 4.

We elected to make the same number of independent observations within all levels. Thus we are conducting a **balanced** experiment in that $n_1 = n_2 = n_3 = n_4 = 3$.

In general, an observation is designated symbolically as x_{ij}, where i is the row index and j is the column index. For our example, x_{32} is 40, for instance. The manner of designating the various totals, means, and number of observations is indicated in the lower portion of Table 12.4. An example of each type of total and mean can be had by referring to the comparable position in Table 12.3.

The Ratio of Variances In Table 12.3, we saw that the mean numbers of packages sold for each of the four posters ($\bar{x}_j$) are 47, 38, 37, and 50. The grand mean ($\bar{x}$) is 43 packages a day. We can establish a null hypothesis by stating that the four sales population distributions for the four posters have equal means–that is,

$$\mu_1 = \mu_2 = \mu_3 = \mu_4,$$

where the μ_j represent the daily package sales population means for the four types of poster. We want to test this hypothesis against the alternative that the population means are not all equal, that at least one of the means differs from the others. The rationale of our test will be to compare the variability among the four level means to the average variability within the levels. We will put the variability among the four means in the numerator of a ratio and the variability within the levels beneath it in the denominator. We can expect the ratio of the variability among means to the variability within levels to be much smaller if the four sales populations have the same mean than if their means differ widely.

If the null hypothesis is true, and we assume that (1) all four sales populations are *normally distributed* and (2) all four populations *have the same variance,* then we have four normally distributed sales populations for which $\mu_1 = \mu_2 = \mu_3 = \mu_4$ and $\sigma_1^2 = \sigma_2^2 = \sigma_3^2 = \sigma_4^2$. It follows that we can consider the 12 sales observations as coming from the *same population.* For simplicity we will use μ and σ^2 for the parameters of this population. Furthermore, it follows that we can view Table 12.3 as being a division of a random sample of 12 observations into four random samples composed of three observations apiece. Under these assumptions and the null hypothesis, the four sample means (47, 38, 37, and 50) belong to the sampling distribution of the mean for $n_j = 3$. From our previous work, we know that the sampling distribution to which the observed values of the column means (the $\bar{x}_j$) belong is a normally distributed statistical population that has a mean and variance of μ and σ^2/n_j, where n_j is equal to 3.

Although we do not know the value of the variance of the sampling distribution of the mean ($\sigma_{\bar{x}_j}^2 = \sigma^2/n_j$), we can form an unbiased estimate of it from the sample information (assuming that the null hypothesis is true):

$$s_{\bar{x}_j}^2 = \frac{\sum (\bar{x}_j - \bar{x})^2}{k - 1}. \qquad \textbf{12.11}$$

For our example,

$$s_{\bar{x}_j}^2 = \frac{(47 - 43)^2 + (38 - 43)^2 + (37 - 43)^2 + (50 - 43)^2}{4 - 1}$$

or

$$s_{\bar{x}_j}^2 = 42.$$

We know that $s_{\bar{x}_j}^2$ is an unbiased estimator of $\sigma_{\bar{x}_j}^2$. We also know that $\sigma_{\bar{x}_j}^2$ is σ^2/n_j. Hence, if we multiply $s_{\bar{x}_j}^2$ by the sample size (n_j), we will have an unbiased estimator of σ^2, the population variance. In symbols, this estimator is

$$(n_j)s_{\bar{x}_j}^2. \qquad \textbf{12.12}$$

In our example,

$$(n_j)s_{\bar{x}_j}^2 = 3(42), \quad \text{or} \quad 126.$$

This is the estimate of population variance based on the variability among means. Note: Because there are four levels in the experiment, 3 degrees of freedom are associated with this estimate of population variance.

If we happened to know the value of the population variance σ^2, we could use the ratio

$$\frac{(n_j)s_{\bar{x}_j}^2}{\sigma^2} \qquad 12.13$$

as the basis for a test of our original null hypothesis (H_0: $\mu_1 = \mu_2 = \mu_3 = \mu_4$). When the hypothesis is true, the numerator of the ratio $[(n_j)s_{\bar{x}_j}^2]$ is an estimate of σ^2 and the ratio will have an expected value of 1. On the other hand, if the four population means are markedly different, the dispersion of the four sample means typically will be much greater than $\sigma_{\bar{x}_j}^2$. In turn, the numerator of the test ratio, Equation 12.13, will be much greater than the denominator, σ^2, and the value of the ratio will be much greater than 1. In this manner, we can use a ratio of variances to test for a difference in means. This is the central concept at the heart of all analysis-of-variance procedures.

Almost never, however, do we know the value of the population variance σ^2 when we want to test for a difference in several population means. To construct a test for the differences in several population means, we must use sample estimates of population variance. We already have one such estimate for the numerator based on variability among level means. Our next step is to get another independent estimate for the denominator based on variability within levels.

We have assumed that the sales population distributions for the four posters in our experiment are identical normal distributions–have the same distribution. As a consequence their variances are assumed to be equal to each other and to that of the composite population ($\sigma^2 = \sigma_1^2 = \sigma_2^2 = \sigma_3^2 = \sigma_4^2$).

Now consider the sample of three observations from the sales population for poster *A*. The variance of this sample (s_1^2) is also an unbiased estimator of the population variance σ^2 if our assumption about variances stated in the previous paragraph is correct. In terms of our present notation,

$$s_1^2 = \frac{\sum (x_{i1} - \bar{x}_1)^2}{n_1 - 1} = \frac{(55 - 47)^2 + (43 - 47)^2 + (43 - 47)^2}{3 - 1}$$

or

$$s_1^2 = 48.$$

Similarly, for the other posters, s_2^2, s_3^2, and s_4^2 are also unbiased estimators of σ^2. Their values are 28, 3, and 19, respectively. We can form a pooled, unbiased estimator ($\bar{s}^2$) of population variance σ^2 by averaging these four sample values. When there are an equal number of observations in all samples,

$$\bar{s}^2 = \frac{\sum (s_j^2)}{k}. \qquad 12.14$$

For our example the estimate is

$$\bar{s}^2 = \tfrac{1}{4}(48 + 28 + 3 + 19)$$

or

$$\bar{s}^2 = 24.5.$$

Note: Since there are 2 degrees of freedom in each of the four samples, a total of 8 degrees of freedom is associated with this estimate.

The pooled sample variance $\bar{s}^2$ is an unbiased estimate of σ^2 based only on the dispersion *within each of the four independent samples.* The four independent sample means ($\bar{x}_j$) constitute the reference points from which the respective dispersions are measured. Consequently, the pooled sample variance within levels is not at all affected by differences among the four sample means.

Given the assumptions, we now have two independent estimates of the population variance σ^2. The first estimate $[(n_j)s^2_{\bar{x}_j}]$ is based on the dispersion of the four sample means $\bar{x}_j$ about the grand sample mean $\bar{x}$. The second estimate ($\bar{s}^2$) is based on the dispersion of the sample observations x_{ij} about their respective level means $\bar{x}_j$.

We form a ratio by placing the estimate of σ^2 based on variance among poster means $[(n_j)s^2_{\bar{x}_j}]$ on top in the numerator and the estimate based on average variance of sales within a poster type ($\bar{s}^2$) underneath in the denominator. The more the four population means differ for the different posters, the greater is the expected value of the numerator of our ratio. Meanwhile, the expected variability in the denominator remains constant. Hence the larger the ratio, the stronger the evidence that the four population means are different. On the other hand, we can expect the value of the ratio to be smaller when the sales populations are identical normal distributions for all four posters than in the former case of unequal means. The latter conditions apply under the null hypothesis. The following statement summarizes the situation under the null hypothesis.

Statement

Consider an experiment to test for differences among the means of k normally distributed populations with the same variance (σ^2). Under the null hypothesis of no differences among means, two independent estimates of σ^2 are $(n_j)s^2_{\bar{x}_j}$ and $\bar{s}^2$ as defined in conjunction with Equations 12.12 and 12.14. Define the ratio of the first of these to the second as an F statistic, that is,

$$F = \frac{(n_j)s^2_{\bar{x}_j}}{\bar{s}^2}. \qquad \textbf{\textit{12.15}}$$

The sampling distribution of this statistic is an F distribution with $(k - 1)$ degrees of freedom in the numerator and $(n - k)$ degrees of freedom in the denominator.

A new type of distribution is mentioned and we will discuss it shortly.

Our objective in one-way analysis of variance is to test the null hypothesis that several level means are equal. The test statistic is the ratio of variability among level means to the variability within levels. Under assumptions just mentioned, the test ratio will follow an F ratio distribution. A given member of the family of F ratio distributions is distinguished from the others by the number of degrees of freedom associated with its numerator (m_1) and with its denominator (m_2).

For one-way analysis of variance the number of degrees of freedom for the numerator (m_1) is one less than the number of level means,

$$m_1 = k - 1,$$

where k is the number of levels. The degrees of freedom for the denominator are the sum of the degrees of freedom within the levels, where the number of observations in a given level less one is the degrees of freedom for that level. Hence the degrees of freedom for the denominator are

$$m_2 = (n_1 - 1) + (n_2 - 1) + \cdots + (n_k - 1)$$

or

$$m_2 = n - k.$$

The Family of F Distributions Several F ratio distributions for different values of m_1 and m_2 are shown in Figure 12.1. It is especially important to note that the distribution with $m_1 = 8$ and $m_2 = 4$ is a *different distribution* from the one with $m_1 = 4$ and $m_2 = 8$. Appendix D identifies different F distributions by showing the degrees of freedom (m_1) for the numerator of an F ratio along the top of the table and the degrees of freedom (m_2) for the denominator along the sides. For each set (m_1, m_2), a pair of numbers appears in the body of the table. The top number is the value of F such that 5 percent of the values in the distribution are greater. The bottom number is the value of F such that 1 percent of the values in the distribution are greater. For example, consider the F distribution for which m_1 is 4 and m_2 is 8. Appendix D shows that 5 percent of this distribution lies above an F value of 3.84 and 1 percent lies above an F value of 7.01. The first of these two points can be located on the horizontal axis in Figure 12.1. The scale has not been extended far enough to locate the second value.

For our example concerned with poster advertising, we have already found that the numerator has the value 126 with 3 degrees of freedom and the denominator has the value 24.5 with 8 degrees of freedom. The value of the test statistic is therefore

$$F = \frac{126}{24.5}, \quad \text{or} \quad 5.14.$$

From Appendix D we see that the value of $F_{.95}$ for 3 and 8 degrees of freedom is 4.07. At the 0.05 level of significance, we can reject the null hypothesis that the long-

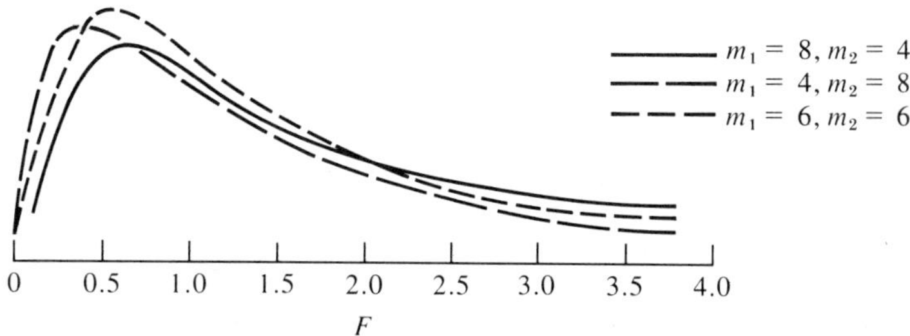

Figure 12.1 *Three F distributions*

run mean numbers of sales from the four posters are equal. There is too much variability among the four sample means ($\bar{x}_j$) as compared with the variability within the four samples for us to accept the hypothesis that all four samples come from populations with the same mean.

One-Way ANOVA Calculations

As early as Chapter 1 we found that calculating variances from deviations usually is very inconvenient unless the deviations are whole numbers. There and in subsequent chapters we developed convenient calculation relationships based on sums and sums of squares of the original observations.

Similar relationships convenient for computation with real data exist for analysis of variance. The three relationships that apply to a one-way analysis of variance with equal numbers of observations in each level are

$$\blacksquare \quad \text{SST} = \sum_j \sum_i (x_{ij}^2) - \frac{T^2}{n}, \quad \blacksquare \qquad \textit{12.16(a)}$$

$$\blacksquare \quad \text{SSA} = \sum_j \frac{C_j^2}{n_j} - \frac{T^2}{n}, \quad \blacksquare \qquad \textit{12.16(b)}$$

$$\blacksquare \quad \text{SSE} = \text{SST} - \text{SSA}. \quad \blacksquare \qquad \textit{12.16(c)}$$

In these equations, C_j is a column (level) total and T is the grand total as illustrated in Table 12.4. Similarly, n_j is the number of observations in a level (column) and n is the total number of observations.

The quantity SST is called the "total sum of squares." To find it we first square all of the observations in the entire ANOVA table and sum the squares. Then we find the sum of all the observations, square the sum, divide by n, and subtract the quotient from the sum of squares. The quantities SSA and SSE are the "sum of squares among level means" and the "sum of squares within levels," respectively. We will discuss how to find them shortly. All these sums of squares are entered into a table for finding an

F ratio. This value will be the same as that obtained by use of Equation 12.15. The general arrangement of this table appears in Table 12.5. The mean squares are the sums of squares divided by the degrees of freedom that apply to each sum. For example, MSA is SSA/$(k - 1)$.

Table 12.5 *Format for One-Way ANOVA*

Source of Variation	Sums of Squares	Degrees of Freedom	Mean Squares	F
Factor	SSA	$k - 1$	MSA*	MSA/MSE
Residual	SSE	$n - k$	MSE†	
Total	SST	$n - 1$		

**MSA = mean square among levels.*
†MSE = mean square within levels.

To repeat, the relationship for the total sum of squares (SST), Equation 12.16(a), states that each observation must first be squared. Then the sum of all n squared observations is found and the correction term (T^2/n) is subtracted from this result. For our example,

$$\sum_j \sum_i (x_{ij}^2) = (55^2 + 43^2 + 43^2) + \cdots + (47^2 + 48^2 + 55^2) = 22{,}762.$$

Then

$$\text{SST} = \sum_j \sum_i (x_{ij}^2) - \frac{T^2}{n} = 22{,}762 - \frac{516^2}{12}$$

or

$$\text{SST} = 574.$$

The relationship for the sum of squares among means (SSA), Equation 12.16(b), states that the sum of observations for each level must first be squared. Then each such square must be divided by the number of observations for that level. The resulting quotients for the separate levels are then added and the correction term is subtracted from this latter total. For our example, we refer to Table 12.3 and see that

$$\sum_j \frac{C_j^2}{n_j} = \frac{141^2}{3} + \frac{114^2}{3} + \frac{111^2}{3} + \frac{150^2}{3} = 22{,}566.$$

Then

$$\text{SSA} = 22{,}566 - \frac{516^2}{12}$$

or

$$SSA = 378.$$

Finally, the relationship for the residual sum of squares (SSE), Equation 12.16(c), states that SSE is the difference between SST and SSA. Hence, for our example,

$$SSE = SST - SSA = 574 - 378$$

or

$$SSE = 196.$$

When the sums of squares have been calculated, they are entered in a table modeled after Table 12.5. The general format applied to our example appears in Table 12.6. The resulting *F* ratio is, of course, the same as we obtained earlier.

Table 12.6 *ANOVA for Poster Experiment*

Source of Variation	Sums of Squares	Degrees of Freedom	Mean Squares	*F*
Posters	378	3	126	5.14
Residual	196	8	24.5	
Total	574	11		

As we pointed out earlier, an *F* value of 5.14 with 3 and 8 degrees of freedom exceeds the *F* value for the 0.05 level of significance. For our poster experiment we conclude that there is a real difference in sales response to the four posters. In Table 12.3, the four means (47, 38, 37, and 50) indicate that posters *A* and *D* are more effective than posters *B* and *C*. Although we will not discuss them, techniques exist for testing to see which differences in level means are most likely to be responsible for the significance of the *F* ratio in a situation such as our poster example.

Summary

One-way analysis of variance provides a technique for testing whether several levels of a given factor result in a different mean response for some variable. The test assumes that the sample observations within the levels are independent random selections from identical normal populations. The null hypothesis of no difference in mean response for the levels is tested by using a statistic which depends on the ratio of the variability among sample level means to the average variability within levels. Standard calculation procedures and standard formats for presenting the results of a one-way analysis have been described.

See Self-Correcting Exercises 12B.

Exercise Set B

1. For an F distribution with 12 degrees of freedom in the numerator and 30 degrees of freedom in the denominator, above what F value will 1 percent of the distribution lie? Five percent?

2. For an F distribution with 30 degrees of freedom in the numerator and 12 degrees of freedom in the denominator, above what F value will 1 percent of the distribution be? Five percent?

3. In a rectangular array of data, in which row and column would each of the following values be found?

   (a) x_{35} (c) x_{2j} (e) x_{ij}

   (b) x_{12} (d) x_{i_1}

4. A study is made of the adverse side effects of drugs used in the treatment of chronic euphoria. In each of four clinics, the same number of prescriptions with the same dosage instructions are dispensed for each drug. Complaints of users over a 6-month period are recorded and totaled as follows. The three complaint populations are assumed to be normally distributed with equal variance.

Clinic	Drug Type *A*	*B*	*C*
1	12	14	13
2	7	6	14
3	9	13	16
4	4	12	12

   (a) What is k?

   (b) What is n? What is n_1?

   (c) Determine the level means.

   (d) What is the grand mean?

(e) Use Equation 12.11 to estimate the variance of the sampling distribution of the mean.

(f) Using your result from part (e), determine an estimate of the value of the variance of the populations.

(g) What is the average variance within levels?

(h) Determine the ratio of your result in (f) to your result in (g).

(i) At the 0.05 level of significance, can you conclude that complaints are of about the same frequency for all three drugs?

5. Four different instructors gave the same quiz to their students. A random sample of three student papers was selected from each instructor's class to see whether there was a difference in mean performance. The scores as listed by instructor are:

Instructor			
A	*B*	*C*	*D*
3	10	6	7
6	6	6	6
3	8	3	8

Use calculations as described in Equations 12.16 to test the hypothesis of no difference in means at a level of significance of 0.05.

6. In a service station location study, three traffic observers are being trained to count and record the number of cars making right turns adjacent to prospective sites. The observers count simultaneously but independently on the same site for four 4-hour periods. The mean number of cars per hour is then found. These results appear in the following table:

Observer		
A	*B*	*C*
51	34	25
33	46	57
32	50	40
65	29	29

Use the method described in connection with Equation 12.15 and a 0.05 level of significance to test whether the means are equal for the three observers.

7. For the data in Exercise 6, test the hypothesis that the means are equal for the three observers at a level of significance of 0.01.

Overview

1. For an F distribution with 9 degrees of freedom in the numerator and 24 degrees of freedom in the denominator, below what F value will 95 percent of the distribution lie?

2. Two types of truck engines are being compared for mileage between major tune-ups. The truckers' trade association selects a sample of 100 trucks with engine A and 121 trucks with engine B. The test results were as follows:

	Engine A	Engine B
Sample mean (miles)	24,800	25,100
Sample variance	1,210,000	1,742,400

With these large samples, assume that the sample variances are equal to the population variances. Conduct a two-tailed test of the hypothesis that the two population means are equal. Use 0.01 as the level of significance.

3. Two sections of Introductory Calculus, one with 25 students and the other with 22, are given an algebra skills test at the beginning of the semester. On the basis of equal test scores, 12 matched pairs are obtained consisting of one student from each section. The larger section is taught in a traditional lecture method; the other is taught as a "programmed" course under the guidance of the same professor. At the end of the semester, the students in both classes are given an achievement test. Test scores (in percent) were recorded for the 12 pairs as follows:

	Student Pair											
	1	2	3	4	5	6	7	8	9	10	11	12
Lecture	38	59	63	92	87	62	74	78	72	71	83	49
Programmed	42	57	65	94	89	67	71	72	76	71	81	40

The professor, with no preconceived notions about the superiority of either method of instruction, wants to test at a level of significance of 0.05 the null hypothesis that both methods are equally effective. Perform the appropriate test and state your conclusion.

4. Several changes designed to improve productivity have been made in the assembly department of an electronics manufacturing firm. "Before and after" data on number of assemblies completed in a day's time have been gathered for eight of the assemblers:

	Assembler							
	1	2	3	4	5	6	7	8
Before	37	42	28	32	38	35	29	34
After	40	40	32	35	41	37	36	38

Find the 0.95 confidence interval estimate of the mean number of additional assemblies made per day after the change. Assume a normal population.

5. Three different makes of electronic calculator to be used for classroom programming instruction are being evaluated for volume purchase by a university. A standard set of problems is created that is representative of the uses to which the calculators will be put. It is estimated that about 10 hours will be required to process the set of problems. Because of variations in the times manufacturers could make demonstrators available, only 12 operators of 15 assigned to the three calculators were able to complete the standard set of problems. The completion times, rounded to the nearest hour, are:

	Calculator	
A	*B*	*C*
7	8	14
6	6	9
8	9	12
	11	10
	7	

(a) At the 0.05 level, test the hypothesis that the mean completion times are the same for all three calculators. Use Equations 12.16 and the format in Table 12.5.

(b) Based only on the outcome of this experiment, which calculator should the university choose?

6. What assumptions permit use of the F table in Appendix D to test the significance of the sample F ratio in Exercise 5? Are these assumptions likely to be met in the experiment described?

7. Why are all the F tests for these exercises upper one-tailed tests? That is, why test only for excessively large values of F?

Glossary of Equations

12.2 $$z = \frac{(\bar{x}_1 - \bar{x}_2) - (\mu_1 - \mu_2)}{\sigma_{\bar{x}_1 - \bar{x}_2}}$$

The difference in sample means from two normally distributed populations can be transformed to the standard normal statistic (z) by subtracting the difference in population means and dividing by the standard error of the difference in sample means. This statistic is used as a basis for hypothesis tests about and interval estimates of the difference in population means; it applies when the two population variances are known or have been estimated from samples of more than 30.

12.7 $$t = \frac{(\bar{x}_1 - \bar{x}_2) - (\mu_1 - \mu_2)}{s_{\bar{x}_1 - \bar{x}_2}}$$

This statistic belongs to a t distribution and is used for tests and estimates concerning the difference in population means when the populations are normal. The statistic is based on the assumption that the two populations have equal variances.

12.10 $$t = \frac{\bar{d} - \mu_d}{s_{\bar{d}}}$$

The mean difference ($\bar{d}$) of a sample set of differences from matched pairs can be used to test hypotheses about the difference in two population means (μ_d). If the differences (d) are normally distributed, then the sample-mean difference can be transformed into a t statistic by means of this equation. The denominator is the estimated standard deviation of the sampling distribution of the variable $\bar{d}$.

12.16(a) $$\text{SST} = \sum_j \sum_i (x_{ij}^2) - \frac{T^2}{n}$$

The total sum of squares is the sum of the squares of all observations less the sum of all observations squared over the number of observations.

12.16(b) $$\text{SSA} = \sum_j \frac{C_j^2}{n_j} - \frac{T^2}{n}$$

The sum of squares among levels is the sum of the ratios of a column total squared to the number of observations in the column less the correction term.

12.16(c) $$\text{SSE} = \text{SST} - \text{SSA}$$

The sum of squares within levels is the total sum of squares less the sum of squares among levels.

13

Sample Survey Methods

The methods of statistical inference that have been discussed in previous chapters have all assumed that sample data were generated by successive independent draws from an underlying statistical population. Only when the sample data are the result of a genuine probability process is a quantitative measure of sampling error valid. The sampling error—for example, the standard error of the mean—is the critical quantity in testing a hypothesis or in constructing a confidence interval.

In this chapter we discuss selection procedures that are used to ensure that sample data are the result of a probability process. Such methods are called **probability sampling** techniques. We will examine some commonly used methods besides simple random sampling. They give rise to different formulas for the standard error of a mean that take account of the different probability processes by which the sample data are selected. We will describe these different selection processes by carrying through an abbreviated example for each. The circumstances that favor the use of each method, and some appreciation of how the methods are combined in practice, are also emphasized.

The map in Figure 13.1 shows all the occupied dwelling units in a hypothetical village called Georges Mills. Table 13.1 indicates whether each dwelling unit is owner-occupied or renter-occupied, and also presents the most recent week's grocery bill for the occupant family. This list, which covers completely the universe of units of observation, will be used to illustrate the principles of several kinds of probability sampling methods.

Simple Random Sampling

Definition | A **simple**, or unrestricted, **random sample** of n units from a universe of N units of observation is defined as a sample resulting from a process in which each possible combination of n out of N units has an equal probability of selection.

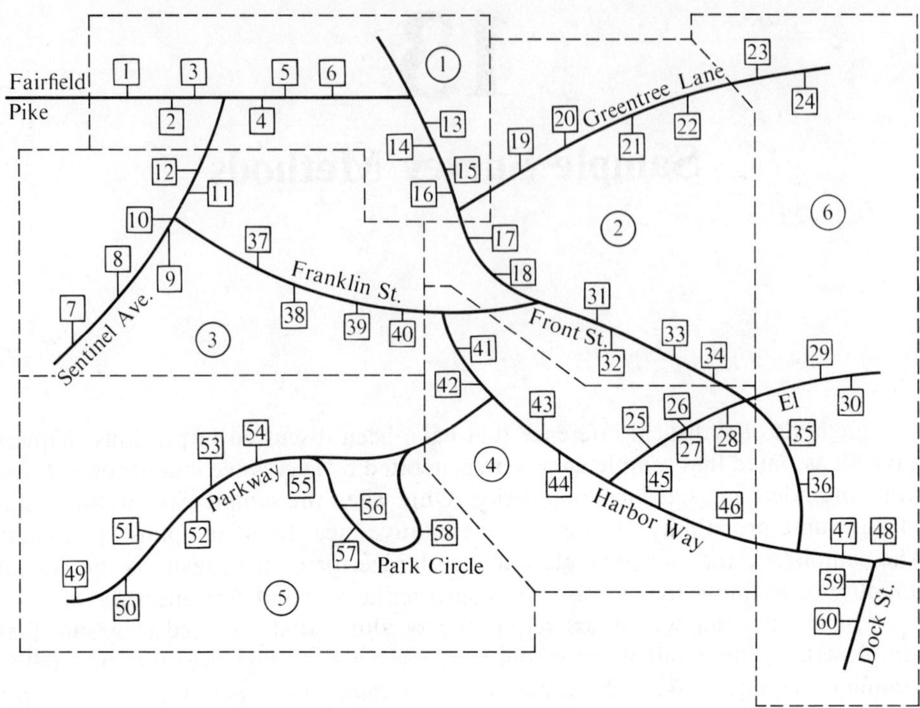

Figure 13.1 Dwelling unit map for village of Georges Mills

This definition is phrased in terms of sampling without replacement. In the village of Georges Mills we have $N = 60$ units of observation, or households. If we were contemplating a sample of $n = 5$ households, a simple random sample would be one in which each of the ${}_{60}C_5 = 5{,}461{,}512$ combinations of 60 households taken five at a time has an equal probability of being the ultimate sample. The formulas we have dealt with so far have applied to sampling with replacement. For example, the basic formula we have used for the standard error of a sample mean is

$$\sigma_{\bar{x}} = \frac{\sigma}{\sqrt{n}}.$$

When we conduct sampling without replacement, we apply a factor called the **finite population multiplier** in finding the standard errors of sample statistics. In sampling without replacement, the standard error of the mean is

$$\blacksquare \quad \sigma_{\bar{x}} = \frac{\sigma}{\sqrt{n}}\sqrt{\frac{N-n}{N-1}}. \quad \blacksquare \qquad \textit{13.1}$$

The fraction under the radical is the finite population multiplier. The effect of sampling without replacement is to reduce the standard error. For example, the variance

Table 13.1 *Dwelling Units with Ownership Status and Weekly Grocery Bill**

Unit	Bill	Unit	Bill
1	$49.6	31	$67.1
2	45.2	32	58.2
3(R)	41.5	33(R)	35.6
4(R)	28.0	34	52.6
5(R)	41.2	35	56.4
6(R)	29.8	36	52.9
7(R)	38.4	37(R)	41.3
8	52.8	38(R)	35.1
9(R)	59.1	39(R)	61.1
10(R)	31.3	40(R)	48.8
11	36.3	41	66.0
12(R)	45.3	42	52.6
13	57.0	43(R)	44.5
14	60.8	44(R)	48.2
15(R)	30.4	45	65.4
16	59.8	46	55.2
17(R)	45.1	47(R)	52.2
18	43.2	48	72.5
19	48.2	49	60.7
20(R)	31.6	50	71.2
21(R)	35.8	51	57.3
22(R)	26.0	52	60.2
23	51.9	53(R)	53.6
24	33.1	54	71.6
25(R)	30.4	55	55.0
26	55.4	56	58.8
27	57.0	57(R)	47.5
28	43.5	58(R)	49.0
29	50.0	59	63.1
30	53.9	60	62.1

**R indicates renter-occupied; all others are owner-occupied.*

of the 60 expenditure figures for the households in Georges Mills is 137.3. If we take a sample of size $n = 12$ *with replacement,* the standard error of the sample mean is

$$\sigma_{\bar{x}} = \frac{\sigma}{\sqrt{n}} = \frac{\sqrt{137.3}}{\sqrt{12}} = \sqrt{11.4} = 3.38.$$

If the sample is taken *without replacement,* we have

$$\sigma_{\bar{x}} = \frac{\sigma}{\sqrt{n}}\sqrt{\frac{N-n}{N-1}} = 3.38\sqrt{\frac{60-12}{60-1}} = 3.38\sqrt{0.814} = 3.05.$$

When a sample drawn without replacement is a small fraction of the units in the universe, the finite population multiplier, while still correct, is often omitted. For a sample of 1000 households in a city of 50,000 households, the multiplier would be

$$\sqrt{\frac{50{,}000 - 1000}{50{,}000 - 1}} = \sqrt{\frac{49{,}000}{49{,}999}} = \sqrt{0.980} = 0.990.$$

For the samples drawn from our small illustrative population, we will include the finite population multiplier. It can be omitted with but small effect when the sample size is less than 10 percent of the population size.

Drawing a Simple Random Sample

To draw a simple random sample we must find a procedure which ensures that every combination of n out of the N units of observation in the universe has an equal probability of selection. One might think to number the addresses of the 60 households referred to earlier and then place slips of paper numbered from 1 to 60 in an urn. If a sample of 12 households were desired, 12 draws without replacement would be made to determine the numbers of the sample households. The idea is sound, but it is difficult to design and conduct a physical drawing that ensures random sampling.

A better procedure is to use a table of random rectangular numbers. Appendix F, a portion of such a table, requires some explanation before we can use it. The table appears to be nothing more than a collection of the digits 0, 1, 2, . . . , 7, 8, 9 arranged in pairs with no discernible order. Indeed, this is exactly the intention. If you machined a cylinder to have 10 balanced faces and labeled the faces with the digits 0, 1, 2, . . . , 7, 8, 9, you could generate a table of random rectangular two-digit numbers by repeatedly rolling the device twice and recording the numbers facing up when it comes to rest. The probability of each two-digit number would be 1/100, and the appearance of any specified two-digit number would be independent of the appearance of other two-digit numbers. A large table of such two-digit numbers would represent an approximation to a population in which each of 100 two-digit numbers (00 through 99) appeared independently and with equal probability. They would form a rectangular probability distribution.

To draw a *simple random sample* with the aid of a table of random numbers, we first number all the units of observation in the universe. On our map the dwelling units are numbered. A list of addresses numbered from 01 to 60 would do as well. We then enter the random number table (Appendix F) at an arbitrary starting point and read successive two-digit numbers to select our sample. We will select a sample of $n = 12$ dwelling units. Only the numbers 01–60 will draw a dwelling unit. Should we encounter 00 or any two-digit number in the range 61–99, we will simply pass along to the next two-digit number. Table 13.2 shows the sequence of random two-digit numbers found in the first row of Appendix F. The third two-digit number, 73, does not draw one of the 60 dwelling units. Notice that the thirteenth number, 35, has already been drawn (the ninth number). We ignore it and continue to read two-digit numbers until we have drawn 12 nonrepeating numbers between 01 and 60.

Table 13.2 *Selection of a Random Sample of 12 out of 60 Numbered Dwelling Units*

Random Number	Dwelling Unit Selected	Random Number	Dwelling Unit Selected
10	10	67	—
09	9	35	—
73	—	48	48
25	25	76	—
33	33	80	—
76	—	95	—
52	52	90	—
01	1	91	—
35	35	17	17
86	—	39	39
34	34	29	29

The mean and variance of the expenditure data for the sample units in Table 13.2 are calculated here:

Unit	x	Unit	x
10	\$31.3	35	\$56.4
9	59.1	34	52.6
25	30.4	48	72.5
33	35.6	17	45.1
52	60.2	39	61.1
01	49.6	29	50.0

$$\sum x = 603.9 \qquad \sum x^2 = 32{,}219.41 \qquad \frac{\left(\sum x\right)^2}{n} = 30{,}391.27$$

$$\bar{x} = \frac{\sum x}{n} = \frac{603.9}{12} = \$50.32$$

$$s^2 = \frac{\sum x^2 - \left(\sum x\right)^2/n}{n-1} = \frac{1828.14}{11} = 166.2$$

We now use the sample data to estimate the standard error of the mean:

$$s_{\bar{x}} = \frac{s}{\sqrt{n}}\sqrt{\frac{N-n}{N-1}} \qquad \textbf{13.2}$$

$$= \frac{\sqrt{166.2}}{\sqrt{12}}\sqrt{\frac{60-12}{60-1}} = 3.36.$$

To form a confidence interval for the population mean based on the sample data, we use the Student t distribution.* This involves an assumption that the population of expenditures is normally distributed. For a 95 percent confidence interval we find the proper t value for $m = n - 1 = 11$ degrees of freedom to be 2.201. The interval then is

$$\$50.32 \pm 2.201\ (\$3.36), \text{ or } \$50.32 \pm \$7.40.$$

The actual mean of the 60 expenditure figures is \$49.79, and our interval estimate is, in fact, correct. However, in taking a sample of size $n = 12$ and basing our estimate wholly on sample information, we have put ourselves in a role where we do not know any parameter values. If the assumption of normality seems unduly restrictive, recall from Chapter 9 that when n is large you may calculate $\bar{x} \pm z_{\alpha/2}s_{\bar{x}}$ without having to be concerned about normality of the population. This would be the more commonly encountered practical situation.

Stratified Random Sampling

In **stratified random sampling**, the universe of units of observation is divided into groups called strata and simple random samples are drawn separately from each stratum. A complete listing of units is necessary as in unrestricted (nonstratified) random sampling. In addition, the characteristic on which the stratification is based must be known for each unit in the universe. The basis for stratification will depend on the nature of the survey; common bases are sex, geographic area, occupational class, political party, and so forth. If the ratio of units drawn to total units is the same for each stratum, the term **proportionate stratified random sampling** is applied. This ratio is called the *sampling fraction,* or *sampling ratio.*

To illustrate proportionate stratified random sampling, let us return to Georges Mills. Before selecting a sample (again of 12 dwelling units) we have determined for each dwelling unit whether it is owner-occupied or rented. It occurs to us that there may well be an association between home ownership and expenditures on groceries.

*The t distribution was first developed by W. S. Gossett in 1908. Gossett was employed by the Guinness Brewery of Dublin and London, who insisted that he use a pseudonym in signing professional articles. Gossett's articles were simply signed "Student," and the distribution came to be called the Student t distribution.

Specifically we think that homeowners may be more affluent and spend more for groceries on the average than do renters. Therefore it would seem that we should take care to have owners and renters included in the sample of the same respective proportions that they represent in the population.

Table 13.1 shows that there are 35 owner-occupied dwelling units and 25 renter-occupied units. A proportionate stratified random sample will require that 35/60 of the units drawn be owner-occupied and 25/60 be renter-occupied. For a 20 percent proportionate stratified random sample this means 7 owner-occupied and 5 renter-occupied units. We now divide the universe into the two strata and select simple random samples of the required size from each stratum.

Definition A **proportionate stratified random sample** is a sample in which the numbers of observation units drawn from each stratum are proportional to their respective numbers in the universe. The resulting subsamples are drawn on a simple random basis from the respective strata.

In Table 13.3 the list of dwelling units from Table 13.1 has been divided into owner- and renter-occupied strata and renumbered within each stratum. Also given are the weekly grocery bill figures, which of course would not be known at the time the sample was selected.

Table 13.3 *Renumbering of Dwelling Units by Stratum*

Owner-Occupied				Renter-Occupied			
New Number	x	New Number	x	New Number	x	New Number	x
1	49.6	19	52.6	1	41.5	19	48.8
2	45.2	20	56.4	2	28.0	20	44.5
3	52.8	21	52.9	3	41.2	21	48.2
4	36.3	22	66.0	4	29.8	22	52.2
5	57.0	23	52.6	5	38.4	23	53.6
6	60.8	24	65.4	6	59.1	24	47.5
7	59.8	25	55.2	7	31.3	25	49.0
8	43.2	26	72.5	8	45.3		
9	48.2	27	60.7	9	30.4		
10	51.9	28	71.2	10	45.1		
11	33.1	29	57.3	11	31.6		
12	55.4	30	60.2	12	35.8		
13	57.0	31	71.6	13	26.0		
14	43.5	32	55.0	14	30.4		
15	50.0	33	58.8	15	35.6		
16	53.9	34	63.1	16	41.3		
17	67.1	35	62.1	17	35.1		
18	58.2			18	61.1		

We now select seven random numbers for the owner-occupied sample and five for the renter-occupied sample. Our random numbers begin with the twenty-third two-digit number in the first row of Appendix F (our previous sample selection ended with the twenty-second number). To avoid having to pass over a large number of nonapplicable two-digit random numbers, we let random numbers in the interval 36–70 as well as 1–35 select owner-occupied units. For rented units numbered 1–25 we let the random numbers 26–50, 51–75, and 76–00 select a numbered dwelling unit as well as the random numbers 01–25. For example, the first random number selected for owners is 27, and the twenty-seventh owner-occupied unit is thereby selected. The second random number drawn is 49. This selects the fourteenth owner-occupied unit in the list, since 49 is the fourteenth number in the interval 36–70. Shown here are the random numbers and the serial numbers from Table 13.3 that they selected, along with the expenditure figures for the sample:

Owners			Renters		
Random Number	Serial Number	x_1	Random Number	Serial Number	x_2
27	27	$60.7	05	5	$38.4
49	14	43.5	64	14	30.4
45	10	51.9	89	14	Repeat
37	2	45.2	47	22	52.2
54	19	52.6	42	17	35.1
20	20	56.4	96	21	48.2
48	13	57.0			

The mean and variance for the sample data drawn from each stratum are:

Owners	Renters
$\bar{x}_1 = \$52.47$	$\bar{x}_2 = \$40.86$
$s_1^2 = 39.56$	$s_2^2 = 82.78$

Estimates from Stratified Samples

Having illustrated the process of drawing a (proportionate) stratified random sample and secured our sample data, we now consider the *why* of stratified sampling.

What do we hope to accomplish? By sampling from individual strata we hope to reduce the standard error of the mean as compared with a simple random sample of equal size. This can be illustrated most clearly by analyzing our population data for the 60 households.

The population stratum means are \$55.90 for the 35 homeowners and \$41.23 for the 25 renters. Thus our intuition that homeowners spend more for groceries than renters is confirmed. Since there are 35 homeowners and 25 renters, the mean of the 60 expenditure values can be obtained by

$$\frac{35}{60}(\$55.90) + \frac{25}{60}(\$41.23) = \$49.79.$$

The $\bar{x}_1 = \$52.47$ and $\bar{x}_2 = \$40.86$ we obtained from our stratified sample are estimates of the population means for owners and renters, respectively. The sample means for the strata will be combined just as above to obtain an estimate of the overall population mean. The sampling error for this estimate depends on the sampling errors for the individual stratum means. These errors will be smaller than for comparable draws from the entire population to the degree that the various strata really do separate the population into layers of x values.

When a stratified sample is taken, the best point estimate of the population mean will be

$$\blacksquare \quad \bar{x} = \sum_j (w_j \bar{x}_j). \quad \blacksquare \qquad \mathit{13.3}$$

In Equation 13.3,

$\bar{x}_j$ = sample mean for the jth stratum,

w_j = N_j/N, the proportion of the universe of units of observation that are in the jth stratum.

Now $\bar{x}$ in Equation 13.3 is a weighted mean of the several stratum means, where the weights are determined by the sizes of the strata. Further, the samples from the several strata are independent of one another. Under these conditions the variance of the mean estimated from a stratified random sample is

$$\blacksquare \quad \sigma_{\bar{x}}^2 = \sum_j (w_j^2 \sigma_{\bar{x}_j}^2). \quad \blacksquare \qquad \mathit{13.4}$$

Using Equation 13.3, the point estimate of the population mean is

$$\bar{x} = \sum_j (w_j \bar{x}_j) = \frac{35}{60}(\$52.47) + \frac{25}{60}(\$40.86) = \$47.63.$$

Then, using the sample data, we estimate the variances of the individual stratum means:

$$s_{\bar{x}_1}^2 = \frac{s_1^2}{n_1}\left(\frac{N_1 - n_1}{N_1 - 1}\right) = \frac{39.56}{7}\left(\frac{35 - 7}{35 - 1}\right) = 4.65,$$

$$s_{\bar{x}_2}^2 = \frac{s_2^2}{n_2}\left(\frac{N_2 - n_2}{N_2 - 1}\right) = \frac{82.78}{5}\left(\frac{25 - 5}{25 - 1}\right) = 13.8.$$

Now we apply Equation 13.4, using the estimated variances of the stratum means:

$$\blacksquare \quad s_{\bar{x}}^2 = \sum_j (w_j^2 s_{\bar{x}_j}^2) \quad \blacksquare \qquad \textit{13.5}$$

$$= (7/12)^2(4.65) + (5/12)^2(13.8)$$

$$= 3.98$$

and

$$s_{\bar{x}} = \sqrt{3.98} = \$1.99.$$

The confidence interval is $\bar{x} \pm t_{\alpha/2}s_{\bar{x}}$. The degrees of freedom for t are $\sum_j (n_j - 1)$, in our case $(7 - 1) + (5 - 1) = 10$. For our 95 percent confidence interval, the t multiple is then 2.228 and we have

$$\bar{x} \pm 2.228 s_{\bar{x}},$$

$$\$47.63 \pm 2.228(\$1.99),$$

$$\$47.63 \pm \$4.43.$$

This confidence interval is narrower than the 95 percent confidence interval from the unrestricted random sample because owning versus renting is related to spending on groceries. Stratified random samples permit more reliable estimates of parameters than unrestricted random samples of the same size when the basis of stratification is related to the variable of interest in the survey. Cutting down on sampling error is our objective in stratification.

Cluster Sampling

The units of observation in a universe often come in convenient groupings, or clusters. Examples of clusters are pages in a listing of phone subscribers, blocks of dwelling units in a city, places of employment of individuals, and so on.

Definition

Cluster sampling is a procedure whereby units of observation in the universe are divided into clusters, or *primary sampling units.* Then sampling proceeds by stages. First a sample of clusters is selected, and then units of observation are drawn from the clusters that have been selected. Units drawn at the final stage are called *elementary units.*

Thus we might draw 50 pages at random from a telephone directory and from each page drawn select four subscribers at random. We would have a cluster sample of $n = 200$ with 50 primary sampling units and 4 elementary units per primary sampling unit. Many other combinations could produce the sample of 200 subscribers. One attractive feature of cluster sampling is the saving in listing and numbering. Instead of having to list and number the entire universe of elementary units, we need list and number only the primary sampling units and then the elementary units within the primary units that are drawn.

On the map of Georges Mills (Figure 13.1) the village has been divided into six geographic areas with 10 dwelling units in each area. We will select a sample of 12 dwelling units by selecting three clusters and then selecting 4 dwelling units from each of the selected clusters. Both stages of selection will be conducted without replacement.

The areas are already numbered from 1 to 6. Our drawing of random numbers to select the stratified sample ended with the tenth two-digit number of the second row in Appendix F. We now draw one-digit numbers, ignoring 0, 7, 8, and 9. The sequence is 2,4,8,0,5. The clusters selected are thus areas 2, 4, and 5. In Table 13.4 we show the selection of four dwelling units from each of these areas and the resulting expenditure data. We continue reading one-digit random numbers and regard the dwelling units in each area as renumbered from 1 to 10 with the lowest-numbered unit reassigned the number 1 and the highest-numbered unit in an area reassigned the number 10. The original numbers are given in Figure 13.1 and the associated data in Table 13.1. We associate the random digit 0 with the tenth, or highest-numbered,

Table 13.4 *Cluster Sample–Selection of Dwelling Units from Geographic Clusters*

Area 2 (Cluster 1)			Area 4 (Cluster 2)			Area 5 (Cluster 3)		
Random Number	Observation Number	x_1	Random Number	Observation Number	x_2	Random Number	Observation Number	x_3
2	18	43.2	7	43	44.5	3	51	57.3
4	20	31.6	2	26	55.4	6	54	71.6
0	34	52.6	0	46	55.2	1	49	60.7
3	19	48.2	6	42	52.6	0	58	49.0
		175.6			207.7			238.6

dwelling unit in an area. The foregoing is just one of a number of ways of achieving a one-to-one relation between each random number and a specific dwelling unit within a cluster. We establish our rules for this before we draw the random numbers.

Estimates from Cluster Samples

Cluster sampling involves a two-stage selection process whereby we first select clusters and then select elementary units from each cluster selected. In cluster sampling, each elementary unit is given an equal probability of selection. We can do this in one of two ways. In the first, each primary sampling unit (cluster) is given an equal probability of selection and then a constant fraction of elementary units within each selected primary unit is drawn. In the second method, primary sampling units are selected with probabilities proportional to the numbers of elementary units they contain and then a constant number of elementary units are drawn from each primary unit. The first method is more common, because often the count of elementary units in the primary units is not known and the constant sampling ratio is often easier to carry out in the survey. In our example the clusters are of equal size, so that the two methods amount to the same thing.

As long as each elementary unit has an equal probability of selection in a cluster sample, the best point estimate of the population mean is the mean of the sample observations:

$$\bar{x} = \frac{\sum_k \sum_i x_{ik}}{n}, \qquad 13.6$$

where x_{ik} is the ith observation in the kth cluster selected in the sample.

The variance of the cluster sample mean depends on two sources of variation. The first is the variance among cluster means and the second is the variance within clusters. The variance *among* cluster means reflects sampling variability introduced in the selection of clusters for the sample; the variance *within* clusters reflects sampling variability introduced by selecting elementary units from clusters. We use the following symbols:

N = number of elementary units in population,

N_j = number of elementary units in jth cluster,

x_{ij} = value of ith observation in jth cluster,

n = sample size of cluster sample,

J = number of clusters in population,

k = number of clusters drawn into sample,

n_k = number of sample observations drawn from kth cluster,

x_{ik} = value of ith sample observation in kth cluster drawn,

μ = population mean,

μ_j = cluster mean (population).

In our example $N = 60$, $N_j = 10$, $n = 12$, $J = 6$, $k = 3$, and $n_k = 4$.

If N_j and N_k are constants,* the variance of the cluster sample mean is

$$\sigma_{\bar{x}}^2 = \frac{n_k\sigma_a^2 + \sigma_w^2}{n}. \qquad 13.7$$

In Equation 13.7, σ_a^2 is the among-cluster variance and σ_w^2 is the within-cluster variance. The true values of these variances are

$$\sigma_a^2 = \frac{\sum_j N_j(\mu_j - \mu)^2}{N}$$

and

$$\sigma_w^2 = \frac{\sum_i \sum_j (x_{ij} - \mu_j)^2}{N}.$$

The sum of the variance among and the variance within clusters is the population variance, that is,

$$\sigma_a^2 + \sigma_w^2 = \sigma^2 = \frac{\sum_{ij} (x_{ij} - \mu)^2}{N}.$$

To see the effect of cluster selection on sampling error, we can rearrange Equation 13.7 as follows:

$$\sigma_{\bar{x}}^2 = \frac{n_k\sigma_a^2 + \sigma_w^2}{n} = \frac{n_k\sigma_a^2 + (\sigma^2 - \sigma_a^2)}{n} = \frac{\sigma^2 + \sigma_a^2(n_k - 1)}{n}$$

$$= \frac{\sigma^2}{n} + \frac{\sigma_a^2(n_k - 1)}{n}.$$

Now we can see what happens. The variance of the mean of an unrestricted random sample is σ^2/n. If one elementary unit is drawn per cluster selected, $(n_k - 1)$

*For other situations the sampling error is more complex, and is discussed in specialized texts such as Leslie Kish *Survey Sampling* (New York: Wiley, 1965).

in the expression above is zero; and the cluster sample will have the same error as a random sample. But if $n_k \geq 2$, the cluster sampling method will yield greater sampling error than a random sample of the same size. How much greater will depend on both the number of units drawn per cluster and the size of the variance among clusters. It is advantageous in cluster sampling to have clusters that are as alike as possible in regard to the characteristics being surveyed.

To establish a confidence interval from the data of a cluster sample, we must first estimate σ_w^2 and σ_a^2. These estimates are furnished by

$$s_w^2 = \frac{\sum_k \sum_i (x_{ik} - \bar{x}_k)^2}{\sum_k (n_k - 1)} \left(\frac{N_j - n_k}{N_j - 1} \right) \tag{13.8}$$

$$s_a^2 = \frac{J - k}{J - 1} \left(\frac{\sum_k (\bar{x}_k - \bar{x})^2}{k - 1} - \frac{s_w^2}{n_k} \right) \tag{13.9}$$

In our example $J = 6$, $k = 3$, $N = 60$, and $n = 12$. The three cluster means from Table 13.4 are

$$\bar{x}_1 = \frac{175.6}{4} = 43.900,$$

$$\bar{x}_2 = \frac{207.7}{4} = 51.925,$$

$$\bar{x}_3 = \frac{238.6}{4} = 59.650.$$

The cluster sample mean, our best point estimate of the population mean, is (following Equation 13.6):

$$\bar{x} = \frac{\sum_k \sum_i x_{ik}}{n} = \frac{175.6 + 207.7 + 238.6}{12} = \$51.825.$$

The numerator sum in Equation 13.9 is

$$\begin{aligned} \sum_k (\bar{x}_k - \bar{x})^2 &= (43.900 - 51.825)^2 + (51.925 - 51.825)^2 + (59.650 - 51.825)^2 \\ &= 62.806 + 0.010 + 61.231 = 124.047. \end{aligned}$$

The individual numerator sums in Equation 13.8 can be calculated from

$$\sum_i (x_{ik} - \bar{x}_k)^2 = \sum_i x_{ik}^2 - \frac{\left(\sum_i x_{ik}\right)^2}{n_k}.$$

These are the sums of squared deviations for the sample observations in each cluster. For our example, the results are:

Cluster	$\sum_i x_{ik}^2$	$\left(\sum_i x_{ik}\right)^2 / n_k$	$\sum_i (x_{ik} - \bar{x}_k)^2$
1	7,954.80	7,708.84	245.96
2	10,863.21	10,784.82	78.39
3	14,495.34	14,232.49	262.85
		$\sum_k \sum_i (x_{ik} - \bar{x}_k)^2 =$	587.20

Substituting in Equations 13.8 and 13.9, we obtain

$$s_w^2 = \frac{\sum_k \sum_i (x_{ik} - \bar{x}_k)^2}{\sum_k (n_k - 1)} \left(\frac{N_j - n_k}{N_j - 1}\right) = \frac{587.20}{9}\left(\frac{10 - 4}{10 - 1}\right) = 43.50,$$

$$s_a^2 = \frac{J - k}{J - 1}\left(\frac{\sum_k (\bar{x}_k - \bar{x})^2}{k - 1} - \frac{s_w^2}{n_k}\right) = \frac{6 - 3}{6 - 1}\left(\frac{124.047}{3 - 1} - \frac{43.50}{4}\right) = 30.69.$$

Then we estimate the variance of the sample mean, using the sample equivalent of Equation 13.7, as

$$s_{\bar{x}}^2 = \frac{n_k s_a^2 + s_w^2}{n} = \frac{4(30.69) + 43.50}{12} = 13.86$$

and

$$s_{\bar{x}} = \sqrt{13.86} = \$3.72.$$

The degrees of freedom are $(k - 1)$ for s_a^2 and $\sum_k (n_k - 1)$ for s_w^2. These are added to obtain the degrees of freedom for $s_{\bar{x}}^2$. In our case the degrees of freedom are $2 + 9 = 11$. The t distribution is used, and for the 95 percent confidence interval we have

$$\bar{x} \pm 2.201 s_{\bar{x}},$$

$$\$51.825 \pm 2.201(\$3.72),$$

$$\$51.825 \pm \$8.188.$$

Summary

We have now constructed 95 percent confidence intervals from three different samples of size 12. First we used simple random sampling, then stratified random sampling, and finally cluster sampling.

In simple random sampling, we select the units of observation from the entire universe without restriction. In stratified sampling, we first separate the units of observation into strata and then draw separate samples of predetermined size from each stratum. In cluster sampling, the clusters are identifiable collections of elementary units of observation. First we select a sample of clusters and then draw elementary units from the selected clusters.

The results we obtained reflect the general sampling error characteristics of the three methods:

Simple random sample:	\$50.32 ± 7.40
Stratified random sample:	\$47.63 ± 4.43
Cluster sample:	\$51.825 ± 8.19

Compared with simple random sampling, stratification tends to reduce sampling error to the extent that the basis of stratification is associated with the variables measured in a survey. Clustering tends to increase sampling error, a disadvantage that may be offset by lower costs per unit sampled. In practice, then, designers of surveys look for ways of stratifying survey populations that will produce strata that are as *different* as possible with respect to variables of interest. When we use cluster selection, it is advantageous to create clusters that promise to be as *similar* as possible in regard to important survey variables.

See Self-Correcting Exercises 13A.

Exercise Set A

1. A safety standards agency randomly selected 50 from the 200 firms under its jurisdiction and found the firms had spent an average of \$760 to meet a recently issued safety standard. The standard deviation s of amounts spent by the firms sampled was \$80. Find the standard error of the mean:
   (a) If the sampling was conducted with replacement.
   (b) If the sampling was conducted without replacement.
2. Use the table of random numbers to select an unrestricted random sample (with replacement) of 10 dwelling units from the list in Table 13.1. Make a systematic record of your selection process.
3. Use the table of random numbers to select an unrestricted random sample (without replacement) of 10 dwelling units from the list in Table 13.1. Make a systematic record of your selection process.
4. Did any of the random numbers drawn in Exercise 2 fail to select a dwelling unit? Did any of the random numbers drawn in Exercise 3 fail to select a dwelling unit? Explain the reason in each case.
5. Construct the 95.0 percent confidence interval for mean expenditures per household from the sample data in Exercise 2.
6. Construct the 95.0 percent confidence interval for mean expenditures per household from the sample data in Exercise 3.
7. A large accounting firm employs 100 auditors—50 men and 50 women. A sample of 10 men and 10 women produced the following data on the number of major audits they had supervised:

Men:	4,5,4,3,5,5,6,0,5,6
Women:	5,3,4,2,4,5,3,1,0,5

   Find the 95.0 percent confidence interval for the average number of audits supervised for the entire staff if the data were the result of an unrestricted random sample conducted without replacement.
8. Find the 95.0 percent confidence interval for the average number of audits supervised for the entire staff if the data were the result of a stratified random sample with selection without replacement.
9. A fast-food franchise operates its units with a fixed table of organization calling for a manager, an assistant manager, and four full-time and four part-time employees in each unit. They operate 200 units. They are considering a sample survey of 200 employees to determine attitudes toward the franchise, its products, customers, and management practices. Personal interviews will be conducted by staff from a management consulting firm. A list of employees with their location (unit) and status (position) is available that can be organized in various ways by the computer. Discuss the relative merits of:
   (a) An unrestricted random sample of 200 employees.
   (b) A proportionate stratified random sample of 20 managers, 20 assistant managers, and 80 full-time and 80 part-time employees.
   (c) A cluster sample of 40 units with five employees randomly selected from each unit.
   (d) A cluster sample of 20 units with every employee in each unit included in the sample.

10. Tabulate the page numbers in Chapters 1 through 11 of this text as indicated:

Chapter	1	2	3	4	5	6	7	8	9	10	11
Beginning page	___	___	___	___	___	___	___	___	___	___	___
Ending page	___	___	___	___	___	___	___	___	___	___	___

Use the table of random numbers to select an unrestricted random sample (without replacement) of 20 pages from the text. Make a systematic record of your selection process.

11. Regard Part One (Chapters 1 to 3), Part Two (Chapters 4 to 7), and Part Three (Chapters 8 to 11) as separate strata and select a proportionate (to the nearest integer) stratified sample of 20 pages from the text. Use random selection without replacement and make a systematic record of your selection process.

12. Regard the chapters as clusters and select four clusters with probability proportionate to the size (number of pages) of the clusters. Do this by selecting four random numbers in the interval covered by the total number of pages in the 11 chapters. The chapters containing the page numbers corresponding to the random numbers are the ones selected. Then select five pages from each of the four chapters with the aid of the random number table. Make a systematic record of your selection process. Sample without replacement.

13. Consider the variable "number of occurrences per page of the word *population.*" Discuss the relative merits of the sample designs in Exercises 10, 11, and 12 for estimating the average value of this variable for the entire set of 11 chapters.

14. In a survey to determine the age of the newest automobile maintained by households in a small community, households were grouped into fifty clusters of 16 households each and 4 households randomly chosen for each of five selected clusters. All sampling was without replacement. All households were found to be maintaining at least one automobile, and the ages (to the nearest whole year) were as follows:

		Cluster		
1	2	3	4	5
4	4	2	3	2
3	4	1	4	2
1	5	0	3	1
0	6	4	4	3

(a) Determine $\bar{x}$ and $\sum (x - \bar{x})^2$ for each cluster. The most convenient way to calculate $\sum (x - \bar{x})^2$ is from $\sum (x^2) - (\sum x)^2/n$.

(b) Use Equation 13.8 to estimate the within-cluster variance.

(c) Use Equation 13.9 to estimate the among-cluster variance.

(d) From Equation 13.6 and the sample equivalent of Equation 13.7, estimate the population mean and the variance of the population mean.

(e) Determine the 95.0 percent confidence interval for the population mean.

15. Would the answers to parts (b), (c), and (e) of Exercise 14 be changed much if the finite population multiplier were ignored? Explain.

Replicated Systematic Sampling

A selection procedure that is easier to carry out than using random number tables is to select every *C*th unit in a listing of units of observation. Suppose there are about 40,000 records in a credit file, but they are not numbered serially. It may be easier to draw every hundredth record from the file to obtain a sample of about 400 records than to number serially all the records so that 400 random numbers can be drawn. A random start would be made by selecting a single random number between 00 and 99. If the record selected by this draw were the fifty-seventh, then records 57, 157, 257, . . . , 39,957 would constitute the sample. These records could then be separated out by hand or machine count.

Systematic sampling is not equivalent to random sampling. If there is a periodic order in the file related to the variables under study, a systematic sample may have larger sampling error than a random sample. Imagine a file of military personnel by battalion, company, and platoon in which the order of listing within each battalion was by company with company officers listed first and within companies by platoon with platoon leaders listed first. Various systematic sampling intervals (values of *C*) with certain starting points would draw chiefly platoon personnel from a list ordered this way. Others could draw samples with a greatly disproportionate number of company officers and platoon leaders. This would be a poor way to draw a sample for a survey of attitudes toward military life.

Many lists or natural orderings of units of observation have a built-in stratification rather than periodic ordering. When this is so, a systematic sample takes on the features of a stratified sample and can have a smaller sampling error than a random sample of the same size. The list of military personnel could easily be organized with all battalion-level officers first, then all company officers, all platoon leaders, and all platoon personnel. Then a systematic sample with a random start will produce a nearly proportionate representation of persons in each of the categories.

Error formulas for random sampling are not really appropriate when the sample is drawn systematically. While it is often difficult to conceive of any way in which certain lists might have a periodic ordering, you can never be sure about it. Listings and natural orderings more often than not have unknown elements of stratification built into them.

Replicated Subsamples

The principle of replicate sampling, a procedure often useful in survey designs, is to divide the total sample survey into *K* independent subsurveys. Each survey is carried out as if it were the only survey—that is, independently of the other subsurveys —and each provides an independent estimate of the parameter of interest. The average of these estimates provides the single best estimate of the parameter, and the variance among the estimates is used to provide the measure of sampling errors. Although this method is used in more complex applied designs, it can be understood in the context of replicated random samples. Suppose we were to regard a random sample of $n = 400$ as 10 independent subsamples of $n_k = 40$ each. We could use $\bar{x}_1, \bar{x}_2, \bar{x}_3, \ldots, \bar{x}_k$ in the following way, where K is the number of subsamples, and $k = 1, 2, \ldots, K$,

$$\bar{x} = \frac{\sum_k \bar{x}_k}{K}, \tag{13.10}$$

$$s_{\bar{x}}^2 = \frac{\sum_k (\bar{x}_k - \bar{x})^2}{K - 1}\left(\frac{1}{K}\right). \tag{13.11}$$

Note that the first term on the right-hand side of Equation 13.11 estimates the variance among sample means of size n_k (40 in the example suggested). But the variance of the mean of 10 subsample means is one-tenth of the variance of the subsample means. Therefore the estimated variance among subsample means is divided by the number of subsamples (multiplied by $1/K$) to estimate the variance of the mean of the entire sample.

The only advantage of replicate sampling apparent in the foregoing discussion is some saving in computational effort, which would be substantial in a large multipurpose survey (with many variables of interest), even in the age of high-speed computers. A more fundamental advantage is that the replicate approach permits computation of sampling errors for surveys where no direct conventional sampling error formula is applicable. Such is the case in systematic sampling.

An Example

Replicated systematic subsampling can be illustrated with the Georges Mills example we used for other sampling methods. We will select four subsamples of three observations each. Each subsample will be chosen systematically. Since there are 60 dwelling units, a systematic sample of 3 units is carried out by selecting every twentieth (60/3) unit with a random starting point between 1 and 20. We use the original listing of dwelling units in Table 13.1.

Our previous sequences of random numbers ended with the thirty-seventh number in the second row of Appendix F. We now decide that 01 to 19 will dictate those starting points and 00 will dictate the twentieth dwelling unit as a starting point. So also will 21 to 39 and 40 (by subtracting 20), 41 to 59 and 60 (by subtracting 40), and so on. The sequence of numbers beginning with the thirty-eighth number in the second row is

40 20 08 22 91.

Table 13.5 Replicated Systematic Subsamples

Subsample 1		Subsample 2		Subsample 3		Subsample 4	
Unit	x_1	Unit	x_2	Unit	x_3	Unit	x_4
20	\$31.6	8	\$52.8	2	\$45.2	11	\$36.3
40	48.8	28	43.5	22	26.0	31	67.1
60	62.1	48	72.5	42	52.6	51	57.3
Totals	\$142.5		\$168.8		\$123.8		\$160.7
Means	47.50		56.27		41.27		53.57

The random number 40 dictates that the first subsample starts with dwelling unit 20. The random number 20 would dictate the same starting point, so we skip that number (starting points are being selected without replacement). The numbers 08, 22, and 91 produce the starting points of 8, 2, and 11 for the second, third, and fourth subsamples. Table 13.5 presents the resulting sample data.

For the sample design, $K = 4$ and the sample mean (following Equation 13.10) is

$$\bar{x} = \frac{\sum_k \bar{x}_k}{K} = \frac{47.50 + 56.27 + 41.27 + 53.57}{4} = \$49.65.$$

To estimate the variance of the sample mean, we modify Equation 13.11 to take account of the establishment of starting points without replacement. The number of possible subsamples under this condition is the selection interval, 20 in our example. We drew 4 of the 20 possible subsamples in the design. The finite multiplier is $(C - K)/(C - 1)$, where C is the systematic selection interval. With this multiplier the estimated variance of the sample mean in a replicated systematic subsample design is

$$s_{\bar{x}}^2 = \frac{\sum_k (\bar{x}_k - \bar{x})^2}{K - 1}\left(\frac{1}{K}\right)\left(\frac{C - K}{C - 1}\right). \qquad \textbf{13.12}$$

We can now calculate

$$\begin{aligned}\sum_k (\bar{x}_k - \bar{x})^2 &= (47.50 - 49.65)^2 + (56.27 - 49.65)^2 + (41.27 - 49.65)^2 \\ &\quad + (53.57 - 49.65)^2 \\ &= 4.6225 + 43.8244 + 70.2244 + 15.3664 \\ &= 134.0377.\end{aligned}$$

Substituting in Equation 13.12,

$$s_{\bar{x}}^2 = \frac{134.04}{4 - 1}\left(\frac{1}{4}\right)\left(\frac{20 - 4}{20 - 1}\right) = 9.406,$$

$$s_{\bar{x}} = \sqrt{9.406} = 3.067.$$

There are $(K - 1)$ degrees of freedom for the estimated variance. The 95 percent confidence interval requires the appropriate Student t value for $(K - 1)$ degrees of freedom. In our case the 95 percent confidence interval is

$$\$49.65 \pm 3.182(3.067),$$

$$\$49.65 \pm 9.759.$$

This is a wider interval than any of the intervals for samples of size 12 selected earlier. However, there is nothing inherent about replicated systematic subsample selection that permits a generalization from this fact. A major attraction of the replicated approach is the ease of calculating sampling error.

So far we have considered some of the basic sampling methods for surveys. A survey may combine several of these methods. For example, a survey to estimate average hourly wages of employees in a community might employ clustering by using the firm as a primary sampling unit. Stratification might be introduced by placing the primary sampling units into standard industrial classification groupings.* Within each stratum, the ordering of firms might be made in accordance with size categories and a systematic selection of firms made. Within the firms selected, it would be convenient to include every kth employee from a prepared list of hourly wage employees. The entire procedure could be replicated by carrying out the survey as five replicated subsurveys and confidence limits set from the five subsample means.

Nonsampling Errors

We complete our discussion of sample surveys by considering sources of error not embodied in the various sampling error formulas. Nonsampling errors are those that would remain in a survey even if the survey method were carried out over and over again. They are errors that are not lessened by increasing the size of the sample. Often they are called **persistent errors** or **bias**. Sampling and nonsampling errors can be likened to the two sides of a right triangle. Total survey error, represented by the hypotenuse, is not reduced much by shortening the height (sampling error) of the triangle if the base (nonsampling error) is large. By the same token, if nonsampling errors can be made small, reduction of sampling errors through good design and increased sample size will be effective in reducing total error. Either way, there is good

*Lists of firms by SIC (standard industrial classification) code are frequently available from state employment commissions.

reason to hold nonsampling error to a low level. Some sources of these persistent errors and measures that can be taken to control them are discussed now.

Faulty Sampling Frame It sometimes happens that the correspondence between the sampling frame and the elementary units of observation is shaky. A classic situation of this kind was the *Literary Digest* poll of 1936. Based on a very large sample, this poll predicted an election victory for Alfred Landon over Franklin Roosevelt. The sampling frame was composed of telephone subscribers* and automobile registration lists. Unfortunately for the *Literary Digest* (which went out of business soon after), many voters were not included in this frame—and these voters without automobiles and telephones voted heavily for Roosevelt.

The only solution to faulty sampling frames is care and persistence in developing the frame. It is dangerous to settle for incomplete frames on the assumption that units not in the frame are like units in the frame. If surveys are worth doing at all, they should be done with care.

Incomplete Response The two major sources of incomplete returns in sampling human populations are respondents not located and refusals. Here the frame may be good but incomplete returns can cause bias in the results. Mail-back questionnaires are particularly a problem. A 30 percent response is high even for a well-prepared mail survey that provides some motivation for a completed response. Can we afford to assume that those not responding have characteristics similar to the responding group? The usual solution is to sample the nonrespondents to find out. Follow-up reminders can be sent out and late returns can be compared with early returns to find the differences for the slower group. If there is much at stake, phone calls or personal interviews can be attempted with a sample of nonrespondents. But if the survey is important, it might be better to have conducted personal or telephone interviews from the beginning.

Refusals are much lower in well-conducted personal interviews. Even in general household interviewing, they can be kept well below 10 percent. Respondents who are not at home are another problem, and the solution is a series of call-backs at different hours to bring them into the returns column. Substitution of next-door families for "not-at-homes" is bad practice. In most surveys, the not-at-homes are different from the at-homes since they tend to be families with several wage earners and younger and older families without children at home. Substitution can be worse than doing nothing, because it just compounds the bias of underrepresentation of certain types of families.

Performance of Survey Personnel Every survey should be designed so that gross errors or persistent variation from planned procedures can be corrected in the normal conduct of the survey. This means adequate training of staff followed by

*Many household surveys today use phone interviews based on a sampling frame of nonbusiness telephone subscribers. Since telephone subscription is much more widespread than 40 years ago, the likelihood of introducing bias is not as great as it once was.

routine checks on their work. Falsifying interviews, making unauthorized substitutions, and so forth can be easily detected (and discouraged initially) by adequate field reporting forms and checks on a sample of the work reported. Accuracy of interview work can be checked by repeat interviews by highly skilled personnel with a small sample of respondents. The U.S. Census Bureau checks the quality of the U.S. Census of Population and Housing through just such a procedure called the Post-enumeration Survey.

In respectable opinion polling and marketing research work, routine sample audits of the performance of interviewers are made by checking back with a proportion of the respondents contacted by each interviewer. The coding and tabulating of completed interviews can be checked through quality control procedures. Usually this involves checking periodic small samples of work and increased inspection when quality is suspect. Errors found are of course rectified. Use of control totals in preparing detailed breakdowns, whether by hand or machine, is standard operating procedure. Mechanical card sorters and counters, mark-sense systems, and electronic computers are not error-free.

Another important area is questionnaire design. How to phrase questions so they are unambiguous and meaningful to respondents is an art compounded of some rules of thumb and much experience. Some books on opinion and marketing research are useful in this regard. In the development of all but the simplest questionnaire forms, pilot testing to find and correct the weak spots in the questionnaire should be carried out. Trying the questions out on a small informal sample of potential respondents quickly reveals the difficulties that need attention.

Summary

Systematic sampling is a selection process in which every kth unit in a list of units of observation is selected for the sample. It is often easier to apply than simple random sampling, but it is not the same. The error in a systematic sample may be larger or smaller than for a random sample of the same size. If the order of listing the units corresponds to a useful stratification of the units, systematic sampling may yield errors comparable to stratified sampling.

To calculate the sampling error for a systematic sample, it is necessary to execute the selection with replicated subsamples. In this procedure, several subsamples are independently selected and processed. The means of the several subsamples provide the information needed for calculating the estimate of sampling error.

Control of nonsampling errors is important in survey work. The chief sources of nonsampling error are faulty sampling frames, incomplete response, and inadequate checks on the performance of survey personnel. The questions in a survey should be as simple and unambiguous as possible.

See Self-Correcting Exercises 13B.

Exercise Set B

1. Use the table of random numbers to select five replicated systematic subsamples of 16 pages each from the pages in the first 11 chapters of this text. (It is more important that every page have a chance of inclusion in the sample than that each subsample have exactly 16 pages.) Be sure to document your selection process.

2. Do you think the selection method of Exercise 1 would yield a more or a less accurate estimate of the average number of occurrences per page of the word *population* in the first 11 chapters than an unrestricted random sample (without replacement) of the same size? Explain.

3. A survey with five systematic replicate subsamples of 50 each was conducted from a listing of dwelling units in an area in order to determine the percentage of units in need of major repairs. The percentages for the subsamples were 40, 45, 41, 53, and 51. Calculate the 95.0 percent confidence interval for the population:
   (a) Given that there were 1000 dwelling units in the list.
   (b) Given that there were 10,000 dwelling units in the list.

4. The basic variable (0 = not in need of repair, 1 = in need of repair) in Exercise 3 is certainly not normally distributed. Why is the Student t distribution nevertheless appropriate in setting the confidence interval?

5. Discuss the appropriateness of the following sampling frames for surveying the target populations indicated. Can you suggest better sampling frames in each case?
   (a) Sampling of persons entering park visitor centers to determine lengths of stay and activities of persons using the national parks.
   (b) Sampling of persons with telephone listings to determine voter opinion on an upcoming school bond referendum in a city.
   (c) Sampling a hotel registry to determine the satisfaction of guests with the services of the hotel.

6. Suppose you are designing a questionnaire for a survey of students at your college. The focus of the survey is the on-campus activities experienced and desired by different students. Criticize the following questions, suggesting improvements where possible:
   (a) Are you a full-time student? ____ Yes ____ No
   (b) Do you usually attend the concerts sponsored by the Associated Students? ____ Yes ____ No
   (c) Why don't you attend more of the lecture series?
   (d) Did you attend a varsity basketball game this past season? ____ Yes ____ No
   (e) Don't you feel that the rock concerts could be better organized? ____ Yes ____ No
   (f) What kind of music do you like most? Check one. ____ Country ____ Rock ____ Jazz ____ Pop

7. A department store with 2000 charge account customers wishes to gather information on which competing stores are patronized by these customers. The alternatives being considered are a mailing to all 2000 customers and a phone survey of 500 customers randomly selected. It is anticipated that the mail survey would yield about 1000 returns. The costs are about the same for both plans. Which plan would you favor? Why?

Overview

1. Compare unrestricted random sampling and stratified random sampling in regard to:
   (a) The work required to select the sample.
   (b) The accuracy of estimates from the sample.
2. Compare simple random and cluster sampling in regard to the criteria in Exercise 1.
3. Suppose in a survey to determine the average age of homes that ages differ by neighborhoods in Georges Mills (Figure 13.1). What are the consequences of using the numbered neighborhoods (a) As strata in a stratified sample? (b) As clusters in a cluster sample?
4. In the situation of Exercise 3, compare the merits of using neighborhoods as strata with replicate systematic samples from the list of dwelling units in numerical order.
5. Suppose the numbering of dwelling units was by alphabetical order of the owners of the units rather than as it now appears in Figure 13.1. Would your answer to Exercise 4 be changed? Explain.
6. Suppose, unknown to an investigator, strata means are in fact unrelated to characteristics under investigation in a survey. What good or harm would be involved in carrying out a stratified sample design?
7. Suppose, unknown to an investigator, cluster means are strongly related to characteristics under investigation in a survey. What good or harm would be involved in carrying out a cluster sample design?
8. Sampling errors measure the variability of repeated applications of the same survey design. Why isn't it enough to consider just this type of error in sample surveys?

Glossary of Equations

13.1 $$\sigma_{\bar{x}} = \frac{\sigma}{\sqrt{n}}\sqrt{\frac{N-n}{N-1}}$$

The fundamental formula for the standard error of a sample mean when an unrestricted random sample is drawn without replacement.

13.2 $s_{\bar{x}} = \frac{s}{\sqrt{n}}\sqrt{\frac{N-n}{N-1}}$

The formula for estimating the standard error of a sample mean from the sample data when an unrestricted random sample is drawn without replacement.

13.3 $\bar{x} = \sum_j (w_j \bar{x}_j)$

The point estimate of the population mean from a stratified random sample is a weighted mean of the stratum means, where the weights are the fractions of the units of observation belonging in each stratum, that is, $w_j = N_j/N$.

13.4 $\sigma_{\bar{x}}^2 = \sum_j (w_j^2 \sigma_{\bar{x}_j}^2)$

The fundamental formula for determining the variance of the mean of a stratified random sample from the variance of the individual stratum means.

13.5 $s_{\bar{x}}^2 = \sum_j (w_j^2 s_{\bar{x}_j}^2)$

The formula for estimating the variance of the mean of a stratified random sample from the estimated variances of the individual stratum means.

13.6 $\bar{x} = \frac{\sum_k \sum_i x_{ik}}{n}$

The point estimate of the population mean from a cluster sample of k clusters is the sum of all the sample observations divided by the sample size.

13.7 $\sigma_{\bar{x}}^2 = \frac{n_k \sigma_a^2 + \sigma_w^2}{n}$

The fundamental formula for the variance of the mean of a cluster sample in which the clusters are of equal size and the number of elementary units sampled is the same within each cluster selected.

13.8

$$s_w^2 = \frac{\sum_k \sum_i (x_{ik} - \bar{x}_k)^2}{\sum_k (n_k - 1)}\left(\frac{N_j - n_k}{N_j - 1}\right)$$

13.9

$$s_a^2 = \frac{J-k}{J-1}\left(\frac{\sum_k (\bar{x}_k - \bar{x})^2}{k-1} - \frac{s_w^2}{n_k}\right)$$

Formulas for estimating σ_w^2 and σ_a^2 for use in the previous formula to obtain $s_{\bar{x}}^2$ entirely from the sample data of a cluster sample. The formulas assume selection without replacement at both sampling stages.

13.10 $\bar{x} = \frac{\sum_k \bar{x}_k}{K}$

The formula for the point estimate of a mean based on K equal-sized replicated subsamples.

13.11 $s_{\bar{x}}^2 = \frac{\sum_k (\bar{x}_k - \bar{x})^2}{K-1}\left(\frac{1}{K}\right)$

The formula for estimating the variance of a mean when the mean is based on K equal-sized replicated subsamples, assuming sampling with replacement.

13.12 $s_{\bar{x}}^2 = \dfrac{\sum_k (\bar{x}_k - \bar{x})^2}{K - 1}\left(\dfrac{1}{K}\right)\left(\dfrac{C - K}{C - 1}\right)$	Modification of the preceding formula when replicated subsamples are selected systematically with selection interval C and starting points are randomly chosen without replacement.

14

Nonparametric Statistics

The hypothesis tests discussed up to now usually have been *parametric* in form. In other words, the hypothesis to be tested was a statement about the value of a parameter such as the population mean. The test used sample information to assess the plausibility of the hypothesized value. Here we will consider some *nonparametric* tests—that is, tests for which the hypotheses are not explicit statements about the value of some population parameter.

In our earlier hypothesis tests we usually made an assumption about the distribution of the population. In most cases the population was assumed to be normally distributed. By contrast, none of the tests we are about to discuss require an assumption about the exact distribution of the population. Such tests are said to be **distribution-free**. Although it is not strictly correct, it is customary to refer to a test as nonparametric if it is either distribution-free or nonparametric or both. Among other purposes, some tests of these types are useful with small samples from populations known to be or suspected of being skewed. Hence they can sometimes bridge the gaps in the summary table at the end of Chapter 9 (page 253).

Our working lives offer many opportunities for applying nonparametric tests. Sales and income distributions measured in money units usually exhibit strong positive skewness. Distribution-free tests often are appropriate in such cases. Behavioral studies frequently permit measuring variables only on nominal or ordinal scales. Judgments of product adequacy and personnel performance are cases in point. As compared with parametric tests, which are designed for use with interval and ratio scales, most nonparametric tests require no more than ordinal scaling. Nevertheless, nonparametric tests sometimes can be substituted for parametric tests.

Typically, nonparametric tests are not so powerful as their parametric counterparts, but a few of them are very nearly as powerful. When there is reasonable doubt about the assumptions needed for a parametric test, a nonparametric test that is nearly as powerful if the assumptions apply, and broadly applicable if they do not, can be a most attractive alternative.

A nonparametric technique we have already encountered is the chi-square (χ^2) test for independence discussed in Chapter 10. A typical use of this test is to compare frequencies observed in the cells of a cross-classified table with those we would expect to find in the cells if the variables used to classify the data were independent. In such a test we do not ask about the value of a population parameter. Hence the test qualifies as being nonparametric.

The chi-square test and those to be described in this chapter are only a few of the many nonparametric tests developed in recent years.

Differences in Location for Two Populations

Two tests that are particularly sensitive to differences in location of two populations are the **Wilcoxon rank-sum test** and the Mann-Whitney U test. Since these are equivalent we will only describe the Wilcoxon test. This test is a distribution-free alternative to the two-sample test we discussed in Chapter 12 for testing the hypothesis that two populations have the same mean (see Equations 12.2 and 12.7 and the related discussion). The rank-sum test, however, requires measurement only up to an ordinal (ranking) scale.

To illustrate the rank-sum test, let us suppose that a new training program for salespeople has been in use for the past 2 years in a certain company. Composite performance ratings for the year following training are available for four salespeople trained under the new plan and six trained under the old. In the past, distributions of composite ratings have had marked positive skewness. Because of this and because ratings are subject to serious question with regard to the specifications for interval measurement, the two-sample t test is not applicable.

We want to test the null hypothesis (H_0) that both composite-score population distributions are identical against the upper one-tailed alternative (H_A) that the distribution of ratings under the new program is located in a higher range of composite scores than is the distribution under the old program. The level of significance α will be 0.025. The rating scores for the 10 salespeople, arranged by training program, are shown below:

New program:	82	79	78	65		
Old program:	75	71	62	59	43	37

Scores under the new program appear to be higher than under the old. We want to see whether this apparent shift is large enough to be statistically significant.

The first step in conducting the rank-sum test is to rank all observations as if they belonged to a single collection of data. For either a one-tailed or two-tailed test, ranking is done in the manner that will result in the minimum total for the smaller of the two samples. The ranks are next assigned to the applicable training program. Finally, the sum of the ranks assigned to the smaller sample is found. This sum constitutes the test statistic W.

The steps just described have been carried out for the rating scores in our example and the results are as follows:

Program	Ranks	Sum
New	1 2 3 6	$w = 12$
Old	4 5 7 8 9 10	

The largest three scores in the set of 10 are 82, 79, and 78. These scores are assigned ranks 1, 2, and 3 respectively (rather than 10, 9, and 8) to make the value of w as small as possible. In some cases the direction in which ranking should be done to minimize the value of W is not obvious. Trying both directions will resolve the issue. The ranking continues through all 10 numbers. The rank numbers are next assigned to the applicable training program. Finally, the rank numbers for the smaller sample are summed to produce the value 12 for W, the test statistic.

We next consider how to interpret the value of the test statistic. Under the null hypothesis both programs are presumed to produce identical distributions of rating scores. Given any sample composed of four observations for the new program and six for the old, we can reason that under the null hypothesis any subset of four of the 10 rank numbers has the same probability of being assigned to the new program as does any other subset of four. There are $({}_{19}C_4)$, or 210, possible combinations of rank numbers, any one of which could be assigned to the smaller sample under the designated conditions. Of these, only rank combinations (1,2,3,4), (1,2,3,5), (1,2,3,6), and (1,2,4,5) would produce values of W (rank sums) equal to or less than 12, the observed value. All combinations are assumed to have equal probability of occurrence under the null hypothesis. Consequently, the probability of getting a value of 12 or less for W under the null hypothesis is 4/210, or 0.019. Since 0.019 is less than our preselected level of significance ($\alpha = 0.025$), we reject the null hypothesis and accept the alternative. The new training program appears to be producing higher performance rating scores.

Appendix G gives critical single-tail values (W_α) of the rank sum for the n_1 observations in the smaller sample classified by the number of observations in the larger sample (n_2) and levels of significance (α). For instance, for a one-tailed test such as ours with $n_1 = 4$, $n_2 = 6$, and $\alpha = 0.025$, the critical value of W is $W = 12$. Under the null hypothesis, the probability is 0.025 *or less* that the rank sum will be 12 or less. Actually, from our earlier analysis we know the exact probability is 0.019, but it would be inconvenient to table exact probabilities for all combinations of n_1 and n_2. For representative values of α, the table shows values of W for which the probabilities are as large as possible without exceeding the selected values of α. By doubling the values of α, Appendix G becomes applicable for two-tailed tests. For a two-tailed test in our salesperson example, the table value of 12 for W with n_1 of 4 and n_2 of 6 would be significant at the 0.05 level.

For values of n_1 and n_2 of 12 or greater, the rank sum is very nearly normally distributed with mean

$$\mu_w = \frac{n_1(n_1 + n_2 + 1)}{2} \qquad \text{14.1(a)}$$

and variance

$$\sigma_w^2 = \frac{n_1 n_2(n_1 + n_2 + 1)}{12},$$

or

$$\sigma_w^2 = \frac{n_2 \mu_w}{6}. \qquad \text{14.1(b)}$$

Consequently, the statistic

$$z = \frac{W - \mu_w}{\sigma_w}$$

can be used in the familiar procedure that employs Appendix A to perform one- and two-tailed tests for differences in location between two populations.

Under rather broad variations in assumptions about the shape of the populations, the rank-sum test performs very effectively as compared with alternatives. For instance, when all conditions are met for the two-sample t test mentioned earlier in this chapter, the rank-sum test has an efficiency of 0.95 compared to 1.00 for the t test. In other words, a sample of 106 observations with the Wilcoxon test is as sensitive as a sample of 100 with the t test. Since the rank-sum test does not require assumptions that are often difficult to make, it is more widely applicable than the two-sample t test.

The theory on which the distributions of rank sums is based assumes that no ties can occur. To prevent tied ranks, effort should be made to measure with enough precision before ranking. When ties do occur, however, one of two situations may arise. Suppose that on page 362 the score in the top row had been 79 rather than 78. The two scores of 79 would account for ranks 2 and 3. The total value of 12 for W will remain the same no matter which of the two scores is assigned either rank. In this first situation, the tie causes no special difficulty.

To illustrate the second type of tie, we can replace the score of 78 in the top row with 75. It is now tied with the highest score in the second row, and these two scores share ranks 3 and 4. In this circumstance, we first assume that the tied score in the top row is larger than its counterpart in the second. Rank 3 would fill in the top row, and W would be 11. We then assume that the reverse is true, so that rank 4 falls in the top row and W is 12. If both values of W fall in the rejection region or if both fall in the acceptance region for the null hypothesis, the issue is resolved. If one value of W leads to acceptance and the other to rejection, we must call the result inconclusive and repeat the experiment or reconsider our judgment about the level of significance.

The Sign Test for Matched Pairs

The **sign test** is valuable when two treatments are to be compared under a variety of conditions. For example, two grievance procedures for dealing with several standard complaints in a work group are to be rated, or two types of lawn seed are to be compared with respect to the qualities of the lawn they produce under a variety of growing conditions. When independent pairs of observations can be made, when each pair is matched with respect to relevant conditions other than the treatment, and when pairs are observed under different conditions, the sign test can be highly effective. Observing pairs under widely varying conditions can invalidate the t test for use with matched pairs (as discussed in Chapter 12). The normality assumption about the distribution of differences may not be justified.

To illustrate the sign test, we will compare two different types of house paint for wearing quality. Adjacent patches of the two types are painted on a number of houses in different parts of the country. Thus the patches of each pair are matched as to color, climate, chemical pollutants in the environment, and many other conditions that could affect wear. The differences in environmental conditions among pairs are great. The months of satisfactory service for each patch are recorded in Table 14.1. In addition to the service lives, the table shows the sign of the differences when the service life of paint B is subtracted from that of paint A.

As in the rank-sum test, the sign test assumes that ties cannot occur within any pair. Should a tie occur, however, the difference for the pair will be zero and no sign can be attached. Exclude such ties from the analysis and reduce the number of paired observations by the number excluded.

The null hypothesis for this example is that the median service lives of the two types of paint are equal under any one of the 16 sets of conditions. Under this hypothesis the probability of observing a negative difference in service lives for any pair is equal to the probability of observing a positive difference and both are equal to 0.5. We can regard the sample as 16 independent draws from a binary population of plus and minus signs, wherein a plus sign (or a minus sign) has a constant probability of occurrence equal to 0.5 for every draw. These are the specifications for a binomial sampling distribution, discussed in Chapter 6 and again in Chapter 10. Let the number of pairs be the number of observations n and let r be the number of plus signs in the

Table 14.1 Service Months for Two Paint Types under Matched Conditions

Pair	A	B	Sign of Difference	Pair	A	B	Sign of Difference
1	19	22	−	9	20	21	−
2	27	26	+	10	38	41	−
3	13	14	−	11	21	23	−
4	8	18	−	12	19	20	−
5	28	30	−	13	8	10	−
6	7	6	+	14	12	16	−
7	27	28	−	15	21	26	−
8	18	19	−	16	13	8	+

sample. The parameter π of the binary population will be 0.5 under the null hypothesis, and we can use Appendix E to find relevant probabilities for 16 observations or fewer.

Suppose that the paint manufacturer in our example wants to continue making paint A unless there is conclusive evidence that paint B is superior. A lower one-tailed test is called for, and an appropriate hypothesis statement is

$$H_0: \quad \pi \geq 0.5,$$

$$H_A: \quad \pi < 0.5,$$

where π is the probability that a plus sign occurs. The selected level of significance α is 0.025.

In the experiment composed of 16 observations, three plus signs occurred. Consulting Appendix E with n and π values of 16 and 0.5, respectively, we find that the probabilities for r values of three or fewer are 0.0085, 0.0018, 0.0002, and 0.0000. The sum of these probabilities is 0.0105. Hence the probability of observing three or fewer plus signs under the null hypothesis is only 0.0105. This is less than the preselected value of α (0.025). Consequently, we accept H_A and conclude that paint B has demonstrated superior service life characteristics.

For values of n larger than 16, the normal approximation to the binomial can be used. This approximation and its application were discussed in Chapter 10.

Sample size is important in the sign test. When the number of observations is very small, there may be no way to reject the null hypothesis for the designated level of significance. If n is 3, for example, the probabilities for possible values of r are all 0.125 or greater. Also, in comparison to the t test the information regarding the signs of differences is used but the information regarding the amounts of differences is ignored. This results in a loss of power that must be compensated with additional observations. Modifications and extensions of the sign test exist. Pairs need not be matched to compare two treatments, for instance.

Summary

Many nonparametric tests have been developed in recent years. The Wilcoxon rank-sum test applies to two sets of data measured on a common ordinal scale. It tests the hypothesis that both sets of data are from the same population and is highly sensitive to differences in central location of the populations. This test is distribution-free and nearly as sensitive as the t test when the latter applies. The Wilcoxon test applies in many more situations than does the t test.

The sign test for matched pairs is a more broadly applicable, but less sensitive, alternative to the t test for matched pairs. The form of the sign test discussed here tests null hypotheses about medians of the populations from which each pair came.

See Self-Correcting Exercises 14A.

Exercise Set A

1. Scores are compared on a reading comprehension test for two groups of randomly selected second-graders, one group from a traditional classroom and the other from an open classroom. Scores were as follows:

Traditional:	68	74	72	75	84	76
Open:	87	71	69	73	70	

   (a) What is n_1? What is n_2?

   (b) Rank numbers for the Wilcoxon rank-sum test should begin with what value? And end with what value?

   (c) What rank should be assigned to the score of 87? Why?

   (d) What is the rank sum for the data from the smaller sample?

   (e) What are the one-tail critical values for these sample sizes for levels of significance of 0.05, 0.025, and 0.01?

   (f) What question might be asked about these data that would require a one-tailed test?

2. Two recipes for beef stew are served in the college cafeteria, one on each of two different nights. On each night, every twentieth student who orders the dish also receives a card on which ratings from 1 ("very poor") to 8 ("excellent") are to be made in several categories concerning the quality of the stew. These ratings are totaled into a composite score to determine which dish, if either, meets with greater student acceptance. Recipe A is rated

by 10 students and recipe B by 9. The ratings are:

Recipe A:	40	31	35	38	36	36	42	44	41	43
Recipe B:	28	46	32	42	38	39	37	45	44	

(a) Use the Wilcoxon rank-sum test to determine which stew, if either, the food service can conclude will meet with greater student satisfaction at a level of significance of 0.05.

(b) Does changing the level of significance to 0.01 in part (a) result in a different conclusion? If so, what is it? If not, why not?

3. An automobile manufacturer plans to introduce incentives to worker productivity. Two plans are suggested, and both are put into effect on a temporary basis in the plant. Plan W consists of payment of 10 percent higher than normal rates for each job completed beyond quota. Plan S carries the same reward, but instead of being paid directly in wages, the extra payment is invested in company stock at slightly reduced prices. To determine which plan should be adopted throughout the plant as the greater stimulus to increased production, the outputs of two workers, one working under plan W and the other under plan S, are recorded from each of 16 of the plant's shops. (Quotas for the shops vary widely, since the job performed in a shop may be anywhere from quite simple to considerably involved.) The following "beyond-quota" results were obtained:

Pair	Plan W	Plan S	Pair	Plan W	Plan S
1	8	6	9	41	34
2	4	0	10	17	22
3	7	9	11	6	4
4	16	22	12	14	12
5	12	7	13	10	10
6	9	8	14	43	36
7	37	34	15	14	16
8	11	9	16	32	31

Suppose that management would rather adopt plan S for what it considers the psychological benefits of "worker ownership," unless the wage incentive plan is really the greater stimulus to production.

(a) State explicitly the null and alternative hypotheses here.

(b) To each pair assign the sign of the difference obtained when the plan W result is subtracted from the plan S value.

(c) What sign is assigned to the pair numbered 13? How does this tie affect the analysis under the matched-pairs test?

(d) What is the sample size you will deal with?

(e) For what value of π are you interested in the binomial distribution?

(f) Determine at a level of significance of 0.05 whether management should put plan W into effect.

4. Suppose pairs of workers, one signed up under plan S and the other under plan W, are selected from each of 40 of the company's shops. When plan W results are subtracted from those of plan S, the differences are as follows: 9 positive differences; 27 negative differences; 4 zeros.

(a) Why is Appendix E of no use here?

(b) What method of attack may be used as an alternative to the binomial distribution?

(c) If the management's attitude and the level of significance are as stated in Exercise 3, determine whether or not plan W should be adopted.

5. How do you determine whether to assign signs of differences based on subtracting plan W results from those of plan S or vice versa?

6. In a matched-pairs sign test consisting of five pairs of values, why would it be impossible to reject a null hypothesis at a 0.01 level of significance? At a 0.02 level of significance? What values for r would permit rejection of H_0 in a lower one-tail test at a level of significance of 0.05? What values for r would permit rejection of H_0 in an upper one-tail test at a level of significance of 0.05?

7. In the "before and after" test of the reducing regimen described in Exercise 6 of Set A in Chapter 12, test the null hypothesis that adult males weigh at least as much after the regimen as before. Use a matched-pairs sign test and a 0.05 level of significance.

8. Discuss the reasons why the matched-pairs sign test in Exercise 7 resulted in a conclusion different from that obtained after performing the test in Exercise 6 of Set A in Chapter 12.

9. Forty automobiles were entered in a matched-pairs signed-rank test of mileages from two brands of gasoline. For each automobile, the mileage with brand A was compared with the mileage when brand B was used. At the 0.01 level, make a two-tailed test of the null hypothesis that there is no difference in mileages if brand B got better mileage in 32 of the automobiles.

10. For Exercise 4 of the Chapter 12 Overview, does the matched-pairs sign test support the conclusion that productivity improved? Use a one-tailed test and let α be 0.05.

Differences in Location for Many Populations

The **Kruskal-Wallis test**, a direct extension of the Wilcoxon rank-sum test for two populations, is sensitive to differences in location for two or more populations. Hence it is a distribution-free counterpart of one-way analysis of variance as described in Chapter 12. If the assumptions applicable to one-way analysis of variance are met, the Kruskal-Wallis test has an efficiency of 0.955 relative to the F test. For marked departures from these assumptions, the Kruskal-Wallis test is to be preferred to the F test. Consequently, it provides a broadly applicable and highly effective test for differences in central locations for treatment levels.

When applied to just two populations, the Kruskal-Wallis test is identical to the Wilcoxon rank-sum test. Therefore, to illustrate it we will consider a situation with three populations. A firm that grows rosebushes on a mass production basis for sale to nurseries wants to determine which of three climates produces the best bushes. Nineteen bushes of comparable ages and varieties are used in the experiment. Six bushes come from region A, five from region B, and eight from C. A panel of qualified judges scores each bush on a composite of characteristics. Then, without regard to the region from which they came, the 19 bushes are ranked on the basis of the judges'

Table 14.2 Composite-Score Ranks of 19 Bushes

	Growing Region		
	A	B	C
	2	10	1
	3	11	5
	4	15	8
	6	16	9
	7	19	13
	12		14
			17
			18
$T_j =$	34	71	85
$T_j^2 =$	1156	5041	7225
$n_j =$	6	5	8
$\frac{T_j^2}{n_j} =$	192.7	1008.2	903.1

scores, the smallest score, 1, being assigned to the poorest bush. The results appear in Table 14.2. Below the rankings are the quantities needed to calculate the value of the test statistic, H. The first row shows the sum of the ranks for each region (T_j); the second row contains the squares of the numbers in the first row (T_j^2); the third row is the number of observations in each treatment (n_j). The total number of observations in the experiment (n) is, of course, the sum of the values in the third row; the final row is the square of the sums in each column divided by the number of observations in that column (T_j^2/n_j). The test statistic H is defined as follows:

$$■ \quad H = \frac{12}{n(n+1)}\left(\sum \frac{T_j^2}{n_j}\right) - 3(n+1). \quad ■ \qquad \textbf{14.2}$$

In this expression, recall that n is the total number of observations in the experiment. For our example,

$$H = \frac{12}{19(20)}(192.7 + 1008.2 + 903.1) - 3(20)$$
$$= 6.44.$$

Before considering a more general argument, we can examine two examples that indicate what to expect. Suppose region A produces very inferior bushes, B is in the middle, and C produces superior bushes. Then 4 bushes from each region might be ranked as follows.

A	B	C
1	5	9
2	6	10
3	7	11
4	8	12

For this pattern, H is 9.85. Next, suppose all three regions are identical. Then the rankings would be mixed something like the following.

A	B	C
1	2	3
6	5	4
9	8	7
10	11	12

For this pattern, H is zero. Hence we can expect large values of H to indicate significant differences in the central location of distributions and small values to indicate no significant differences.

Sound procedure requires that we state the null hypothesis before we look at this or any set of observations. Then, given the hypothesis statement, we must consider how the test statistic H will be distributed under the null hypothesis. Our null hypothesis will be that the populations of scores for each of the three growing regions have identical distributions. They are then, in effect, all the same population. Now imagine that we repeat our experiment a very great number of times when the null hypothesis is true. We repeatedly select 19 ranked scores at random from the population and always assign six of the rank numbers at random to the first column, five to the second, and the remaining eight to the third.

Consider any one column. Each of the n_j rank numbers in that column can be viewed as a random selection from a rectangular population of the numbers 1 through 19, sampled without replacement between draws. The mean μ of a rectangular population composed of consecutive integers 1 through n is $(n + 1)/2$ and the variance σ^2 is $(n - 1)(n + 1)/12$.

Now consider the sum (T_j) of the n_j rank numbers in a given column. The mean (μ_T) of the distribution of this sum will be the product $n_j\mu$, where μ is $(n + 1)/2$. The variance (σ_T^2) will be $n_j\sigma^2\ (n - n_j)/(n - 1)$, where σ^2 is $(n - 1)(n + 1)/12$ and $(n - n_j)/(n - 1)$ is the correction for sampling without replacement. By the central limit theorem, as n_j grows large the distribution of the variable T_j approaches normality. Furthermore, the distribution of

$$z = \frac{T_j - \mu_T}{\sigma_T}$$

approaches the standard normal distribution. Suppose we square z for each of the k columns in the experiment. Then it can be shown that the sum of the squared z values is distributed very nearly as chi square (χ^2) with $(k - 1)$ degrees of freedom—provided there are at least five observations in each of three treatments (columns). The statistic H in Equation 14.2 is an algebraic transformation of the sum just described. This transformation is a convenient form of the test statistic for calculation purposes.

What happens if the null hypothesis is not true? Specifically, consider the case in which one of the treatment means is very much smaller than the other means. The other treatments will, on the average, contain more large rank numbers than they would under the null hypothesis, and their values of T_j also will be large. Conversely, T_j will tend to be unusually small for the one treatment we mentioned. When we square the values of T_j, those that are unusually large will produce squares that more than compensate for the square of the unusually small sum. Consequently, the value of H will tend to be much larger than it would be under the null hypothesis. The clue to a departure from the null hypothesis is an unusually large value of H.

Returning to our example, we have a sample value of H equal to 6.44. There are five or more observations in each of the minimum of three treatments. We can, therefore, employ the chi-square approximation. Since there are three treatment levels, there will be 2 degrees of freedom. If we select the 0.05 level of significance, Appendix C shows that 5.99 is the critical value. Because our sample value of H exceeds the critical value, we reject the null hypothesis of equal growing effectiveness for the three regions.

When we examine the mean squared sums (T_j^2/n_j) in the last row of Table 14.2, the quantity for region A is very small relative to the other two regions, which seem to be about on a par with one another. A preponderance of low ranks in column A stems from low scores and poor rosebushes. Hence it appears that region A produces considerably poorer bushes than do the other two regions.

In some situations there will be fewer than five observations for at least one of the minimal three treatments needed for the chi-square approximation. Tables based on exact distributions of the H statistic for small samples can be used in such situations.

This test, like the Wilcoxon rank-sum test, assumes that the underlying variable is continuous, so that ties cannot occur. In practice, of course, ties do occur. When they do, they should be handled in the same manner described for the Wilcoxon test earlier in this chapter.

Rank-Difference Correlation

In Chapter 11 we discussed the Pearson correlation coefficients ρ and r for a population and for a sample. This was seen to be an effective measure of association between pairs of values for two variables measured on interval or ratio scales. When the statistical population of paired values has a bivariate normal distribution, the sampling distribution of r is known and inferences about population values of ρ are possible.

When observations are judgments, attitudes, preferences, or aspects of behavior, only ordinal scales may apply. In other instances, we may not be able to support the assumption of bivariate normality. The **Spearman rank-difference correlation statistic** (r_s) often is an effective alternative in these circumstances.

Let us consider an industrial training program in which employees are selected and trained at company expense. In an effort to improve the program, a test designed to measure motivation was given to 12 trainees before they entered the program. At the end of the program, the comprehensive course scores were collected for these same people. If there is a strong relationship between these two measures–motivation and scores–the test may be helpful in selecting trainees. The null hypothesis is that there is no correlation between motivation upon entering the course and performance upon leaving it. The distribution of scores for several hundred trainees exhibits rather marked negative skewness. Hence rank-difference correlation is chosen to test for a relationship between motivation and performance.

After ranking each set of scores for the 12 trainees, we have recorded the results in Table 14.3. The fourth column shows the difference between the ranks of the pairs. The fifth column shows the squared differences and the sum of the squared differences ($\sum d^2$). We can find the value of the rank-difference correlation coefficient r_s from $\sum d^2$ and n, the number of paired observations. In Table 14.3, n is 12. The computation equation for the Spearman rank-difference correlation coefficient is

$$\blacksquare \quad r_s = 1 - \frac{6\sum d^2}{n(n^2 - 1)}. \quad \blacksquare \qquad \textit{14.3}$$

For our example,

$$r_s = 1 - \frac{6(82)}{12(143)}, \quad \text{or} \quad +0.713.$$

As with the Pearson correlation coefficient, the Spearman correlation coefficient can range from -1 through zero to $+1$. A value of zero indicates no relationship, and sample values of either $+1$ or -1 suggest a strong relationship in the population. In our example, low numbers were assigned to trainees with poor motivation and poor performance. Hence $+0.713$ suggests that poor motivation at the outset of the program is followed by poor performance and vice versa.

The null hypothesis states that there is no association between motivation and performance in the population. Then for a given sample set of n rank scores for motivation there are $n!$ equally likely arrangements of n rank scores for performance. Each arrangement will produce a value for the rank-difference correlation coefficient r_s. The number of such arrangements that produce a given value of r_s can be divided by $n!$ to obtain the probability for that value under the null hypothesis. When this

Table 14.3 Ranks of Trainees on Two Tests

Trainee	Rank: Motivation	Rank: Performance	Difference (d)	d^2
1	3	7	−4	16
2	7	2	5	25
3	1	4	−3	9
4	11	9	2	4
5	2	1	1	1
6	9	12	−3	9
7	6	5	1	1
8	4	6	−2	4
9	5	3	2	4
10	10	8	2	4
11	12	11	1	1
12	8	10	−2	4
			$\Sigma d^2 =$	82

is done for all possible values of r_s, the result is the sampling distribution of the test statistic r_s when there is no correlation in the population. The distribution of r_s ranges from -1 to $+1$ and is symmetrical for any sample size. As sample size increases, values of r_s tend to become more concentrated around zero and the probabilities of values in the vicinity of $+1$ and -1 become very small.

For small numbers of paired observations, special tables are available. When there are 10 or more pairs, however, and when we want to test the hypothesis of no correlation in the population, the statistic

$$t = r_s \sqrt{\frac{n-2}{1-r_s^2}} \qquad \textbf{14.4}$$

can be assumed to come from a t distribution with $(n - 2)$ degrees of freedom.

For our example, we obtained a sample rank-difference correlation coefficient r_s of $+0.713$ from 12 pairs of ranks. We want to test the null hypothesis that the population rank-difference correlation coefficient ρ_s is zero against the two-tailed alternative that ρ_s is not zero. The level of significance α will be 0.02. In Appendix B we read that $t_{0.01}$ is -2.764 and that $t_{0.99}$ is $+2.764$ for 10 degrees of freedom $(n - 2)$. These are the critical values for the test statistic. From Equation 14.4 we find that

$$t = r_s \sqrt{\frac{n-2}{1-r_s^2}}$$

$$= 0.713 \sqrt{\frac{10}{1-0.713^2}}, \quad \text{or} \quad +3.22.$$

We can reject the null hypothesis and accept this result as evidence of association between motivation and performance in the population.

When the bivariate normality assumptions described in Chapter 11 are satisfied, the efficiency of the Spearman coefficient applied to ranked data is 0.91 relative to the Pearson coefficient applied to the data as measured on the original interval or ratio scale.

As with the other ranking techniques in this chapter, the assumptions underlying rank-difference correlation exclude ties in ranks. Since ties do occur in practice, however, the procedure here is the same as that discussed in conjunction with the Wilcoxon rank-sum test.

Summary

The Kruskal-Wallis test applies to two or more collections of data measured on a common ordinal scale. It tests the hypothesis that all the collections come from populations with the same central location and can substitute for one-way analysis of variance when the necessary conditions for the latter do not hold. Even when the conditions hold, the Kruskal-Wallis test is almost as sensitive as the F test. When there are only two collections of data, the Kruskal-Wallis test is equivalent to the Wilcoxon rank-sum test discussed earlier in this chapter.

Rank-difference correlation can be used to test the hypothesis that a sample of paired rank numbers came from a bivariate population in which there is no correlation. The test is more broadly applicable than the one based on the Pearson coefficient and is nearly as sensitive as the latter when the necessary conditions are met.

It should be reemphasized that this chapter considers only four from a very large number of nonparametric tests.

See Self-Correcting Exercises 14B.

Exercise Set B

Exercises 1 through 5 and Exercise 7 refer to the following: Most members of the surgical nursing staff at a small suburban hospital are graduates of four nursing schools. In an effort to determine whether hiring preferences should be given to graduates of one or more of the nursing schools on the basis of adequacy of preparation in operating room procedures, 12 surgeons affiliated with the hospital are asked to rate each newly hired recent nursing graduate after 6 weeks on the job in various areas of competency. Over a year's time, such ratings are obtained for 24 graduates of the four schools. The following composite scores are obtained, with higher ratings given to those nurses considered more capable:

School *A*	School *B*	School *C*	School *D*
41	17	87	84
54	36	84	24
56	52	28	74
62	58	19	63
60	71	46	31
20		78	18
			76

1. Rank these nurses on the basis of these scores.
2. Is the chi-square approximation of the H statistic applicable here? Justify your answer.
3. (a) How many degrees of freedom are present?
   (b) For your result in part (a), what is the critical value of chi square at the 0.025 level of significance?
4. Use the Kruskal-Wallis test to determine whether these schools should be considered at a level of significance of 0.025 to be equal in adequacy of preparation in surgical procedures. If preference to hiring graduates of one or more schools is indicated, to what schools should preference be given on the basis of these results?
5. If the last score for graduates of school D were 78 instead of 76, how would this affect the applicability of the Kruskal-Wallis test? Would the outcome of the test be affected? If so, how? If not, why not?
6. In a Kruskal-Wallis test with four levels of treatments, what is the minimum sample size from each treatment level to permit use of the chi-square approximation of the H distribution? Why might $n = 29$ not be adequate to ensure the applicability of the chi-square approximation?
7. Why would the chi-square approximation to the H distribution not be applicable if the last score in the second column were omitted, that is, if ratings had only been made of four graduates of school B?
8. Suppose that in the motivation-achievement testing program discussed in the text we had found that $r_s = -0.82$ rather than 0.713. What sort of relationship, if any, would this have indicated about the population? Would such a result be considered plausible by the training program coordinators? What would a value for r_s of 0.05 indicate about a relationship between motivation and achievement in the training program?
9. The personnel department at Amalgamated United Corporation, in an effort to formulate criteria for hiring electrical engineers, reviews the company's files on 16 electrical engineering graduates of Siwash University it has hired in the past 5 years. Each such employee is ranked according to IQ and job competency ratings. The rankings are given below:

Employee	Rank IQ	Rank Competency	Employee	Rank IQ	Rank Competency
1	15	5	9	16	7
2	13	4	10	10	9
3	7	2	11	11	8
4	3	1	12	5	15
5	8	3	13	6	13
6	9	11	14	4	16
7	1	12	15	12	10
8	2	14	16	14	6

(a) What is the number of pairs of observations?

(b) In determining r_s, does it matter whether you find d by subtracting competency ranks from IQ ranks, or may you equally well subtract IQ ranks from competency ranks?

(c) Determine r_s.

(d) Use your result from part (c) and Equation 14.4 to test at a level of significance of 0.02 the null hypothesis that there is no correlation in the population. What is the number of degrees of freedom? What is the appropriate value of t for $1 - \alpha = 0.98$. What conclusion do you draw from performing this test?

(e) How might the personnel department interpret the results of performing this test? Should these results influence their hiring practices? If so, in what way?

10. Suppose Amalgamated's personnel department wants to gain more information in formulating hiring standards by studying the grade point averages at graduation of the 16 engineers referred to in Exercise 9. Suppose the averages were as follows:

Employee	Grade Point Average	Competency Ranking	Employee	Grade Point Average	Competency Ranking
1	2.84	5	9	2.92	7
2	2.80	4	10	3.00	9
3	3.62	2	11	2.98	8
4	3.65	1	12	3.22	15
5	2.86	3	13	2.42	13
6	2.65	11	14	2.38	16
7	3.92	12	15	3.10	10
8	2.58	14	16	3.14	6

(a) What must be done before the value of d for each pair of values can be determined?

(b) What value of d is obtained for employee 3? For employee 16?

(c) Determine r_s. If any significant correlation should be found to exist, would it be positive or negative?

(d) Test, at a level of significance of 1 percent, the null hypothesis of no correlation. State your conclusion.

(e) How, if at all, should the personnel office implement the results of this test for correlation?

Overview

1. Give five examples of nonparametric tests referred to in this chapter.
2. Give two examples of parametric tests previously studied.
3. How do "distribution-free" tests differ from parametric tests? From nonparametric tests?
4. Where both are applicable, parametric tests are generally preferred to nonparametric ones. Why? Under what conditions must we settle for the results of nonparametric tests?
5. The impact strengths of two materials being considered for milk cartons are given below. At the 0.01 level, test the hypothesis that B is at least as strong as A against the alternative that A is stronger. Use the rank-sum test.

Material A:	0.95	0.98	0.99	0.96	0.90	0.89	0.92	0.94
Material B:	0.93	0.88	0.91	0.83	0.87	0.86	0.82	0.79

6. Measurements of the abrasive materials per tube contained in two brands of toothpaste yielded the following results (in milligrams):

Brand A:	26	23	25	24	28
Brand B:	19	22	24	21	20

   At the 0.05 level, test the null hypothesis that brand B has at least as much abrasive material per tube as does brand A. Use the rank-sum test.
7. The clerical force in a large office was divided into two groups to test the effectiveness of two new makes of typewriter. Fifteen typists used typewriter brand A and 18 used brand B. When the productivities of all 33 machines were ranked, with the highest productivity being given rank 1, the sum of the ranks for brand A was 171. At the 0.01 level, make a two-tailed test of the hypothesis that there is no difference in productivity for the two brands.
8. Transparent sacks, paper bags, and bulk display were the three methods used to market oranges in 20 supermarkets, all belonging to the same chain. The same grade of oranges was sold at the same price in all markets during the test period. Sales in pounds per day for each type of display were:

Transparent Sacks	Paper Bags	Bulk Display
36	29	16
40	30	26
47	45	17
49	39	22
23	13	28
41	32	
38	18	
	31	

Using the Kruskal-Wallis procedure, test the null hypothesis that all three displays result in the same sales per day. Use 0.05 as the level of significance. What display would you select as a result of this test?

9. Suppose that the first observation in the left column in Exercise 8 is 39 instead of 36. What change in procedure is indicated to conduct the same test? Discuss fully.

10. A panel of housewives and a panel of college students rated 12 television programs. The mean ratings for the program were:

Program	Housewives' Rating	Students' Rating
1	61	86
2	55	70
3	83	58
4	42	87
5	31	52
6	64	48
7	69	78
8	78	92
9	92	67
10	94	82
11	70	75
12	28	59

Calculate the Spearman rank-difference correlation coefficient as a measure of the consistency of the two ranking efforts.

11. Test the null hypothesis that there is no consistency in the ratings by the two groups in Exercise 10 at a level of significance of 0.05.

12. A group of party workers and a group of independent voters were asked to rank four candidates for office. The two sets of ranks are:

	Candidate			
	A	*B*	*C*	*D*
Party workers	2	1	4	3
Independents	1	3	4	2

Calculate the Spearman correlation coefficient.

Glossary of Equations

14.1(a) $$\mu_w = \frac{n_1(n_1 + n_2 + 1)}{2}$$

14.1(b) $$\sigma_w^2 = \frac{n_2\mu_w}{6}$$

The mean μ_w of the Wilcoxon rank-sum statistic for two identical populations is a function of the number of observations in the smaller (n_1) and larger (n_2) samples. The variance σ_w^2 can be found from the mean. The distribution is essentially normally distributed when n_1 and n_2 are at least 12.

14.2 $$H = \frac{12}{n(n + 1)}\left(\sum \frac{T_j^2}{n_j}\right) - 3(n + 1)$$

The test statistic H for the Kruskal-Wallis test for differences in location of three or more populations is a function of the total number of observations in the experiment (n), the sum of the ranks (T_j) in each treatment, and the numbers of observations in each treatment (n_j).

14.3 $$r_s = 1 - \frac{6\sum d^2}{n(n^2 - 1)}$$

The Spearman rank-difference correlation coefficient r_s is found from the sum of the squared rank differences (Σd^2) for the paired observations and the number of pairs (n).

14.4 $$t = r_s\sqrt{\frac{n - 2}{1 - r_s^2}}$$

When there is no association between the two variables in the population, the sample rank-difference correlation coefficient r_s is distributed approximately as t with $(n - 2)$ degrees of freedom for 10 or more paired observations.

15

Index Numbers

The importance of index numbers in describing changes in the nation's economy is highlighted in the following description of the beginning of the 1973–1975 recession: "By the end of 1973, just as prices entered a period of almost unbridled growth, the industrial expansion ground to a halt. The index of industrial production for all industry had hit its peak in November 1973. Then it declined and ran virtually flat through all of 1974."*

The passage refers to the behavior of prices and industrial production in very general terms. Of course, these changes are made up of price and production changes in scores of industries and thousands of products. The author based his description on **index numbers** of prices and production. This chapter outlines the construction and uses of price and quantity indexes. Moreover, several of the most widely used indexes in the United States are described.

Price Indexes

Figure 15.1 shows the recent behavior of the Consumer Price Index (CPI) of the Bureau of Labor Statistics. Here the rise since 1972 in the index for all items can be seen, and the even more rapid rise for food items is obvious. The index for food items stood at 159.5 in the second quarter of 1974 compared with 123.5 in 1972. But what do these indexes mean? In this section we introduce, by way of an example, the fundamental idea behind price index construction. This is followed by a brief indication of how this idea is carried out in the Consumer Price Index.

Table 15.1, derived from reports of the Bureau of Labor Statistics, shows the average retail prices in the United States for six selected food items in 1967, May 1973, and May 1974. We can see that all the prices have risen since 1967. From 1973 to 1974 the average prices for bacon and eggs decline somewhat, while prices of the other foods increase. But how can we summarize the changing prices of the six food items as a group? What we seek is an index of the behavior of prices for these six items as a group. A meaningful index is provided by considering the quantities of the food items

**World Almanac and Book of Facts—1975,* Newspaper Enterprise Association, Inc., November 1975, p. 83.

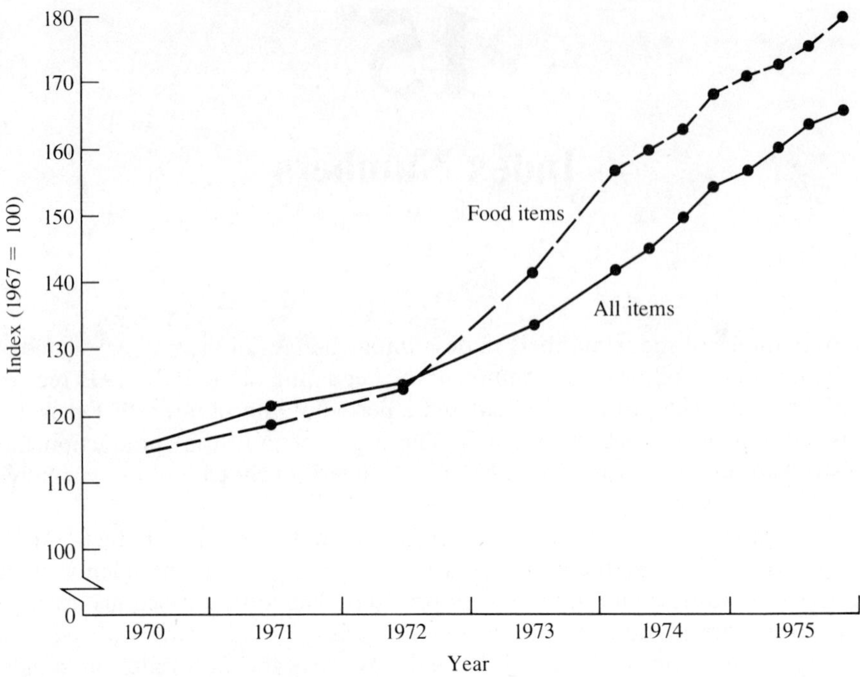

Figure 15.1 *The Consumer Price Index—all items and food items, 1970–1975*

that are purchased or consumed by a relevant economic unit or collection of units. If a family does not use coffee, for example, the price of coffee is of no direct personal concern to them.

Suppose a family consumed only these six foods for breakfast and determined that they used 2 pounds of bread, ½ pound of butter, 1 pound of bacon, 2 dozen eggs, ½ pound of coffee, and 4 quarts of milk weekly for breakfast. A useful index could then be constructed of the price (cost) of providing this particular *market basket* of goods. We need only multiply the unit prices in each period by the quantities of concern to us. The result of summing the price × quantity extensions, shown in Table 15.2, is the dollar expenditure needed to purchase the market basket in the different periods. The weekly market basket could be purchased in 1967 for about $4.21. In

Table 15.1 *Average Prices (in cents) of Selected Food Items in the United States*

Item	1967	May 1973	May 1974
White bread (lb)	22.4	27.0	35.3
Butter (lb)	83.0	85.5	94.3
Bacon (lb)	83.7	121.5	120.2
Eggs (dozen)	49.1	68.1	65.4
Coffee (lb)	76.9	102.6	120.4
Milk (qt)	28.7	35.0	44.9

May 1973 the same market basket cost about \$5.46, and in May 1974 the cost was up to nearly \$6.09.

Definition A fixed-weight **price index** measures the relative price of a fixed bill (or market basket) of goods. At any point in time the index compares the total price of the bill of goods to the total price of the same bill of goods in the base year:

$$\blacksquare \quad PI_n = \frac{\sum(p_n q)}{\sum(p_0 q)} \times 100. \quad \blacksquare \qquad \textit{15.1}$$

In this equation, the summation is understood to be over the several commodities; q refers to the quantity weights for each item; p refers to prices, with p_n indicating prices for any given period and p_0 referring to prices in the period used as the **base** for the index.

The aggregates involved in Equation 15.1 are the total costs that we have already identified. For example, $\Sigma(p_{1973}q) = \$(0.5400 + 0.4275 + 1.2150 + 1.3620 + 0.5130 + 1.4000) = \$5.4575$, the cost of our market basket in May 1973. If we select 1967 as the base period, the indexes are

$$\text{Price index (1967)} = \frac{\sum(p_{1967}q)}{\sum(p_{1967}q)} \times 100 = \frac{4.2145}{4.2145} \times 100 = 100.0,$$

$$\text{Price index (1973)} = \frac{\sum(p_{1973}q)}{\sum(p_{1967}q)} \times 100 = \frac{5.4575}{4.2145} \times 100 = 129.49,$$

$$\text{Price index (1974)} = \frac{\sum(p_{1974}q)}{\sum(p_{1967}q)} \times 100 = \frac{6.0855}{4.2145} \times 100 = 144.39.$$

The quantity weights represented a weekly bill of goods, or market basket, to provide breakfasts for our illustrative family of four persons. In 1967 (the base period) the market basket cost roughly \$4.21. In May 1973 the same market basket cost about \$5.46. The index relates the two costs, indicating that the cost in May 1973 is 129.5 percent of the cost in 1967. By May 1974 the cost had risen to 144.4 percent of the base period cost. From 1967 to May of 1974 the price of the fixed market basket of breakfast items increased by 44.4 percent.

Table 15.2 Calculation of Breakfast Price Index for a Family of Four

		Price ($)			Cost ($) to Purchase Quantity in		
Item	Quantity	1967	May 1973	May 1974	1967	May 1973	May 1974
White bread (lb)	2.0	$0.224	0.270	0.353	$0.4480	0.5400	0.7060
Butter (lb)	0.5	0.830	0.855	0.943	0.4150	0.4275	0.4715
Bacon (lb)	1.0	0.837	1.215	1.202	0.8370	1.2150	1.2020
Eggs (dozen)	2.0	0.491	0.681	0.654	0.9820	1.3620	1.3080
Coffee (lb)	0.5	0.769	1.026	1.204	0.3845	0.5130	0.6020
Milk (quart)	4.0	0.287	0.350	0.449	1.1480	1.4000	1.7960
Total					$4.2145	5.4575	6.0855
Index (1967 = 100)					100.00	129.49	144.39

The BLS Consumer Price Index

The index of breakfast prices calculated by the method described above could also have been obtained as a weighted mean of price relatives. The formula would be

$$\mathrm{PI}_n = \frac{\sum[(p_n/p_0 \times 100) \times (p_0 q)]}{\sum (p_0 q)}.$$

Using figures from Table 15.2, the index for May 1973 using 1967 as a base is calculated below according to the weighted mean formula. The subscript 1 is used for May 1973 and the subscript 0 for the base period of 1967.

	p_0	p_1	$\frac{p_1}{p_0} \times 100$	$p_0 q$	$\left(\frac{p_1}{p_0} \times 100\right) \times p_0 q$
White bread	$0.224	$0.270	120.54	$0.4480	54.0019
Butter	0.830	0.855	103.01	0.4150	42.7492
Bacon	0.837	1.215	145.16	0.8370	121.4989
Eggs	0.491	0.681	138.70	0.9820	136.2034
Coffee	0.769	1.026	133.42	0.3845	51.3000
Milk	0.287	0.350	121.95	1.1480	139.9986
Total				4.2145	545.7520

$$\mathrm{PI}_1 = \frac{545.7520}{4.2145} = 129.49$$

In our example, this formula appears (correctly) to involve unnecessary calculations, because we multiply individual percentage price relatives by p_0q products only to have the p_0 element in the denominator of the price relatives cancel into the p_0 element in the p_0q products; that is,

$$\sum\left[\left(\frac{p_n}{\not{p}_0} \times 100\right) \times \left(\not{p}_0 q\right)\right] = \left[\sum (p_n q)\right] \times 100.$$

In a comprehensive price index, this form of calculation *is* useful because quantity weights are not available. Instead, a number of specific items are priced within each category of goods and the *average* price relative for these items is weighted by dollar expenditures on the entire category of items in a fixed period. For example, in 1974 the Bureau of Labor Statistics (BLS) actually priced 11 fresh vegetable items for the CPI. About 105 food items representing 17 different categories were priced. Fifteen items represented men's apparel, and 20 items represented drugs and prescriptions. In all, approximately 400 items were priced in about 57 categories of goods and services. In the CPI calculation there are, then, intermediate indexes for fairly narrow groups of goods and services which are weighted by proportionate dollar expenditures on all items in the group. The expenditure pattern is determined by a separate study of the purchasing habits of consumer units. The expenditure weights used in the CPI through 1976 were based on a study of 4912 urban wage-earner and clerical-worker families and 585 single workers in 72 urban areas in the United States in 1960–1961. Beginning in 1977 the weights will be based on a larger study for 1972–1973. The focus of the expenditure study gives meaning to the index —as a measure of changes in prices of goods and services bought by urban wage earners and clerical workers.

Quantity Indexes

It would seem at first that the main difficulty in constructing quantity indexes is the different units in which quantities are measured. How, for example, do we add tons of steel, gallons of gasoline, and kilowatts of electric power in constructing an index of industrial production? Consider, however, a situation in which all the quantity units are the same. Such a situation might prevail for production of selected grain crops, which are commonly measured in bushels. Why not simply find the total bushels of output in each year for all the crops included and construct from this an index of agricultural production for these crops? The answer is that one bushel of corn and one bushel of wheat may not be of equal importance. The accepted way of measuring the importance of quantity units in a money exchange economy is to refer to the judgment of that economy in terms of unit prices. The unit prices will become weights in the quantity index in the same way that quantities become the weights in

Table 15.3 *Average Prices and Average Weekly Quantities Purchased for Discussion Group Meetings, 1974 and 1975*

	1974 (year 0)		1975 (year 1)			
Item	p_0	q_0	p_1	q_1	p_0q_0	p_0q_1
Coffee (lb)	$0.97	0.6	$1.19	0.4	$0.582	$0.388
Tea (lb)	2.18	0.2	2.29	0.3	0.436	0.654
Bread (lb)	0.37	2.5	0.53	2.0	0.925	0.740
Coffee cake (lb)	0.87	3.2	1.09	3.0	2.784	2.610
Margarine (lb)	0.59	0.4	0.75	0.2	0.236	0.118
Butter (lb)	0.85	0.2	0.92	0.2	0.170	0.170
					$5.133	$4.680

aggregate price indexes. The resulting weighted aggregate index will measure the relative change in *value* of production associated with changes in quantities produced, prices being held constant at levels prevailing at a particular time. Use of a simple aggregate of quantities produced would amount to the application of equal unit price weights for all commodities.

To illustrate the construction of a quantity index, Table 15.3 presents prices and quantities of refreshment items purchased for weekly meetings of a discussion group.

We find that the group spent $5.133 per week to obtain refreshments for meetings in 1974, that is, $\Sigma(p_0q_0) = \$5.133$. In 1975 the quantities purchased changed somewhat from 1974. If the quantities purchased in 1975 had been bought at 1974 prices, the average weekly refreshment bill would have been $4.680. The effect of the changes in quantities purchased would have been to reduce the weekly refreshment bill had prices remained the same. Notice the p_1 column was not used.

$$\text{QI} = \frac{\$4.680}{\$5.133} \times 100 = 91.2$$

The effect of changes in quantities purchased would have been to reduce the weekly refreshment bill by 8.8 percent had prices remained constant at 1974 levels.

Definition

A fixed-weight **quantity index** measures the effect of changes in quantities between the base year and the given year. In measuring this effect, prices are held constant.

$$\blacksquare \quad \text{QI}_n = \frac{\sum (pq_n)}{\sum (pq_0)} \times 100 \quad \blacksquare \qquad \textit{15.2}$$

An alternative method for calculating a fixed-weight quantity index is the average-of-relatives formula using quantity relatives and base-year pq_0 weights. The computation would be

$$\mathrm{QI}_n = \frac{\sum [(q_n/q_0 \times 100) \times (pq_0)]}{\sum (pq_0)}.$$

Clearly, this leads to the same index as Equation 15.2. In a comprehensive quantity index, such as the Federal Reserve Board's Index of Industrial Production, the average-of-relatives computational form has certain advantages. We will discuss this index in the second half of this chapter.

Shifting the Base of an Index

The base given in an index is not the one we want to use in all comparisons. For example, the following table shows an index of farm output and an index of farm prices for selected years. The base of the output index is 1957–1959, but the base of the price index is 1910–1914. The historic base is used because of its connection with farm price support programs.

Year	Index of Farm Output (1957–1959 = 100)	Index of Farm Prices (1910–1914 = 100)
1950	86	258
1955	96	232
1960	106	238
1965	116	248

Source: U.S. Census Bureau, Pocket Data Book, USA 1967, p. 213.

Suppose we want to compare the changes in output with the price changes from 1950 on. It would be best to put both indexes on the same base. To change the base of each index to 1950, the indexes for each series are divided by the existing index for 1950, the desired new base year. For example, 86 is the 1950 index of the first series and the remaining three numbers in the series are divided by 86 to shift the base to 1950.

Statement | To change the base of an index series, divide the existing indexes by the existing index for the desired new base period and state the result in percentage form.

The series with 1950 as a base is:

Year	Index of Farm Output	Index of Farm Prices
1950	100.0	100.0
1955	111.6	89.9
1960	123.3	92.2
1965	134.9	96.1

From these figures we can get a quick appreciation of how changes in prices since 1950 compare with changes in output. We see, for example, that from 1950 to 1955 the drop in farm prices was of the same relative magnitude as the rise in farm output. Over the entire period from 1950 to 1965, however, output rose by 34.9 percent while prices fell only 3.9 percent.

Summary

A fundamental concept in price index construction is the change in price of a fixed collection of goods and services. The collection of goods and services that are priced depends on the purpose of the index. The BLS Consumer Price Index, for example, measures the change in price of goods and services purchased by urban wage earners and clerical workers. Just as quantities are fixed in determining a price index, prices are held constant in calculating a quantity index. A quantity index measures the change in value of an economic aggregate with prices held constant. Examples are farm output, industrial production, and mineral production.

The base year for an index is the year whose index level is equated to 100.0. You can change the base of a published index by dividing the values of the index series by the published index for the desired new base period.

See Self-Correcting Exercises 15A.

Exercise Set A

1. The basic ingredients for a pound cake (excluding a teaspoon of vanilla and a half teaspoon of salt) are 1 pound of butter, 1 pound of sugar, nine eggs, and 2 pounds of flour. Butter cost \$0.73 a pound in 1950 and \$0.87 in 1972; sugar cost \$0.10 a pound in 1950 and \$0.14 in 1972. Eggs declined in price from \$0.60 to \$0.52 a dozen from 1950 to 1972, while flour increased from \$0.98 to \$1.18 per 10-pound bag. Find the cost of making the cake in 1950 and in 1972.

2. Construct a price index for 1972, using 1950 as a base, from your totals in Exercise 1. Give a one-sentence statement of the meaning of the index.

3. Average prices (dollars per bushel) received by farmers in the United States for corn and wheat are given below for selected years:

	1960	1965	1969	1970	1971	1972
Corn	1.00	1.16	1.16	1.33	1.08	1.29
Wheat	1.74	1.35	1.25	1.33	1.34	1.67

   Annual average production from 1966 to 1970 was 4400 million bushels for corn and 1400 million bushels for wheat. Using these figures as quantity weights, construct price indexes for the selected years, using 1970 as the base year.

4. A certain farmer harvested and sold about 8000 bushels of wheat and 4000 bushels of corn in both 1970 and 1972. Does the index for 1972 in Exercise 3 represent a good index of prices received for this farmer? Explain.

5. Calculate a price index for 1972, using 1970 as a base, that is specifically applicable to the farmer in Exercise 4.

6. From the data on cake ingredients in Exercise 1, calculate price relatives for 1972 compared to 1950 for each ingredient. Calculate also the total cost for each ingredient in the recipe in 1950.

7. For each ingredient in the cake recipe, multiply the price relative (from Exercise 6) by the 1950 total cost for the ingredient. How does the sum of these products relate to your answers in Exercise 1?

8. The following table gives the average price (per carat) and quantity (millions of carats) of gem and industrial diamonds imported into the United States in 1965 and 1970. Using 1965 price weights, construct a quantity index for 1970 using 1965 as the base year.

	1965		1970	
	Price	Quantity	Price	Quantity
Gem	\$95.90	3.2	\$98.80	4.3
Industrial	\$ 4.40	12.2	\$ 3.90	11.2

9. The total quantity of gem and industrial diamonds imported barely changed from 15.4 million carats in 1965 to 15.5 million in 1970. How do you explain your result in Exercise 8 in the light of this fact?

10. Data are given below on domestic U.S. production of petroleum (billions of barrels) and natural gas (trillions of cubic feet) for three selected years:

	1960	1965	1970
Petroleum	2.6	2.8	3.5
Natural gas	12.8	16.0	21.9

In 1965 the average petroleum price at the well was \$2.86 a barrel, and the average natural gas price at the well was \$0.156 per thousand cubic feet. Construct a production index for these two fuels for 1960, 1965, and 1970, using 1965 as the base year.

11. Shift the base of your price index series in Exercise 3 to 1960. Show by algebra or by calculation that the new index for 1965 is the same as you would have obtained had you used 1965 as the base year originally.

12. The figures below are Consumer Price Indexes for the United States and the United Kingdom:

	1968	1969	1970	1971	1972
United States (1967 = 100)	104.2	109.8	116.3	121.3	125.3
United Kingdom (1970 = 100)	89.2	94.0	100.0	109.5	117.2

Shift the United States series to a 1970 base so that the two series may be more readily compared.

13. Do your results from Exercise 12 indicate that consumer prices were higher in the United Kingdom than in the United States in 1972? If not, what do the two figures for 1972 indicate?

Price, Quantity, and Value

We have emphasized that a price index measures the relative change in a value aggregate that can be attributed to changing prices. Comparable quantity indexes measure the relative change in a value aggregate attributable to changes in quantities. Since value is the product of price and quantity, we might expect that the change in a value aggregate could be factored into two parts—the price effect and the quantity effect.

Arranged below are the entries of Table 15.3 with two value extensions added, the p_1q_0 and p_1q_1 columns:

Item	p_0	q_0	p_1	q_1	p_0q_0	p_0q_1	p_1q_0	p_1q_1
Coffee	\$0.97	0.6	\$1.19	0.4	\$0.582	\$0.388	\$0.714	\$0.476
Tea	2.18	0.2	2.29	0.3	0.436	0.654	0.458	0.687
Bread	0.37	2.5	0.53	2.0	0.925	0.740	1.325	1.060
Coffee cake	0.87	3.2	1.09	3.0	2.784	2.610	3.488	3.270
Margarine	0.59	0.4	0.75	0.2	0.236	0.118	0.300	0.150
Butter	0.85	0.2	0.92	0.2	0.170	0.170	0.184	0.184
					\$5.133	\$4.680	\$6.469	\$5.827

We find that the average weekly refreshment bill has increased from \$5.133 in year 0 (1974) to \$5.827 in year 1 (1975). In general, the **value index** for any period (n) is defined as

$$\blacksquare \quad VI_n = \frac{\sum (p_n q_n)}{\sum (p_0 q_0)} \times 100. \quad \blacksquare \qquad \textit{15.3}$$

The *value index* for weekly refreshments in 1975 is

$$VI_1 = \frac{\sum (p_1 q_1)}{\sum (p_0 q_0)} \times 100 = \frac{\$5.827}{\$5.133} \times 100 = 113.5,$$

or 113.5 percent of the value of weekly refreshments in 1974.

Factoring a Value Change

Earlier, we constructed a quantity index in which we kept prices constant at their 1974 levels. This index was

$$QI_1 = \frac{\sum (p_0 q_1)}{\sum (p_0 q_0)} \times 100 = \frac{\$4.680}{\$5.133} \times 100 = 91.2.$$

The price index with fixed 1974 quantity weights is

$$PI_1 = \frac{\sum (p_1 q_0)}{\sum (p_0 q_0)} \times 100 = \frac{\$6.469}{\$5.133} \times 100 = 126.0.$$

The effect of quantity changes (holding prices constant) on the value of weekly purchases of refreshments is to reduce value by 8.8 percent, while the effect of price changes (holding quantities constant) is to increase value by 26.0 percent. In combination these effects would produce a value relative of

$$0.912 \times 1.260 = 1.149,$$

which is close but not exactly equal to the value index of 113.5 (percent).

Fixed-weight price and quantity indexes do a reasonable job of "factoring out" the effects of price and quantity changes on total value. As we shall see, price indexes are frequently used to adjust value series so that they become indicators of change in quantities. The procedure used for this adjustment assumes that $VI_n = PI_n \times QI_n$.

Real Income and Purchasing Power of the Dollar

A common use of price indexes is to adjust money values for the changing purchasing power of the monetary unit. For example, Table 15.4 shows average weekly earnings of production workers in manufacturing industries in the United States for selected years, along with the Consumer Price Index for the same periods. Also shown is the result of dividing the average weekly earnings by the relative CPI prevailing for each year. For example, \$133.73/1.163 = \$114.99 and indicates that \$133.73 in 1970 buys only \$114.99 of goods and services (CPI market basket) valued at 1967 (the base year for the CPI) prices. The other "real" earnings figures similarly indicate the value of goods and services, in 1967 dollars, that could be purchased by the average weekly earnings of the remaining periods. We see that, because of rising prices, "real" earnings have risen more modestly than dollar earnings.

Table 15.4 *Average Weekly Earnings in Manufacturing in the United States Adjusted for Purchasing Power of the Dollar*

	1955	1960	1965	1970
Average weekly earnings	\$75.70	\$89.72	\$107.53	\$133.73
Consumer Price Index*	80.2	88.7	94.5	116.3
"Real" earnings in 1967 dollars	\$94.39	\$101.15	\$113.79	\$114.99
Purchasing power of the dollar†	\$1.247	\$1.127	\$1.058	\$0.860

**1967 = 100.*
†1967 = \$1.00.

In the bottom row are shown the reciprocals (times 100) of the price indexes, called the purchasing power of the dollar. For example, the 1955 figure of \$1.247 indicates that \$1 in 1955 purchased (CPI market basket) what required \$1.247 to purchase in 1967. Similarly, \$1 in 1970 purchased what could be had in 1967 for

\$0.860. Thus we say that the 1970 dollar is worth only \$0.86 when compared with the 1967 (base year) dollar.

Laspeyres and Paasche Price Indexes

Our earlier discussion emphasized that fixed quantity weights are basic to the construction of price indexes. But should these quantities be those of the base year or those of the current year? In a consumer price index, should the index embody the consumption pattern of the earlier or of the current year?

A price index involving base-year weights is called a **Laspeyres index**, and a price index with current-year weights is called a **Paasche index**, after the economists who proposed them. The Laspeyres price index is the type we calculated earlier from the refreshment data:

$$\text{PI}_n \text{ (Laspeyres)} = \frac{\sum (p_n q_0)}{\sum (p_0 q_0)} \times 100.$$

The Laspeyres index of refreshment prices for 1975 with 1974 as a base was 126.0. The Paasche price index is

$$\text{PI}_n \text{ (Paasche)} = \frac{\sum (p_n q_n)}{\sum (p_0 q_n)} \times 100.$$

For the refreshment data, the Paasche index is

$$\text{PI}_1 \text{ (Paasche)} = \frac{\sum (p_1 q_1)}{\sum (p_0 q_1)} \times 100 = \frac{\$5.827}{\$4.680} \times 100 = 124.5.$$

The Paasche calculation tells us that quantities of refreshments actually purchased for \$5.827 in 1975 could have been purchased for \$4.680 in 1974. The effect of price changes has been, then, to increase the value (cost) of purchases by 24.5 percent. As measured by the Laspeyres index, the effect of price changes was 26.0 percent. The differences between the two results are emphasized in Table 15.5.

The Laspeyres index, viewed as a weighted average of relatives, is

$$\text{PI}_n \text{ (Laspeyres)} = \frac{\sum [(p_n/p_0 \times 100) \times (p_0 q_0)]}{\sum (p_0 q_0)}$$

and the Paasche index is

Table 15.5 Calculation of Laspeyres and Paasche Indexes as Weighted Averages of Relatives

Item	p_0	p_1	$\frac{p_1}{p_0} \times 100$	Relative Importance p_0q_0	p_0q_1	Index Products Laspeyres	Paasche
Coffee	$0.97	$1.19	122.7	0.113	0.083	13.87	10.18
Tea	2.18	2.29	105.0	0.085	0.140	8.92	14.70
Bread	0.37	0.53	143.2	0.180	0.158	25.78	22.63
Coffee cake	0.87	1.09	125.2	0.543	0.558	67.98	69.86
Margarine	0.59	0.75	127.1	0.046	0.025	5.85	3.18
Butter	0.85	0.92	108.2	0.033	0.036	3.57	3.90
				1.000	1.000	125.97	124.45

$$\text{PI}_n \text{ (Paasche)} = \frac{\sum[(p_n/p_0 \times 100) \times (p_0q_n)]}{\sum(p_0q_n)}.$$

These are the calculations carried out in Table 15.5, except that the value weights have been reduced to proportions. The Laspeyres formula, for example, becomes

$$\text{PI}_n \text{ (Laspeyres)} = \sum\left[\left(\frac{p_n}{p_0} \times 100\right) \times \frac{p_0q_0}{\sum(p_0q_0)}\right].$$

As averages of relatives, the difference between the Laspeyres and Paasche indexes is seen then as a difference in the relative importance attached to different price relatives by the two sets of expenditure weights.

The price relatives show that the price of coffee has risen more than tea, the price of bread has risen more than coffee cake, and the price of margarine has risen more than butter. The original quantity data suggest that the group may have switched some from coffee to tea, bread to coffee cake, and margarine to butter over the period. The result, in the two indexes, is that the relative importance of the items switched away from is greater in the Laspeyres than in the Paasche index. In general, if consumers tend to change consumption patterns in response to relative changes in price, the Laspeyres index will exceed the Paasche index. The Laspeyres index with fixed base-period weights acts as if these changes did not occur by imposing the fixed weights on the current period. The Paasche index imposes the current quantities on the base period, which is equally arbitrary.

These considerations led Professor Irving Fisher to propose the square root of the product of a Laspeyres and a Paasche index as a solution to measuring changes in the cost of living. Fisher's suggestion, called **Fisher's ideal index**, is

$$\text{PI}_n \text{ (Fisher)} = \sqrt{\frac{\sum(p_nq_0)}{\sum(p_0q_0)} \times \frac{\sum(p_nq_n)}{\sum(p_0q_n)}}.$$

Fisher's approach has the further advantage that it will perfectly factor out the price and quantity elements from the value change. His quantity index would be

$$\text{QI}_n\ (\text{Fisher}) = \sqrt{\frac{\sum (p_0 q_n)}{\sum (p_0 q_0)} \times \frac{\sum (p_n q_n)}{\sum (p_n q_0)}}.$$

It is easy to see that the product of PI_n (Fisher) and QI_n (Fisher) is the value relative $\Sigma(p_n q_n)/\Sigma(p_0 q_0)$. While they are an ideal theoretical solution, Fisher's formulas are not used in any comprehensive index today.

Fixed weighting schemes deny the economic reality of interdependent prices arising from alternative means of satisfying given wants. On the other hand, the concept of the relative change in price of a fixed bill of goods is a useful one. Because it is difficult to divorce the connotation of cost of living from price indexes, keep in mind that Laspeyres fixed-weight indexes are likely to overstate increases and understate decreases in the relative cost of maintaining a prescribed level of real income. Although fixed weights are perhaps necessary to free an index from ambiguity, they also constitute an inherent limitation in a changing economy of interrelated prices.

Most of the price indexes prepared by government agencies use a historic set of quantity weights, though the period they represent is not necessarily that of the base year. For example, during the 1960s the BLS Consumer Price Index had a set of weights for 1960–1961, but the base year was the period 1957–1959. The CPI weights are based on an extensive survey of consumer expenditure patterns, and it is not practical to repeat this expensive survey each year in order to obtain weights for a Paasche index. To prevent the CPI from reflecting in time an obsolete consumption pattern, a new expenditure study is carried out about every 10 years. The indexes for the late 1970s will be based on a set of consumption weights for 1972–1973.

Important United States Indexes

In this section brief descriptions of several of the most important indexes in the United States are given. Additional facts about the Consumer Price Index are also given.

Wholesale Prices

The Bureau of Labor Statistics of the Department of Labor has prepared monthly indexes of wholesale prices since 1902. The indexes are designed to measure general price movements in primary markets—that is, the first important commercial transaction for each commodity. The index is a weighted average of price relatives where the weights are net sales values in the weight-base reference period. Periodic revisions of weights are made in accordance with information from the U.S. Censuses of Agriculture, Mineral Production, and Business—generally at 5-year intervals. Monthly indexes are prepared for 14 major groups, 81 subgroups, and 241 product classes. In addition, indexes are prepared reflecting price movements of special groups of commodities and the outputs of certain industries. An example is the index of wholesale prices of building materials. The major groups and the price indexes for 1972 and 1974 (1967 = 100) are shown in Table 15.6.

Table 15.6 *Wholesale Price Indexes for Commodity Groups, 1972 and 1974*

Commodity Group	Index (1967 = 100) 1972	1974
Farm products	125.0	187.7
Processed foods and feeds	120.8	170.9
Textile products and apparel	113.6	139.9
Hides, skins, leather, related products	131.3	145.1
Fuels, related products, power	118.6	208.3
Chemicals and allied products	104.2	146.8
Rubber and plastic products	109.3	136.2
Lumber and wood products	144.3	183.6
Pulp, paper, allied products	113.4	151.7
Metals and metal products	123.5	171.9
Machinery and equipment	117.9	139.4
Furniture and household durables	111.4	127.9
Nonmetallic mineral products	118.0	153.2

Source: Federal Reserve Bulletin, March 1975.

Consumer Prices

The Consumer Price Index of the Bureau of Labor Statistics was initiated in World War I for use in wage negotiations. In its current form it is designed to reflect changes in price of a fixed quantity of goods and services purchased by urban wage earners and salaried clerical workers.* Weights are revised by periodic surveys of expenditure patterns of these families. Prices of goods and services are obtained monthly and quarterly from establishments in 56 cities. The cities include the 22 largest metropolitan areas in the United States and a sample of smaller areas. Five major group indexes and 23 subgroup indexes are published monthly for the nation

*Beginning in 1977, the Bureau of Labor Statistics will also publish an index for *all* urban families.

Table 15.7 Relative Importance of Major Groups of the CPI and Indexes for 1975

Group	Relative Importance* 1950	Relative Importance* 1963	Index for 1975† (1967 = 100)
All items	100.0	100.0	161.2
Food	33.3	22.5	175.4
Housing	25.1	33.4	166.8
Apparel and upkeep	12.8	10.6	142.3
Transportation	11.4	13.9	150.6
Health and recreation	17.4	19.6	153.5

Source: Handbook of Labor Statistics, Bureau of Labor Statistics Bulletin 1790, 1973.
†*Source: Monthly Labor Review, February 1976.*

and quarterly for each of the 22 largest metropolitan areas and Honolulu. The overall index is published monthly for the nation and the five largest metropolitan areas. Table 15.7 gives the relative importance of major categories of expenditure in 1950 and 1963 and presents the average monthly index for each group in 1975.

Industrial Production

The Board of Governors of the Federal Reserve System compiles a monthly index of industrial production–the FRB index–based on over 200 separate series collected by government agencies and trade organizations. The 200 series cover quantities of output in manufacturing, mining, and utilities. In forming quantity indexes, individual quantity relatives for various outputs are weighted by value added in the product or industry sector as determined from periodic census materials. Indexes are determined for various product groups as well as for industry groupings. Table 15.8 indicates the major industry groups in the manufacturing sector and a recent index for each group.

GNP Implicit Price Deflators

The Office of Business Economics of the U.S. Department of Commerce prepares estimates of gross national product, national income, personal income, and related series. The series, known collectively as the *national income accounts,* are published monthly in the *Survey of Current Business.* Briefly, *gross national product (GNP)* is the market value of all goods and services produced by the labor and property of the nation's residents. *National income* is the sum of labor and property

Table 15.8 *Major Manufacturing Industries in the FRB Index and Indexes for 1974*

Industry Group	Index for 1974 (1967 = 100)
All manufacturing	124.8
Durable manufactures	120.8
Primary metals	127.8
Fabricated metal products	131.4
Machinery	116.3
Transportation equipment	96.9
Instruments and related products	143.8
Clay, glass, stone products	125.9
Lumber and products	120.1
Furniture and fixtures	136.2
Nondurable manufactures	129.7
Textile mill products	123.0
Apparel products	105.0
Leather and products	77.6
Paper and printing	133.9
Chemicals and products	154.3
Petroleum products	124.1
Rubber and plastic products	164.6
Food products	126.1

Source: Federal Reserve Bulletin, March 1975.

earnings from current production of goods and services in the nation, and *personal income* is the income currently received by individuals and noncorporate enterprises in the form of wages and salaries, proprietors' and rental income, dividends, interest, and other (transfer) payments from business and government.

Gross national product and other major series in the national income accounts are estimated in current-dollar and constant-dollar terms. The constant-dollar

Table 15.9 *GNP in Current and Constant Dollars*

	Current Dollars (billions)	1958 Dollars (billions)	Implicit Price Deflator (1958 = 100)
1950	285	355	80.3
1955	398	438	90.9
1960	504	488	103.3
1965	685	618	110.8
1970	976	722	135.2

Source: U.S. Census Bureau, Pocket Data Book, USA 1973, p. 194.

estimates are obtained by dividing components of current-dollar GNP, using as fine a product breakdown as possible, by appropriate price indexes. Various components of the price indexes previously discussed are used in these adjustments. Then the constant-dollar figures for all components are totaled to obtain GNP in constant dollars. If the current-dollar GNP is divided by the constant-dollar figure, the result is the **implicit GNP price deflator**. Table 15.9 shows GNP in current and in constant (1958) dollars for selected years. The final column shows the implicit price deflator. It represents a comprehensive price index for the national economy.

Summary

Changes in the value of an economic aggregate can be viewed as the product of the effects of price changes and quantity changes. Fixed-weight price and quantity indexes are approximate measures of these separate effects. When a value series is divided by an appropriately related price index, the result is a measure of quantity changes in constant dollars. An example is "real" income, income adjusted for price changes. The purchasing power of the dollar can be expressed by dividing $100 by the price index.

Fixed-weight price indexes do not provide for changes in consumption that may be made when some commodities and services become expensive compared to available alternatives. Long-term changes in spending patterns are incorporated in periodic revisions of the weights, however.

The BLS Consumer Price Index and the BLS Wholesale Price Index reflect changing price levels at different stages of the U.S. economy. In addition, various components of these indexes measure changing price levels in different economic sectors. The GNP implicit price deflator reflects changing price levels of all goods and services included in the gross national product. The FRB Index of Industrial Production is a comprehensive index of the quantity of output in basic industries in the United States.

See Self-Correcting Exercises 15B.

Exercise Set B

1. The following data refer to the numbers (in million head) and market price per head of livestock on farms in the United States:

	Numbers		Market Price	
	1960	1970	1960	1970
Cattle	96	115	$137	$184
Hogs	19	39	$ 59	$ 57
Sheep	29	17	$ 17	$ 25

Find the index of total market value of livestock on farms for 1970 using 1960 as a base.

2. Find the price index and the quantity index for 1970, using 1960 as a base, from the data of Exercise 1. Compare the value change implied by these indexes with the value change found in Exercise 1.

3. The following table gives the average price (per carat) and quantity (millions of carats) of gem and industrial diamonds imported into the United States in 1965 and 1970:

	1965		1970	
	Price	Quantity	Price	Quantity
Gem	$95.90	3.2	$98.80	4.3
Industrial	$ 4.40	12.2	$ 3.90	11.2

Find the value of imports in 1970 and in 1965.

4. From the data in Exercise 3, find the price index for 1970, using 1965 as a base and using base-year weights.

5. Adjust the 1970 value of imports from Exercise 3 to the 1965 price level. Should the resulting relative change in imports valued at 1965 prices agree closely with a quantity index for 1970 on 1965 as a base using base-year price weights? Why?

6. Average gross weekly earnings in contract construction are given below along with the Consumer Price Index for selected years:

	1960	1965	1970
Average earnings	$113	$138	$196
CPI (1967 = 100)	88.7	94.5	116.3

Adjust the average weekly earnings figures to reflect the 1967 price level. What do your results indicate?

7. Below are the value of manufacturers' shipments and the wholesale price index for industrial commodities for selected years:

	1960	1965	1970
Shipments (billions)	$370	$492	$631
Price index (1967 = 100)	95.3	96.4	110.0

Find the percentage changes from 1960 to 1965 and from 1965 to 1970 in the quantity of shipments.

8. The following data show per capita consumption (in pounds) of beef and pork products in the United States and the Consumer Price Index (1967 = 100) for beef and for pork products:

	Per Capita Consumption		Price Index	
	1960	1970	1960	1970
Beef	85.1	113.7	92.1	117.6
Pork	64.9	66.4	81.7	115.9

Calculate the price index for 1970 on 1960 as a base using base-year weights (Laspeyres) and using current-year weights (Paasche).

9. The text offers some reasons why the Laspeyres index might show a greater rise in prices than the Paasche index—namely, substitution of one product for another in response to different relative changes in price. Does that appear to be the case between beef and pork in Exercise 8? Or can you offer a different explanation in this case?

10. The following data show gold and silver production (in fine ounces) in the United States in 1970 (q_0) and 1972 (q_1) and the average price in dollars per fine ounce in 1970 (p_0) and 1972 (p_1). Also shown are various price × quantity extensions and their totals.

	q_0	q_1	p_0	p_1	p_0q_0	p_1q_0	p_0q_1	p_1q_1
Gold	1,743	1,428	36	59	62,748	102,832	51,408	84,252
Silver	45,000	37,000	1.771	1.685	79,695	75,825	65,527	62,345
					142,443	178,662	116,935	146,597

Calculate Laspeyres and Paasche price indexes for 1972 on 1970 as a base.

11. What is it about the quantity weights in Exercise 10 that explains the comparison between the Laspeyres and Paasche price indexes?

12. Personal income in the United States in current and constant (1958) dollars is shown here for selected years in billions of dollars:

	1960	1965	1970
Current dollars	401	539	806
Constant dollars	390	495	624

Find the implicit price deflator for personal income.

13. Per capita personal income in the United States for selected years is 1960: $2219; 1965: $2773; 1970: $3935. Adjust the per capita personal income figures for the implicit price deflator found in Exercise 12. What do the resulting figures measure?

14. Why is it better to use the implicit price deflator rather than the Consumer Price Index to adjust per capita personal income? If you were adjusting median family income, which index would you prefer for deflation?

Overview

The following data refer to shellfish catch (in millions of pounds) and average prices per pound received by fishermen in the United States:

Variety	Quantity			Price		
	1950	1965	1970	1950	1965	1970
Clams	40	70	100	$0.25	$0.23	$0.29
Crabs	160	320	280	0.06	0.09	0.14
Lobsters	20	30	30	0.32	0.64	0.97
Oysters	80	50	50	0.24	0.50	0.54
Shrimp	200	240	370	0.18	0.32	0.35

The price × quantity sums that might be used in various indexes are given below. For example, $101.50 = $\Sigma(p_{1950}q_{1965})$.

	Quantity		
Price	1950	1965	1970
1950	$ 81.20	$101.50	$130.00
1965	140.40	165.90	210.80
1970	166.60	205.20	253.80

1. Construct value indexes for 1950, 1965, and 1970 using 1965 as the base year. Give a one-sentence statement of the meaning of the value index for 1970.
2. Calculate price indexes for the three years, using 1950 as the base year and using (a) 1950, (b) 1965, and (c) 1970 quantity weights.
3. Give a short statement of the meaning of the price index for 1970, using 1950 weights, found in Exercise 2.
4. Note that the price index with 1950 weights rises faster than the index with 1970 weights in Exercise 2. Can you identify the variety of shellfish whose price-quantity behavior is primarily responsible for this difference?
5. Find the quantity index for 1970 on 1950 as a base, using 1965 prices as weights. Give a short statement of the meaning of the index.
6. Deflate the value of shellfish catch in 1950, 1965, and 1970 for prices as measured by the index from Exercise 2 using 1965 weights. Interpret your deflated value figure for 1970.
7. The following data refer to purchases, prices, and expenditures on new and used cars in the United States:

	Purchases per 100 Households New	Used	Average Price New	Used	Annual Expenditures
1970	11.1	19.6	\$3018	\$ 959	\$33.3 billion
1972	12.8	22.5	\$3380	\$1056	\$44.8 billion

Compute an index of new and used car prices for 1972 on 1970 as a base, using 1970 weights.

8. Deflate the 1972 expenditure figure for price change in Exercise 7, and interpret the adjusted expenditures.
9. Refer to Exercise 7: Construct an index of new and used car purchases for 1972 on 1970 as a base, using 1970 price weights.
10. Multiply the 1970 annual expenditures by your index (in relative form) from Exercise 9, and interpret the resulting figure.

Glossary of Equations

15.1 $$\mathrm{PI}_n = \frac{\sum (p_n q)}{\sum (p_0 q)} \times 100$$

A fixed-weight price index is the percentage relative of the aggregate price of a fixed bill of goods in the given year to the aggregate price of the same bill of goods in the base year.

15.2 $$\mathrm{QI}_n = \frac{\sum (p q_n)}{\sum (p q_0)} \times 100$$

A fixed-weight quantity index is the percentage relative of the aggregate value of given-year quantities at a fixed set of prices to the aggregate value of base-year quantities at the same prices.

15.3 $$\mathrm{VI}_n = \frac{\sum (p_n q_n)}{\sum (p_0 q_0)} \times 100$$

The value index is the percentage relative of aggregate value in the given year to aggregate value in the base year.

Appendixes

A. Areas under the Normal Distribution

*For **Negative** Values of z*

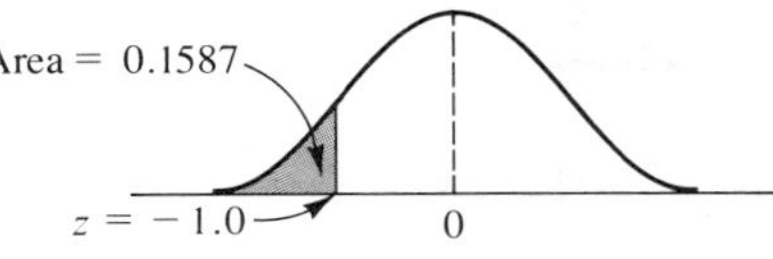

z To 1st Decimal	.00	.01	.02	.03	.04	.05	.06	.07	.08	.09
	Area									
−3.0	.0014	.0013	.0013	.0012	.0012	.0011	.0011	.0011	.0010	.0010
−2.9	.0019	.0018	.0018	.0017	.0016	.0016	.0015	.0015	.0014	.0014
−2.8	.0026	.0025	.0024	.0023	.0023	.0022	.0021	.0021	.0020	.0019
−2.7	.0035	.0034	.0033	.0032	.0031	.0030	.0029	.0028	.0027	.0026
−2.6	.0047	.0045	.0044	.0043	.0041	.0040	.0039	.0038	.0037	.0036
−2.5	.0062	.0060	.0059	.0057	.0055	.0054	.0052	.0051	.0049	.0048
−2.4	.0082	.0080	.0078	.0075	.0073	.0071	.0069	.0068	.0066	.0064
−2.3	.0107	.0104	.0102	.0099	.0096	.0094	.0091	.0089	.0087	.0084
−2.2	.0139	.0136	.0132	.0129	.0126	.0122	.0119	.0116	.0113	.0110
−2.1	.0179	.0174	.0170	.0166	.0162	.0158	.0154	.0150	.0146	.0143
−2.0	.0228	.0222	.0217	.0212	.0207	.0202	.0197	.0192	.0188	.0183
−1.9	.0287	.0281	.0274	.0268	.0262	.0256	.0250	.0244	.0238	.0233
−1.8	.0359	.0352	.0344	.0336	.0329	.0322	.0314	.0307	.0300	.0294
−1.7	.0446	.0436	.0427	.0418	.0409	.0401	.0392	.0384	.0375	.0367
−1.6	.0548	.0537	.0526	.0516	.0505	.0495	.0485	.0475	.0465	.0455
−1.5	.0668	.0655	.0643	.0630	.0618	.0606	.0594	.0582	.0570	.0559
−1.4	.0808	.0793	.0778	.0764	.0749	.0735	.0722	.0708	.0694	.0681
−1.3	.0968	.0951	.0934	.0918	.0901	.0885	.0869	.0853	.0838	.0823
−1.2	.1151	.1131	.1112	.1093	.1075	.1056	.1038	.1020	.1003	.0985
−1.1	.1357	.1335	.1314	.1292	.1271	.1251	.1230	.1210	.1190	.1170
−1.0	.1587	.1562	.1539	.1515	.1492	.1469	.1446	.1423	.1401	.1379
−0.9	.1841	.1814	.1788	.1762	.1736	.1711	.1685	.1660	.1635	.1611
−0.8	.2119	.2090	.2061	.2033	.2005	.1977	.1949	.1922	.1894	.1867
−0.7	.2420	.2389	.2358	.2327	.2297	.2266	.2236	.2206	.2177	.2148
−0.6	.2743	.2709	.2676	.2643	.2611	.2578	.2546	.2514	.2483	.2451
−0.5	.3085	.3050	.3015	.2981	.2946	.2912	.2877	.2843	.2810	.2776
−0.4	.3446	.3409	.3372	.3336	.3300	.3264	.3228	.3192	.3156	.3121
−0.3	.3821	.3783	.3745	.3707	.3669	.3632	.3594	.3557	.3520	.3483
−0.2	.4207	.4168	.4129	.4090	.4052	.4013	.3974	.3936	.3897	.3859
−0.1	.4602	.4562	.4522	.4483	.4443	.4404	.4364	.4325	.4286	.4247
−0.0	.5000	.4960	.4920	.4880	.4840	.4801	.4761	.4721	.4681	.4641

(Column headings .00–.09: Second Decimal)

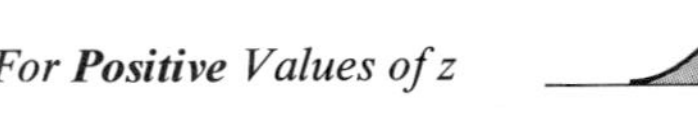

*For **Positive** Values of z*

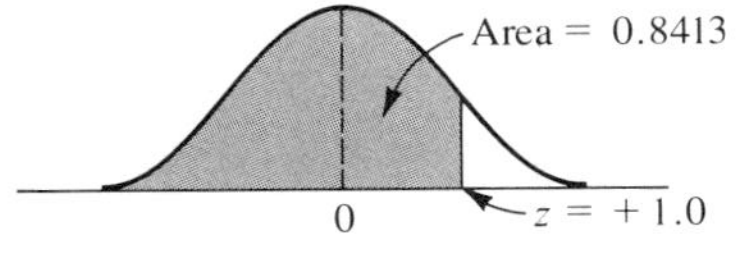

z To 1st Decimal	.00	.01	.02	.03	.04	.05	.06	.07	.08	.09
	Area									
0.0	.5000	.5040	.5080	.5120	.5160	.5199	.5239	.5279	.5319	.5359
0.1	.5398	.5438	.5478	.5517	.5557	.5596	.5636	.5675	.5714	.5753
0.2	.5793	.5832	.5871	.5910	.5948	.5987	.6026	.6064	.6103	.6141
0.3	.6179	.6217	.6255	.6293	.6331	.6368	.6406	.6443	.6480	.6517
0.4	.6554	.6591	.6628	.6664	.6700	.6736	.6772	.6808	.6844	.6879
0.5	.6915	.6950	.6985	.7019	.7054	.7088	.7123	.7157	.7190	.7224
0.6	.7257	.7291	.7324	.7357	.7389	.7422	.7454	.7486	.7517	.7549
0.7	.7580	.7611	.7642	.7673	.7703	.7734	.7764	.7794	.7823	.7852
0.8	.7881	.7910	.7939	.7967	.7995	.8023	.8051	.8078	.8106	.8133
0.9	.8159	.8186	.8212	.8238	.8264	.8289	.8315	.8340	.8365	.8389
1.0	.8413	.8438	.8461	.8485	.8508	.8531	.8554	.8577	.8599	.8621
1.1	.8643	.8665	.8686	.8708	.8729	.8749	.8770	.8790	.8810	.8830
1.2	.8849	.8869	.8888	.8907	.8925	.8944	.8962	.8980	.8997	.9015
1.3	.9032	.9049	.9066	.9082	.9099	.9115	.9131	.9147	.9162	.9177
1.4	.9192	.9207	.9222	.9236	.9251	.9265	.9278	.9292	.9306	.9319
1.5	.9332	.9345	.9357	.9370	.9382	.9394	.9406	.9418	.9430	.9441
1.6	.9452	.9463	.9474	.9485	.9495	.9505	.9515	.9525	.9535	.9545
1.7	.9554	.9564	.9573	.9582	.9591	.9599	.9608	.9616	.9625	.9633
1.8	.9641	.9649	.9656	.9664	.9671	.9678	.9686	.9693	.9700	.9706
1.9	.9713	.9719	.9726	.9732	.9738	.9744	.9750	.9756	.9762	.9767
2.0	.9772	.9778	.9783	.9788	.9793	.9798	.9803	.9808	.9812	.9817
2.1	.9821	.9826	.9830	.9834	.9838	.9842	.9846	.9850	.9854	.9857
2.2	.9861	.9865	.9868	.9871	.9874	.9878	.9881	.9884	.9887	.9890
2.3	.9893	.9896	.9898	.9901	.9904	.9906	.9909	.9911	.9913	.9916
2.4	.9918	.9920	.9922	.9924	.9926	.9928	.9930	.9932	.9934	.9936
2.5	.9938	.9940	.9941	.9943	.9944	.9946	.9948	.9949	.9951	.9952
2.6	.9953	.9955	.9956	.9957	.9958	.9960	.9961	.9962	.9963	.9964
2.7	.9965	.9966	.9967	.9968	.9969	.9970	.9971	.9972	.9973	.9974
2.8	.9974	.9975	.9976	.9977	.9977	.9978	.9979	.9979	.9980	.9981
2.9	.9981	.9982	.9982	.9983	.9984	.9984	.9985	.9985	.9986	.9986
3.0	.9986	.9987	.9987	.9988	.9988	.9988	.9989	.9989	.9990	.9990

(Column headings .00–.09: Second Decimal)

B. Areas under the *t* Distributions

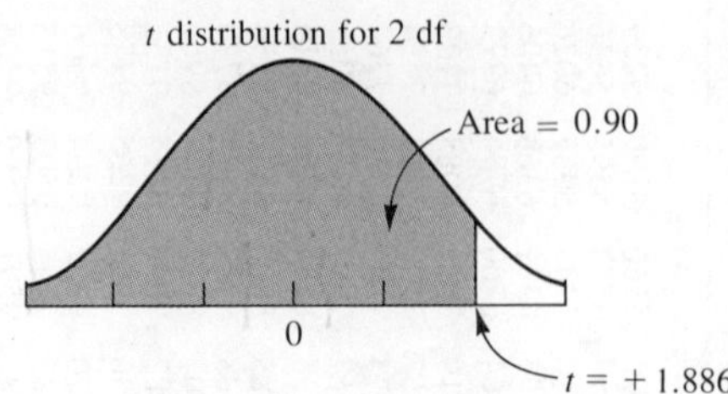

df	$t_{.005}$	$t_{.01}$	$t_{.025}$	$t_{.05}$	$t_{.10}$	$t_{.90}$	$t_{.95}$	$t_{.975}$	$t_{.99}$	$t_{.995}$
1	−63.657	−31.821	−12.706	−6.314	−3.078	3.078	6.314	12.706	31.821	63.657
2	− 9.925	− 6.965	− 4.303	−2.920	−1.886	1.886	2.920	4.303	6.965	9.925
3	− 5.841	− 4.541	− 3.182	−2.353	−1.638	1.638	2.353	3.182	4.541	5.841
4	− 4.604	− 3.747	− 2.776	−2.132	−1.533	1.533	2.132	2.776	3.747	4.604
5	− 4.032	− 3.365	− 2.571	−2.015	−1.476	1.476	2.015	2.571	3.365	4.032
6	− 3.707	− 3.143	− 2.447	−1.943	−1.440	1.440	1.943	2.447	3.143	3.707
7	− 3.499	− 2.998	− 2.365	−1.895	−1.415	1.415	1.895	2.365	2.998	3.499
8	− 3.355	− 2.896	− 2.306	−1.860	−1.397	1.397	1.860	2.306	2.896	3.355
9	− 3.250	− 2.821	− 2.262	−1.833	−1.383	1.383	1.833	2.262	2.821	3.250
10	− 3.169	− 2.764	− 2.228	−1.812	−1.372	1.372	1.812	2.228	2.764	3.169
11	− 3.106	− 2.718	− 2.201	−1.796	−1.363	1.363	1.796	2.201	2.718	3.106
12	− 3.055	− 2.681	− 2.179	−1.782	−1.356	1.356	1.782	2.179	2.681	3.055
13	− 3.012	− 2.650	− 2.160	−1.771	−1.350	1.350	1.771	2.160	2.650	3.012
14	− 2.977	− 2.624	− 2.145	−1.761	−1.345	1.345	1.761	2.145	2.624	2.977
15	− 2.947	− 2.602	− 2.131	−1.753	−1.341	1.341	1.753	2.131	2.602	2.947
16	− 2.921	− 2.583	− 2.120	−1.746	−1.337	1.337	1.746	2.120	2.583	2.921
17	− 2.898	− 2.567	− 2.110	−1.740	−1.333	1.333	1.740	2.110	2.567	2.898
18	− 2.878	− 2.552	− 2.101	−1.734	−1.330	1.330	1.734	2.101	2.552	2.878
19	− 2.861	− 2.539	− 2.093	−1.729	−1.328	1.328	1.729	2.093	2.539	2.861
20	− 2.845	− 2.528	− 2.086	−1.725	−1.325	1.325	1.725	2.086	2.528	2.845
21	− 2.831	− 2.518	− 2.080	−1.721	−1.323	1.323	1.721	2.080	2.518	2.831
22	− 2.819	− 2.508	− 2.074	−1.717	−1.321	1.321	1.717	2.074	2.508	2.819
23	− 2.807	− 2.500	− 2.069	−1.714	−1.319	1.319	1.714	2.069	2.500	2.807
24	− 2.797	− 2.492	− 2.064	−1.711	−1.318	1.318	1.711	2.064	2.492	2.797
25	− 2.787	− 2.485	− 2.060	−1.708	−1.316	1.316	1.708	2.060	2.485	2.787
26	− 2.779	− 2.479	− 2.056	−1.706	−1.315	1.315	1.706	2.056	2.479	2.779
27	− 2.771	− 2.473	− 2.052	−1.703	−1.314	1.314	1.703	2.052	2.473	2.771
28	− 2.763	− 2.467	− 2.048	−1.701	−1.313	1.313	1.701	2.048	2.467	2.763
29	− 2.756	− 2.462	− 2.045	−1.699	−1.311	1.311	1.699	2.045	2.462	2.756
30	− 2.750	− 2.457	− 2.042	−1.697	−1.310	1.310	1.697	2.042	2.457	2.750
40	− 2.704	− 2.423	− 2.021	−1.684	−1.303	1.303	1.684	2.021	2.423	2.704
60	− 2.660	− 2.390	− 2.000	−1.671	−1.296	1.296	1.671	2.000	2.390	2.660
120	− 2.617	− 2.358	− 1.980	−1.658	−1.289	1.289	1.658	1.980	2.358	2.617
∞	− 2.576	− 2.326	− 1.960	−1.645	−1.282	1.282	1.645	1.960	2.326	2.576

C. Areas under the Chi-Square Distributions

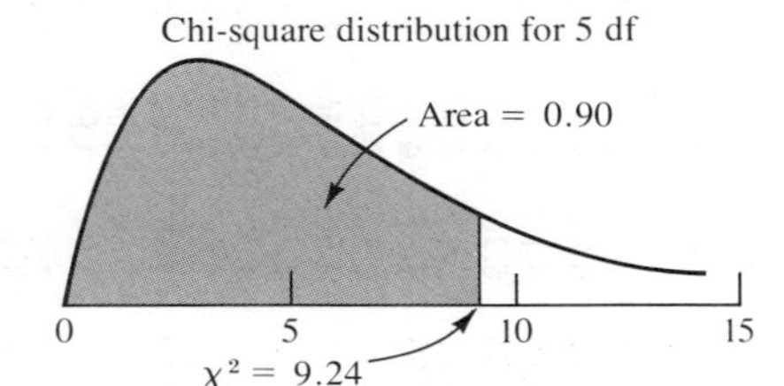

df	$\chi^2_{.005}$	$\chi^2_{.01}$	$\chi^2_{.025}$	$\chi^2_{.05}$	$\chi^2_{.10}$	$\chi^2_{.90}$	$\chi^2_{.95}$	$\chi^2_{.975}$	$\chi^2_{.99}$	$\chi^2_{.995}$
1	0.000039	0.00016	0.00098	0.0039	0.0158	2.71	3.84	5.02	6.63	7.88
2	0.0100	0.0201	0.0506	0.1026	0.2107	4.61	5.99	7.38	9.21	10.60
3	0.0717	0.115	0.216	0.352	0.584	6.25	7.81	9.35	11.34	12.84
4	0.207	0.297	0.484	0.711	1.064	7.78	9.49	11.14	13.28	14.86
5	0.412	0.554	0.831	1.15	1.61	9.24	11.07	12.83	15.09	16.75
6	0.676	0.872	1.24	1.64	2.20	10.64	12.59	14.45	16.81	18.55
7	0.989	1.24	1.69	2.17	2.83	12.02	14.07	16.01	18.48	20.28
8	1.34	1.65	2.18	2.73	3.49	13.36	15.51	17.53	20.09	21.96
9	1.73	2.09	2.70	3.33	4.17	14.68	16.92	19.02	21.67	23.59
10	2.16	2.56	3.25	3.94	4.87	15.99	18.31	20.48	23.21	25.19
11	2.60	3.05	3.82	4.57	5.58	17.28	19.68	21.92	24.73	26.76
12	3.07	3.57	4.40	5.23	6.30	18.55	21.03	23.34	26.22	28.30
13	3.57	4.11	5.01	5.89	7.04	19.81	22.36	24.74	27.69	29.82
14	4.07	4.66	5.63	6.57	7.79	21.06	23.68	26.12	29.14	31.32
15	4.60	5.23	6.26	7.26	8.55	22.31	25.00	27.49	30.58	32.80
16	5.14	5.81	6.91	7.96	9.31	23.54	26.30	28.85	32.00	34.27
18	6.26	7.01	8.23	9.39	10.86	25.99	28.87	31.53	34.81	37.16
20	7.43	8.26	9.59	10.85	12.44	28.41	31.41	34.17	37.57	40.00
24	9.89	10.86	12.40	13.85	15.66	33.20	36.42	39.36	42.98	45.56
30	13.79	14.95	16.79	18.49	20.60	40.26	43.77	46.98	50.89	53.67
40	20.71	22.16	24.43	26.51	29.05	51.81	55.76	59.34	63.69	66.77
60	35.53	37.48	40.48	43.18	46.46	74.40	79.08	83.30	88.38	91.95
120	83.85	86.92	91.58	95.70	100.62	140.23	146.57	152.21	158.95	163.64

D. Values on the *F* Distributions

F distribution for 50 and 5 df

Area = 0.05

Values of *F*

F = 4.44

Right tail of the distribution for $P = 0.05$ (lightface type), 0.01 (boldface type)

m_2 = Degrees of Freedom for Denominator	m_1 = Degrees of Freedom for Numerator											
m_2	1	2	3	4	5	6	7	8	9	10	11	12
1	161	200	216	225	230	234	237	239	241	242	243	244
	4052	**4999**	**5403**	**5625**	**5764**	**5859**	**5928**	**5981**	**6022**	**6056**	**6082**	**6106**
2	18.51	19.00	19.16	19.25	19.30	19.33	19.36	19.37	19.38	19.39	19.40	19.41
	98.49	**99.01**	**99.17**	**99.25**	**99.30**	**99.33**	**99.34**	**99.36**	**99.38**	**99.40**	**99.41**	**99.42**
3	10.13	9.55	9.28	9.12	9.01	8.94	8.88	8.84	8.81	8.78	8.76	8.74
	34.12	**30.81**	**29.46**	**28.71**	**28.24**	**27.91**	**27.67**	**27.49**	**27.34**	**27.23**	**27.13**	**27.05**
4	7.71	6.94	6.59	6.39	6.26	6.16	6.09	6.04	6.00	5.96	5.93	5.91
	21.20	**18.00**	**16.69**	**15.98**	**15.52**	**15.21**	**14.98**	**14.80**	**14.66**	**14.54**	**14.45**	**14.37**
5	6.61	5.79	5.41	5.19	5.05	4.95	4.88	4.82	4.78	4.74	4.70	4.68
	16.26	**13.27**	**12.06**	**11.39**	**10.97**	**10.67**	**10.45**	**10.27**	**10.15**	**10.05**	**9.96**	**9.89**
6	5.99	5.14	4.76	4.53	4.39	4.28	4.21	4.15	4.10	4.06	4.03	4.00
	13.74	**10.92**	**9.78**	**9.15**	**8.75**	**8.47**	**8.26**	**8.10**	**7.98**	**7.87**	**7.79**	**7.72**
7	5.59	4.74	4.35	4.12	3.97	3.87	3.79	3.73	3.68	3.63	3.60	3.57
	12.25	**9.55**	**8.45**	**7.85**	**7.46**	**7.19**	**7.00**	**6.84**	**6.71**	**6.62**	**6.54**	**6.47**
8	5.32	4.46	4.07	3.84	3.69	3.58	3.50	3.44	3.39	3.34	3.31	3.28
	11.26	**8.65**	**7.59**	**7.01**	**6.63**	**6.37**	**6.19**	**6.03**	**5.91**	**5.82**	**5.74**	**5.67**
9	5.12	4.26	3.86	3.63	3.48	3.37	3.29	3.23	3.18	3.13	3.10	3.07
	10.56	**8.02**	**6.99**	**6.42**	**6.06**	**5.80**	**5.62**	**5.47**	**5.35**	**5.26**	**5.18**	**5.11**
10	4.96	4.10	3.71	3.48	3.33	3.22	3.14	3.07	3.02	2.97	2.94	2.91
	10.04	**7.56**	**6.55**	**5.99**	**5.64**	**5.39**	**5.21**	**5.06**	**4.95**	**4.85**	**4.78**	**4.71**
11	4.84	3.98	3.59	3.36	3.20	3.09	3.01	2.95	2.90	2.86	2.82	2.79
	9.65	**7.20**	**6.22**	**5.67**	**5.32**	**5.07**	**4.88**	**4.74**	**4.63**	**4.54**	**4.46**	**4.40**
12	4.75	3.88	3.49	3.26	3.11	3.00	2.92	2.85	2.80	2.76	2.72	2.69
	9.33	**6.93**	**5.95**	**5.41**	**5.06**	**4.82**	**4.65**	**4.50**	**4.39**	**4.30**	**4.22**	**4.16**
13	4.67	3.80	3.41	3.18	3.02	2.92	2.84	2.77	2.72	2.67	2.63	2.60
	9.07	**6.70**	**5.74**	**5.20**	**4.86**	**4.62**	**4.44**	**4.30**	**4.19**	**4.10**	**4.02**	**3.96**
14	4.60	3.74	3.34	3.11	2.96	2.85	2.77	2.70	2.65	2.60	2.56	2.53
	8.86	**6.51**	**5.56**	**5.03**	**4.69**	**4.46**	**4.28**	**4.14**	**4.03**	**3.94**	**3.86**	**3.80**
15	4.54	3.68	3.29	3.06	2.90	2.79	2.70	2.64	2.59	2.55	2.51	2.48
	8.68	**6.36**	**5.42**	**4.89**	**4.56**	**4.32**	**4.14**	**4.00**	**3.89**	**3.80**	**3.73**	**3.67**
16	4.49	3.63	3.24	3.01	2.85	2.74	2.66	2.59	2.54	2.49	2.45	2.42
	8.53	**6.23**	**5.29**	**4.77**	**4.44**	**4.20**	**4.03**	**3.89**	**3.78**	**3.69**	**3.61**	**3.55**
17	4.45	3.59	3.20	2.96	2.81	2.70	2.62	2.55	2.50	2.45	2.41	2.38
	8.40	**6.11**	**5.18**	**4.67**	**4.34**	**4.10**	**3.93**	**3.79**	**3.68**	**3.59**	**3.52**	**3.45**
18	4.41	3.55	3.16	2.93	2.77	2.66	2.58	2.51	2.46	2.41	2.37	2.34
	8.28	**6.01**	**5.09**	**4.58**	**4.25**	**4.01**	**3.85**	**3.71**	**3.60**	**3.51**	**3.44**	**3.37**
19	4.38	3.52	3.13	2.90	2.74	2.63	2.55	2.48	2.43	2.38	2.34	2.31
	8.18	**5.93**	**5.01**	**4.50**	**4.17**	**3.94**	**3.77**	**3.63**	**3.52**	**3.43**	**3.36**	**3.30**
20	4.35	3.49	3.10	2.87	2.71	2.60	2.52	2.45	2.40	2.35	2.31	2.28
	8.10	**5.85**	**4.94**	**4.43**	**4.10**	**3.87**	**3.71**	**3.56**	**3.45**	**3.37**	**3.30**	**3.23**
21	4.32	3.47	3.07	2.84	2.68	2.57	2.49	2.42	2.37	2.32	2.28	2.25
	8.02	**5.78**	**4.87**	**4.37**	**4.04**	**3.81**	**3.65**	**3.51**	**3.40**	**3.31**	**3.24**	**3.17**
22	4.30	3.44	3.05	2.82	2.66	2.55	2.47	2.40	2.35	2.30	2.26	2.23
	7.94	**5.72**	**4.82**	**4.31**	**3.99**	**3.76**	**3.59**	**3.45**	**3.35**	**3.26**	**3.18**	**3.12**
23	4.28	3.42	3.03	2.80	2.64	2.53	2.45	2.38	2.32	2.28	2.24	2.20
	7.88	**5.66**	**4.76**	**4.26**	**3.94**	**3.71**	**3.54**	**3.41**	**3.30**	**3.21**	**3.14**	**3.07**
24	4.26	3.40	3.01	2.78	2.62	2.51	2.43	2.36	2.30	2.26	2.22	2.18
	7.82	**5.61**	**4.72**	**4.22**	**3.90**	**3.67**	**3.50**	**3.36**	**3.25**	**3.17**	**3.09**	**3.03**
25	4.24	3.38	2.99	2.76	2.60	2.49	2.41	2.34	2.28	2.24	2.20	2.16
	7.77	**5.57**	**4.68**	**4.18**	**3.86**	**3.63**	**3.46**	**3.32**	**3.21**	**3.13**	**3.05**	**2.99**
26	4.22	3.37	2.98	2.74	2.59	2.47	2.39	2.32	2.27	2.22	2.18	2.15
	7.72	**5.53**	**4.64**	**4.14**	**3.82**	**3.59**	**3.42**	**3.29**	**3.17**	**3.09**	**3.02**	**2.96**

D. Values on the *F* Distributions (*continued*)

m_1 = Degrees of Freedom for Numerator												m_2 = Degrees of Freedom for Denominator
14	16	20	24	30	40	50	75	100	200	500	∞	m_2
245	246	248	249	250	251	252	253	253	254	254	254	1
6142	**6169**	**6208**	**6234**	**6258**	**6286**	**6302**	**6323**	**6334**	**6352**	**6361**	**6366**	
19.42	19.43	19.44	19.45	19.46	19.47	19.47	19.48	19.49	19.49	19.50	19.50	2
99.43	**99.44**	**99.45**	**99.46**	**99.47**	**99.48**	**99.48**	**99.49**	**99.49**	**99.49**	**99.50**	**99.50**	
8.71	8.69	8.66	8.64	8.62	8.60	8.58	8.57	8.56	8.54	8.54	8.53	3
26.92	**26.83**	**26.69**	**26.60**	**26.50**	**26.41**	**26.35**	**26.27**	**26.23**	**26.18**	**26.14**	**26.12**	
5.87	5.84	5.80	5.77	5.74	5.71	5.70	5.68	5.66	5.65	5.64	5.63	4
14.24	**14.15**	**14.02**	**13.93**	**13.83**	**13.74**	**13.69**	**13.61**	**13.57**	**13.52**	**13.48**	**13.46**	
4.64	4.60	4.56	4.53	4.50	4.46	4.44	4.42	4.40	4.38	4.37	4.36	5
9.77	**9.68**	**9.55**	**9.47**	**9.38**	**9.29**	**9.24**	**9.17**	**9.13**	**9.07**	**9.04**	**9.02**	
3.96	3.92	3.87	3.84	3.81	3.77	3.75	3.72	3.71	3.69	3.68	3.67	6
7.60	**7.52**	**7.39**	**7.31**	**7.23**	**7.14**	**7.09**	**7.02**	**6.99**	**6.94**	**6.90**	**6.88**	
3.52	3.49	3.44	3.41	3.38	3.34	3.32	3.29	3.28	3.25	3.24	3.23	7
6.35	**6.27**	**6.15**	**6.07**	**5.98**	**5.90**	**5.85**	**5.78**	**5.75**	**5.70**	**5.67**	**5.65**	
3.23	3.20	3.15	3.12	3.08	3.05	3.03	3.00	2.98	2.96	2.94	2.93	8
5.56	**5.48**	**5.36**	**5.28**	**5.20**	**5.11**	**5.06**	**5.00**	**4.96**	**4.91**	**4.88**	**4.86**	
3.02	2.98	2.93	2.90	2.86	2.82	2.80	2.77	2.76	2.73	2.72	2.71	9
5.00	**4.92**	**4.80**	**4.73**	**4.64**	**4.56**	**4.51**	**4.45**	**4.41**	**4.36**	**4.33**	**4.31**	
2.86	2.82	2.77	2.74	2.70	2.67	2.64	2.61	2.59	2.56	2.55	2.54	10
4.60	**4.52**	**4.41**	**4.33**	**4.25**	**4.17**	**4.12**	**4.05**	**4.01**	**3.96**	**3.93**	**3.91**	
2.74	2.70	2.65	2.61	2.57	2.53	2.50	2.47	2.45	2.42	2.41	2.40	11
4.29	**4.21**	**4.10**	**4.02**	**3.94**	**3.86**	**3.80**	**3.74**	**3.70**	**3.66**	**3.62**	**3.60**	
2.64	2.60	2.54	2.50	2.46	2.42	2.40	2.36	2.35	2.32	2.31	2.30	12
4.05	**3.98**	**3.86**	**3.78**	**3.70**	**3.61**	**3.56**	**3.49**	**3.46**	**3.41**	**3.38**	**3.36**	
2.55	2.51	2.46	2.42	2.38	2.34	2.32	2.28	2.26	2.24	2.22	2.21	13
3.85	**3.78**	**3.67**	**3.59**	**3.51**	**3.42**	**3.37**	**3.30**	**3.27**	**3.21**	**3.18**	**3.16**	
2.48	2.44	2.39	2.35	2.31	2.27	2.24	2.21	2.19	2.16	2.14	2.13	14
3.70	**3.62**	**3.51**	**3.43**	**3.34**	**3.26**	**3.21**	**3.14**	**3.11**	**3.06**	**3.02**	**3.00**	
2.43	2.39	2.33	2.29	2.25	2.21	2.18	2.15	2.12	2.10	2.08	2.07	15
3.56	**3.48**	**3.36**	**3.29**	**3.20**	**3.12**	**3.07**	**3.00**	**2.97**	**2.92**	**2.89**	**2.87**	
2.37	2.33	2.28	2.24	2.20	2.16	2.13	2.09	2.07	2.04	2.02	2.01	16
3.45	**3.37**	**3.25**	**3.18**	**3.10**	**3.01**	**2.96**	**2.89**	**2.86**	**2.80**	**2.77**	**2.75**	
2.33	2.29	2.23	2.19	2.15	2.11	2.08	2.04	2.02	1.99	1.97	1.96	17
3.35	**3.27**	**3.16**	**3.08**	**3.00**	**2.92**	**2.86**	**2.79**	**2.76**	**2.70**	**2.67**	**2.65**	
2.29	2.25	2.19	2.15	2.11	2.07	2.04	2.00	1.98	1.95	1.93	1.92	18
3.27	**3.19**	**3.07**	**3.00**	**2.91**	**2.83**	**2.78**	**2.71**	**2.68**	**2.62**	**2.59**	**2.57**	
2.26	2.21	2.15	2.11	2.07	2.02	2.00	1.96	1.94	1.91	1.90	1.88	19
3.19	**3.12**	**3.00**	**2.92**	**2.84**	**2.76**	**2.70**	**2.63**	**2.60**	**2.54**	**2.51**	**2.49**	
2.23	2.18	2.12	2.08	2.04	1.99	1.96	1.92	1.90	1.87	1.85	1.84	20
3.13	**3.05**	**2.94**	**2.86**	**2.77**	**2.69**	**2.63**	**2.56**	**2.53**	**2.47**	**2.44**	**2.42**	
2.20	2.15	2.09	2.05	2.00	1.96	1.93	1.89	1.87	1.84	1.82	1.81	21
3.07	**2.99**	**2.88**	**2.80**	**2.72**	**2.63**	**2.58**	**2.51**	**2.47**	**2.42**	**2.38**	**2.36**	
2.18	2.13	2.07	2.03	1.98	1.93	1.91	1.87	1.84	1.81	1.80	1.78	22
3.02	**2.94**	**2.83**	**2.75**	**2.67**	**2.58**	**2.53**	**2.46**	**2.42**	**2.37**	**2.33**	**2.31**	
2.14	2.10	2.04	2.00	1.96	1.91	1.88	1.84	1.82	1.79	1.77	1.76	23
2.97	**2.89**	**2.78**	**2.70**	**2.62**	**2.53**	**2.48**	**2.41**	**2.37**	**2.32**	**2.28**	**2.26**	
2.13	2.09	2.02	1.98	1.94	1.89	1.86	1.82	1.80	1.76	1.74	1.73	24
2.93	**2.85**	**2.74**	**2.66**	**2.58**	**2.49**	**2.44**	**2.36**	**2.33**	**2.27**	**2.23**	**2.21**	
2.11	2.06	2.00	1.96	1.92	1.87	1.84	1.80	1.77	1.74	1.72	1.71	25
2.89	**2.81**	**2.70**	**2.62**	**2.54**	**2.45**	**2.40**	**2.32**	**2.29**	**2.23**	**2.19**	**2.17**	
2.10	2.05	1.99	1.95	1.90	1.85	1.82	1.78	1.76	1.72	1.70	1.69	26
2.86	**2.77**	**2.66**	**2.58**	**2.50**	**2.41**	**2.36**	**2.28**	**2.25**	**2.19**	**2.15**	**2.13**	

D. Values on the *F* Distributions (*continued*)

m_2 = Degrees of Freedom for Denominator; m_1 = Degrees of Freedom for Numerator

m_2	1	2	3	4	5	6	7	8	9	10	11	12
27	4.21 **7.68**	3.35 **5.49**	2.96 **4.60**	2.73 **4.11**	2.57 **3.79**	2.46 **3.56**	2.37 **3.39**	2.30 **3.26**	2.25 **3.14**	2.20 **3.06**	2.16 **2.98**	2.13 **2.93**
28	4.20 **7.64**	3.34 **5.45**	2.95 **4.57**	2.71 **4.07**	2.56 **3.76**	2.44 **3.53**	2.36 **3.36**	2.29 **3.23**	2.24 **3.11**	2.19 **3.03**	2.15 **2.95**	2.12 **2.90**
29	4.18 **7.60**	3.33 **5.42**	2.93 **4.54**	2.70 **4.04**	2.54 **3.73**	2.43 **3.50**	2.35 **3.33**	2.28 **3.20**	2.22 **3.08**	2.18 **3.00**	2.14 **2.92**	2.10 **2.87**
30	4.17 **7.56**	3.32 **5.39**	2.92 **4.51**	2.69 **4.02**	2.53 **3.70**	2.42 **3.47**	2.34 **3.30**	2.27 **3.17**	2.21 **3.06**	2.16 **2.98**	2.12 **2.90**	2.09 **2.84**
32	4.15 **7.50**	3.30 **5.34**	2.90 **4.46**	2.67 **3.97**	2.51 **3.66**	2.40 **3.42**	2.32 **3.25**	2.25 **3.12**	2.19 **3.01**	2.14 **2.94**	2.10 **2.86**	2.07 **2.80**
34	4.13 **7.44**	3.28 **5.29**	2.88 **4.42**	2.65 **3.93**	2.49 **3.61**	2.38 **3.38**	2.30 **3.21**	2.23 **3.08**	2.17 **2.97**	2.12 **2.89**	2.08 **2.82**	2.05 **2.76**
36	4.11 **7.39**	3.26 **5.25**	2.86 **4.38**	2.63 **3.89**	2.48 **3.58**	2.36 **3.35**	2.28 **3.18**	2.21 **3.04**	2.15 **2.94**	2.10 **2.86**	2.06 **2.78**	2.03 **2.72**
38	4.10 **7.35**	3.25 **5.21**	2.85 **4.34**	2.62 **3.86**	2.46 **3.54**	2.35 **3.32**	2.26 **3.15**	2.19 **3.02**	2.14 **2.91**	2.09 **2.82**	2.05 **2.75**	2.02 **2.69**
40	4.08 **7.31**	3.23 **5.18**	2.84 **4.31**	2.61 **3.83**	2.45 **3.51**	2.34 **3.29**	2.25 **3.12**	2.18 **2.99**	2.12 **2.88**	2.07 **2.80**	2.04 **2.73**	2.00 **2.66**
42	4.07 **7.27**	3.22 **5.15**	2.83 **4.29**	2.59 **3.80**	2.44 **3.49**	2.32 **3.26**	2.24 **3.10**	2.17 **2.96**	2.11 **2.86**	2.06 **2.77**	2.02 **2.70**	1.99 **2.64**
44	4.06 **7.24**	3.21 **5.12**	2.82 **4.26**	2.58 **3.78**	2.43 **3.46**	2.31 **3.24**	2.23 **3.07**	2.16 **2.94**	2.10 **2.84**	2.05 **2.75**	2.01 **2.68**	1.98 **2.62**
46	4.05 **7.21**	3.20 **5.10**	2.81 **4.24**	2.57 **3.76**	2.42 **3.44**	2.30 **3.22**	2.22 **3.05**	2.14 **2.92**	2.09 **2.82**	2.04 **2.73**	2.00 **2.66**	1.97 **2.60**
48	4.04 **7.19**	3.19 **5.08**	2.80 **4.22**	2.56 **3.74**	2.41 **3.42**	2.30 **3.20**	2.21 **3.04**	2.14 **2.90**	2.08 **2.80**	2.03 **2.71**	1.99 **2.64**	1.96 **2.58**
50	4.03 **7.17**	3.18 **5.06**	2.79 **4.20**	2.56 **3.72**	2.40 **3.41**	2.29 **3.18**	2.20 **3.02**	2.13 **2.88**	2.07 **2.78**	2.02 **2.70**	1.98 **2.62**	1.95 **2.56**
55	4.02 **7.12**	3.17 **5.01**	2.78 **4.16**	2.54 **3.68**	2.38 **3.37**	2.27 **3.15**	2.18 **2.98**	2.11 **2.85**	2.05 **2.75**	2.00 **2.66**	1.97 **2.59**	1.93 **2.53**
60	4.00 **7.08**	3.15 **4.98**	2.76 **4.13**	2.52 **3.65**	2.37 **3.34**	2.25 **3.12**	2.17 **2.95**	2.10 **2.82**	2.04 **2.72**	1.99 **2.63**	1.95 **2.56**	1.92 **2.50**
65	3.99 **7.04**	3.14 **4.95**	2.75 **4.10**	2.51 **3.62**	2.36 **3.31**	2.24 **3.09**	2.15 **2.93**	2.08 **2.79**	2.02 **2.70**	1.98 **2.61**	1.94 **2.54**	1.90 **2.47**
70	3.98 **7.01**	3.13 **4.92**	2.74 **4.08**	2.50 **3.60**	2.35 **3.29**	2.23 **3.07**	2.14 **2.91**	2.07 **2.77**	2.01 **2.67**	1.97 **2.59**	1.93 **2.51**	1.89 **2.45**
80	3.96 **6.96**	3.11 **4.88**	2.72 **4.04**	2.48 **3.56**	2.33 **3.25**	2.21 **3.04**	2.12 **2.87**	2.05 **2.74**	1.99 **2.64**	1.95 **2.55**	1.91 **2.48**	1.88 **2.41**
100	3.94 **6.90**	3.09 **4.82**	2.70 **3.98**	2.46 **3.51**	2.30 **3.20**	2.19 **2.99**	2.10 **2.82**	2.03 **2.69**	1.97 **2.59**	1.92 **2.51**	1.88 **2.43**	1.85 **2.36**
125	3.92 **6.84**	3.07 **4.78**	2.68 **3.94**	2.44 **3.47**	2.29 **3.17**	2.17 **2.95**	2.08 **2.79**	2.01 **2.65**	1.95 **2.56**	1.90 **2.47**	1.86 **2.40**	1.83 **2.33**
150	3.91 **6.81**	3.06 **4.75**	2.67 **3.91**	2.43 **3.44**	2.27 **3.14**	2.16 **2.92**	2.07 **2.76**	2.00 **2.62**	1.94 **2.53**	1.89 **2.44**	1.85 **2.37**	1.82 **2.30**
200	3.89 **6.76**	3.04 **4.71**	2.65 **3.88**	2.41 **3.41**	2.26 **3.11**	2.14 **2.90**	2.05 **2.73**	1.98 **2.60**	1.92 **2.50**	1.87 **2.41**	1.83 **2.34**	1.80 **2.28**
400	3.86 **6.70**	3.02 **4.66**	2.62 **3.83**	2.39 **3.36**	2.23 **3.06**	2.12 **2.85**	2.03 **2.69**	1.96 **2.55**	1.90 **2.46**	1.85 **2.37**	1.81 **2.29**	1.78 **2.23**
1000	3.85 **6.66**	3.00 **4.62**	2.61 **3.80**	2.38 **3.34**	2.22 **3.04**	2.10 **2.82**	2.02 **2.66**	1.95 **2.53**	1.89 **2.43**	1.84 **2.34**	1.80 **2.26**	1.76 **2.20**
∞	3.84 **6.64**	2.99 **4.60**	2.60 **3.78**	2.37 **3.32**	2.21 **3.02**	2.09 **2.80**	2.01 **2.64**	1.94 **2.51**	1.88 **2.41**	1.83 **2.32**	1.79 **2.24**	1.75 **2.18**

D. Values on the *F* Distributions (*continued*)

m_1 = Degrees of Freedom for Numerator												
14	16	20	24	30	40	50	75	100	200	500	∞	m_2
2.08	2.03	1.97	1.93	1.88	1.84	1.80	1.76	1.74	1.71	1.68	1.67	27
2.83	**2.74**	**2.63**	**2.55**	**2.47**	**2.38**	**2.33**	**2.25**	**2.21**	**2.16**	**2.12**	**2.10**	
2.06	2.02	1.96	1.91	1.87	1.81	1.78	1.75	1.72	1.69	1.67	1.65	28
2.80	**2.71**	**2.60**	**2.52**	**2.44**	**2.35**	**2.30**	**2.22**	**2.18**	**2.13**	**2.09**	**2.06**	
2.05	2.00	1.94	1.90	1.85	1.80	1.77	1.73	1.71	1.68	1.65	1.64	29
2.77	**2.68**	**2.57**	**2.49**	**2.41**	**2.32**	**2.27**	**2.19**	**2.15**	**2.10**	**2.06**	**2.03**	
2.04	1.99	1.93	1.89	1.84	1.79	1.76	1.72	1,69	1.66	1.64	1.62	30
2.74	**2.66**	**2.55**	**2.47**	**2.38**	**2.29**	**2.24**	**2.16**	**2.13**	**2.07**	**2.03**	**2.01**	
2.02	1.97	1.91	1.86	1.82	1.76	1.74	1.69	1.67	1.64	1.61	1.59	32
2.70	**2.62**	**2.51**	**2.42**	**2.34**	**2.25**	**2.20**	**2.12**	**2.08**	**2.02**	**1.98**	**1.96**	
2.00	1.95	1.89	1.84	1.80	1.74	1.71	1.67	1.64	1.61	1.59	1.57	34
2.66	**2.58**	**2.47**	**2.38**	**2.30**	**2.21**	**2.15**	**2.08**	**2.04**	**1.98**	**1.94**	**1.91**	
1.98	1.93	1.87	1.82	1.78	1.72	1.69	1.65	1.62	1.59	1.56	1.55	36
2.62	**2.54**	**2.43**	**2.35**	**2.26**	**2.17**	**2.12**	**2.04**	**2.00**	**1.94**	**1.90**	**1.87**	
1.96	1.92	1.85	1.80	1.76	1.71	1.67	1.63	1.60	1.57	1.54	1.53	38
2.59	**2.51**	**2.40**	**2.32**	**2.22**	**2.14**	**2.08**	**2.00**	**1.97**	**1.90**	**1.86**	**1.84**	
1.95	1.90	1.84	1.79	1.74	1.69	1.66	1.61	1.59	1.55	1.53	1.51	40
2.56	**2.49**	**2.37**	**2.29**	**2.20**	**2.11**	**2.05**	**1.97**	**1.94**	**1.88**	**1.84**	**1.81**	
1.94	1.89	1.82	1.78	1.73	1.68	1.64	1.60	1.57	1.54	1.51	1.49	42
2.54	**2.46**	**2.35**	**2.26**	**2.17**	**2.08**	**2.02**	**1.94**	**1.91**	**1.85**	**1.80**	**1.78**	
1.92	1.88	1.81	1.76	1.72	1.66	1.63	1.58	1.56	1.52	1.50	1.48	44
2.52	**2.44**	**2.32**	**2.24**	**2.15**	**2.06**	**2.00**	**1.92**	**1.88**	**1.82**	**1.78**	**1.75**	
1.91	1.87	1.80	1.75	1.71	1.65	1.62	1.57	1.54	1.51	1.48	1.46	46
2.50	**2.42**	**2.30**	**2.22**	**2.13**	**2.04**	**1.98**	**1.90**	**1.86**	**1.80**	**1.76**	**1.72**	
1.90	1.86	1.79	1.74	1.70	1.64	1.61	1.56	1.53	1.50	1.47	1.45	48
2.48	**2.40**	**2.28**	**2.20**	**2.11**	**2.02**	**1.96**	**1.88**	**1.84**	**1.78**	**1.73**	**1.70**	
1.90	1.85	1.78	1.74	1.69	1.63	1.60	1.55	1.52	1.48	1.46	1.44	50
2.46	**2.39**	**2.26**	**2.18**	**2.10**	**2.00**	**1.94**	**1.86**	**1.82**	**1.76**	**1.71**	**1.68**	
1.88	1.83	1.76	1.72	1.67	1.61	1.58	1.52	1.50	1.46	1.43	1.41	55
2.43	**2.35**	**2.23**	**2.15**	**2.06**	**1.96**	**1.90**	**1.82**	**1.78**	**1.71**	**1.66**	**1.64**	
1.86	1.81	1.75	1.70	1.65	1.59	1.56	1.50	1.48	1.44	1.41	1.39	60
2.40	**2.32**	**2.20**	**2.12**	**2.03**	**1.93**	**1.87**	**1.79**	**1.74**	**1.68**	**1.63**	**1.60**	
1.85	1.80	1.73	1.68	1.63	1.57	1.54	1.49	1.46	1.42	1.39	1.37	65
2.37	**2.30**	**2.18**	**2.09**	**2.00**	**1.90**	**1.84**	**1.76**	**1.71**	**1.64**	**1.60**	**1.56**	
1.84	1.79	1.72	1.67	1.62	1.56	1.53	1.47	1.45	1.40	1.37	1.35	70
2.35	**2.28**	**2.15**	**2.07**	**1.98**	**1.88**	**1.82**	**1.74**	**1.69**	**1.62**	**1.56**	**1.53**	
1.82	1.77	1.70	1.65	1.60	1.54	1.51	1.45	1.42	1.38	1.35	1.32	80
2.32	**2.24**	**2.11**	**2.03**	**1.94**	**1.84**	**1.78**	**1.70**	**1.65**	**1.57**	**1.52**	**1.49**	
1.79	1.75	1.68	1.63	1.57	1.51	1.48	1.42	1.39	1.34	1.30	1.28	100
2.26	**2.19**	**2.06**	**1.98**	**1.89**	**1.79**	**1.73**	**1.64**	**1.59**	**1.51**	**1.46**	**1.43**	
1.77	1.72	1.65	1.60	1.55	1.49	1.45	1.39	1.36	1.31	1.27	1.25	125
2.23	**2.15**	**2.03**	**1.94**	**1.85**	**1.75**	**1.68**	**1.59**	**1.54**	**1.46**	**1.40**	**1.37**	
1.76	1.71	1.64	1.59	1.54	1.47	1.44	1.37	1.34	1.29	1.25	1.22	150
2.20	**2.12**	**2.00**	**1.91**	**1.83**	**1.72**	**1.66**	**1.56**	**1.51**	**1.43**	**1.37**	**1.33**	
1.74	1.69	1.62	1.57	1.52	1.45	1.42	1.35	1.32	1.26	1.22	1.19	200
2.17	**2.09**	**1.97**	**1.88**	**1.79**	**1.69**	**1.62**	**1.53**	**1.48**	**1.39**	**1.33**	**1.28**	
1.72	1.67	1.60	1.54	1.49	1.42	1.38	1.32	1.28	1.22	1.16	1.13	400
2.12	**2.04**	**1.92**	**1.84**	**1.74**	**1.64**	**1.57**	**1.47**	**1.42**	**1.32**	**1.24**	**1.19**	
1.70	1.65	1.58	1.53	1.47	1.41	1.36	1.30	1.26	1.19	1.13	1.08	1000
2.09	**2.01**	**1.89**	**1.81**	**1.71**	**1.61**	**1.54**	**1.44**	**1.38**	**1.28**	**1.19**	**1.11**	
1.69	1.64	1.57	1.52	1.46	1.40	1.35	1.28	1.24	1.17	1.11	1.00	∞
2.07	**1.99**	**1.87**	**1.79**	**1.69**	**1.59**	**1.52**	**1.41**	**1.36**	**1.25**	**1.15**	**1.00**	

m_2 = Degrees of Freedom for Denominator

E. Binomial Distributions

n	r	π = .05	.10	.15	.20	.25	.30	.35	.40	.45	.50	.55	.60	.65	.70	.75	.80	.85	.90	.95
1	0	.9500	.9000	.8500	.8000	.7500	.7000	.6500	.6000	.5500	.5000	.4500	.4000	.3500	.3000	.2500	.2000	.1500	.1000	.0500
	1	.0500	.1000	.1500	.2000	.2500	.3000	.3500	.4000	.4500	.5000	.5500	.6000	.6500	.7000	.7500	.8000	.8500	9000	.9500
2	0	.9025	.8100	.7225	.6400	.5625	.4900	.4225	.3600	.3025	.2500	.2025	.1600	.1225	.0900	.0625	.0400	.0225	.0100	.0025
	1	.0950	.1800	.2550	.3200	.3750	.4200	.4550	.4800	.4950	.5000	.4950	.4800	.4550	.4200	.3750	.3200	.2550	.1800	.0950
	2	.0025	.0100	.0225	.0400	.0625	.0900	.1225	.1600	.2025	.2500	.3025	.3600	.4225	.4900	.5625	.6400	.7225	.8100	.9025
3	0	.8574	.7290	.6141	.5120	.4219	.3430	.2746	.2160	.1664	.1250	.0911	.0640	.0429	.0270	.0156	.0080	.0034	.0010	.0001
	1	.1354	.2430	.3251	.3840	.4219	.4410	.4436	.4320	.4084	.3750	.3341	.2880	.2389	.1890	.1406	.0960	.0574	.0270	.0071
	2	.0071	.0270	.0574	.0960	.1406	.1890	.2389	.2880	.3341	.3750	.4084	.4320	.4436	.4410	.4219	.3840	.3251	.2430	.1354
	3	.0001	.0010	.0034	.0080	.0156	.0270	.0429	.0640	.0911	.1250	.1664	.2160	.2746	.3430	.4219	.5120	.6141	.7290	.8574
4	0	.8145	.6561	.5220	.4096	.3164	.2401	.1785	.1296	.0915	.0625	.0410	.0256	.0150	.0081	.0039	.0016	.0005	.0001	.0000
	1	.1715	.2916	.3685	.4096	.4219	.4116	.3845	.3456	.2995	.2500	.2005	.1536	.1115	.0756	.0469	.0256	.0115	.0036	.0005
	2	.0135	.0486	.0975	.1536	.2109	.2646	.3105	.3456	.3675	.3750	.3675	.3456	.3105	.2646	.2109	.1536	.0975	.0486	.0135
	3	.0005	.0036	.0115	.0256	.0469	.0756	.1115	.1536	.2005	.2500	.2995	.3456	.3845	.4116	.4219	.4096	.3865	.2916	.1715
	4	.0000	.0001	.0005	.0016	.0039	.0081	.0150	.0256	.0410	.0625	.0915	.1296	.1785	.2401	.3164	.4096	.5220	.6561	.8145
5	0	.7738	.5905	.4437	.3277	.2373	.1681	.1160	.0778	.0503	.0312	.0185	.0102	.0053	.0024	.0010	.0003	.0001	.0000	.0000
	1	.2036	.3280	.3915	.4096	.3955	.3602	.3124	.2592	.2059	.1562	.1128	.0768	.0488	.0284	.0146	.0064	.0022	.0004	.0000
	2	.0214	.0729	.1382	.2048	.2637	.3087	.3364	.3456	.3369	.3125	.2757	.2304	.1811	.1323	.0879	.0512	.0244	.0081	.0011
	3	.0011	.0081	.0244	.0512	.0879	.1323	.1811	.2304	.2757	.3125	.3369	.3456	.3364	.3087	.2637	.2048	.1382	.0729	.0214
	4	.0000	.0004	.0022	.0064	.0146	.0284	.0488	.0768	.1128	.1562	.2059	.2592	.3124	.3602	.3955	.4096	.3915	.3280	.2036
	5	.0000	.0000	.0001	.0003	.0010	.0024	.0053	.0102	.0185	.0312	.0503	.0778	.1160	.1681	.2373	.3277	.4437	.5905	.7738
6	0	.7351	.5314	.3771	.2621	.1780	.1176	.0754	.0467	.0277	.0156	.0083	.0041	.0018	.0007	.0002	.0001	.0000	.0000	.0000
	1	.2321	.3543	.3993	.3932	.3560	.3025	.2437	.1866	.1359	.0938	.0609	.0369	.0205	.0102	.0044	.0015	.0004	.0001	.0000
	2	.0305	.0984	.1762	.2458	.2966	.3241	.3280	.3110	.2780	.2344	.1861	.1382	.0951	.0595	.0330	.0154	.0055	.0012	.0001
	3	.0021	.0146	.0415	.0819	.1318	.1852	.2355	.2765	.3032	.3125	.3032	.2765	.2355	.1852	.1318	.0819	.0415	.0146	.0021
	4	.0001	.0012	.0055	.0154	.0330	.0595	.0951	.1382	.1861	.2344	.2780	.3110	.3280	.3241	.2966	.2458	.1762	.0984	.0305
	5	.0000	.0001	.0004	.0015	.0044	.0102	.0205	.0369	.0609	.0938	.1359	.1866	.2437	.3025	.3560	.3932	.3993	.3543	.2321
	6	.0000	.0000	.0000	.0001	.0002	.0007	.0018	.0041	.0083	.0156	.0277	.0467	.0754	.1176	.1780	.2621	.3771	.5314	.7351
7	0	.6983	.4783	.3206	.2097	.1335	.0824	.0490	.0280	.0152	.0078	.0037	.0016	.0006	.0002	.0001	.0000	.0000	.0000	.0000
	1	.2573	.3720	.3960	.3670	.3115	.2471	.1848	.1306	.0872	.0547	.0320	.0172	.0084	.0036	.0013	.0004	.0001	.0000	.0000
	2	.0406	.1240	.2097	.2753	.3115	.3177	.2985	.2613	.2140	.1641	.1172	.0774	.0466	.0250	.0115	.0043	.0012	.0002	.0000
	3	.0036	.0230	.0617	.1147	.1730	.2269	.2679	.2903	.2918	.2734	.2388	.1935	.1442	.0972	.0577	.0287	.0109	.0026	.0002
	4	.0002	.0026	.0109	.0287	.0577	.0972	.1442	.1935	.2388	.2734	.2918	.2903	.2679	.2269	.1730	.1147	.0617	.0230	.0036
	5	.0000	.0002	.0012	.0043	.0115	.0250	.0466	.0774	.1172	.1641	.2140	.2613	.2985	.3177	.3115	.2753	.2097	.1240	.0406
	6	.0000	.0000	.0001	.0004	.0013	.0036	.0084	.0172	.0320	.0547	.0872	.1306	.1848	.2471	.3115	.3670	.3960	.3720	.2573
	7	.0000	.0000	.0000	.0000	.0001	.0002	.0006	.0016	.0037	.0078	.0152	.0280	.0490	.0824	.1335	.2097	.3206	.4783	.6983

E. Binomial Distributions (*continued*)

n	r	π = .05	.10	.15	.20	.25	.30	.35	.40	.45	.50	.55	.60	.65	.70	.75	.80	.85	.90	.95
8	0	.6634	.4305	.2725	.1678	.1001	.0576	.0319	.0168	.0084	.0039	.0017	.0007	.0002	.0001	.0000	.0000	.0000	.0000	.0000
	1	.2793	.3826	.3847	.3355	.2670	.1977	.1373	.0896	.0548	.0312	.0164	.0079	.0033	.0012	.0004	.0001	.0000	.0000	.0000
	2	.0515	.1488	.2376	.2936	.3115	.2965	.2587	.2090	.1569	.1094	.0703	.0413	.0217	.0100	.0038	.0011	.0002	.0000	.0000
	3	.0054	.0331	.0839	.1468	.2076	.2541	.2786	.2787	.2568	.2188	.1719	.1239	.0808	.0467	.0231	.0092	.0026	.0004	.0000
	4	.0004	.0046	.0185	.0459	.0865	.1361	.1875	.2322	.2627	.2734	.2627	.2322	.1875	.1361	.0865	.0459	.0185	.0046	.0004
	5	.0000	.0004	.0026	.0092	.0231	.0467	.0808	.1239	.1719	.2188	.2568	.2787	.2786	.2541	.2076	.1468	.0839	.0331	.0054
	6	.0000	.0000	.0002	.0011	.0038	.0100	.0217	.0413	.0703	.1094	.1569	.2090	.2587	.2965	.3115	.2936	.2376	.1488	.0515
	7	.0000	.0000	.0000	.0001	.0004	.0012	.0033	.0079	.0164	.0312	.0548	.0896	.1373	.1977	.2670	.3355	.3847	.3826	.2793
	8	.0000	.0000	.0000	.0000	.0000	.0001	.0002	.0007	.0017	.0039	.0084	.0168	.0319	.0576	.1001	.1678	.2725	.4305	.6634
9	0	.6302	.3874	.2316	.1342	.0751	.0404	.0207	.0101	.0046	.0020	.0008	.0003	.0001	.0000	.0000	.0000	.0000	.0000	.0000
	1	.2985	.3874	.3679	.3020	.2253	.1556	.1004	.0605	.0339	.0176	.0083	.0035	.0013	.0004	.0001	.0000	.0000	.0000	.0000
	2	.0629	.1722	.2597	.3020	.3003	.2668	.2162	.1612	.1110	.0703	.0407	.0212	.0098	.0039	.0012	.0003	.0000	.0000	.0000
	3	.0077	.0446	.1069	.1762	.2336	.2668	.2716	.2508	.2119	.1641	.1160	.0743	.0424	.0210	.0087	.0028	.0006	.0001	.0000
	4	.0006	.0074	.0283	.0661	.1168	.1715	.2194	.2508	.2600	.2461	.2128	.1672	.1181	.0735	.0389	.0165	.0050	.0008	.0000
	5	.0000	.0008	.0050	.0165	.0389	.0735	.1181	.1672	.2128	.2461	.2600	.2508	.2194	.1715	.1168	.0661	.0283	.0074	.0006
	6	.0000	.0001	.0006	.0028	.0087	.0210	.0424	.0743	.1160	.1641	.2119	.2508	.2716	.2668	.2336	.1762	.1069	.0446	.0077
	7	.0000	.0000	.0000	.0003	.0012	.0039	.0098	.0212	.0407	.0703	.1110	.1612	.2162	.2668	.3003	.3020	.2597	.1722	.0629
	8	.0000	.0000	.0000	.0000	.0001	.0004	.0013	.0035	.0083	.0176	.0339	.0605	.1004	.1556	.2253	.3020	.3679	.3874	.2985
	9	.0000	.0000	.0000	.0000	.0000	.0000	.0001	.0003	.0008	.0020	.0046	.0101	.0207	.0404	.0751	.1342	.2316	.3874	.6302
10	0	.5987	.3487	.1969	.1074	.0563	.0282	.0135	.0060	.0025	.0010	.0003	.0001	.0000	.0000	.0000	.0000	.0000	.0000	.0000
	1	.3151	.3874	.3474	.2684	.1877	.1211	.0725	.0403	.0207	.0098	.0042	.0016	.0005	.0001	.0000	.0000	.0000	.0000	.0000
	2	.0746	.1937	.2759	.3020	.2816	.2335	.1757	.1209	.0763	.0439	.0229	.0106	.0043	.0014	.0004	.0001	.0000	.0000	.0000
	3	.0105	.0574	.1298	.2013	.2503	.2668	.2522	.2150	.1665	.1172	.0746	.0425	.0212	.0090	.0031	.0008	.0001	.0000	.0000
	4	.0010	.0112	.0401	.0881	.1460	.2001	.2377	.2508	.2384	.2051	.1596	.1115	.0689	.0368	.0162	.0055	.0012	.0001	.0000
	5	.0001	.0015	.0085	.0264	.0584	.1029	.1536	.2007	.2340	.2461	.2340	.2007	.1536	.1029	.0584	.0264	.0085	.0015	.0001
	6	.0000	.0001	.0012	.0055	.0162	.0368	.0689	.1115	.1596	.2051	.2384	.2508	.2377	.2001	.1460	.0881	.0401	.0112	.0010
	7	.0000	.0000	.0001	.0008	.0031	.0090	.0212	.0425	.0746	.1172	.1665	.2150	.2522	.2668	.2503	.2013	.1298	.0574	.0105
	8	.0000	.0000	.0000	.0001	.0004	.0014	.0043	.0106	.0229	.0439	.0763	.1209	.1757	.2335	.2816	.3020	.2759	.1937	.0746
	9	.0000	.0000	.0000	.0000	.0000	.0001	.0005	.0016	.0042	.0098	.0207	.0403	.0725	.1211	.1877	.2684	.3474	.3874	.3151
	10	.0000	.0000	.0000	.0000	.0000	.0000	.0000	.0001	.0003	.0010	.0025	.0060	.0135	.0282	.0563	.1074	.1969	.3487	.5987
11	0	.5688	.3138	.1673	.0859	.0422	.0198	.0088	.0036	.0014	.0005	.0002	.0000	.0000	.0000	.0000	.0000	.0000	.0000	.0000
	1	.3293	.3835	.3248	.2362	.1549	.0932	.0518	.0266	.0125	.0054	.0021	.0007	.0002	.0000	.0000	.0000	.0000	.0000	.0000
	2	.0867	.2131	.2866	.2953	.2581	.1998	.1395	.0887	.0513	.0269	.0126	.0052	.0018	.0005	.0001	.0000	.0000	.0000	.0000
	3	.0137	.0710	.1517	.2215	.2581	.2568	.2254	.1774	.1259	.0806	.0462	.0234	.0102	.0037	.0011	.0002	.0000	.0000	.0000
	4	.0014	.0158	.0536	.1107	.1721	.2201	.2428	.2365	.2060	.1611	.1128	.0701	.0379	.0173	.0064	.0017	.0003	.0000	.0000
	5	.0001	.0025	.0132	.0388	.0803	.1321	.1830	.2207	.2360	.2256	.1931	.1471	.0985	.0566	.0268	.0097	.0023	.0003	.0000
	6	.0000	.0003	.0023	.0097	.0268	.0566	.0985	.1471	.1931	.2256	.2360	.2207	.1830	.1321	.0803	.0388	.0132	.0025	.0001
	7	.0000	.0000	.0003	.0017	.0064	.0173	.0379	.0701	.1128	.1611	.2060	.2365	.2428	.2201	.1721	.1107	.0536	.0158	.0014
	8	.0000	.0000	.0000	.0002	.0011	.0037	.0102	.0234	.0462	.0806	.1259	.1774	.2254	.2568	.2581	.2215	.1517	.0710	.0137
	9	.0000	.0000	.0000	.0000	.0001	.0005	.0018	.0052	.0126	.0269	.0513	.0887	.1395	.1998	.2581	.2953	.2866	.2131	.0867
	10	.0000	.0000	.0000	.0000	.0000	.0000	.0002	.0007	.0021	.0054	.0125	.0266	.0518	.0932	.1549	.2362	.3248	.3835	.3293
	11	.0000	.0000	.0000	.0000	.0000	.0000	.0000	.0000	.0002	.0005	.0014	.0036	.0088	.0198	.0422	.0859	.1673	.3138	.5688

E. Binomial Distributions (*continued*)

n	r	π = .05	.10	.15	.20	.25	.30	.35	.40	.45	.50	.55	.60	.65	.70	.75	.80	.85	.90	.95
12	0	.5404	.2824	.1422	.0687	.0317	.0138	.0057	.0022	.0008	.0002	.0001	.0000	.0000	.0000	.0000	.0000	.0000	.0000	.0000
	1	.3413	.3766	.3012	.2062	.1267	.0712	.0368	.0174	.0075	.0029	.0010	.0003	.0001	.0000	.0000	.0000	.0000	.0000	.0000
	2	.0988	.2301	.2924	.2835	.2323	.1678	.1088	.0639	.0339	.0161	.0068	.0025	.0008	.0002	.0000	.0000	.0000	.0000	.0000
	3	.0173	.0852	.1720	.2362	.2581	.2397	.1954	.1419	.0923	.0537	.0277	.0125	.0048	.0015	.0004	.0001	.0000	.0000	.0000
	4	.0021	.0213	.0683	.1329	.1936	.2311	.2367	.2128	.1700	.1208	.0762	.0420	.0199	.0078	.0024	.0005	.0001	.0000	.0000
	5	.0002	.0038	.0193	.0532	.1032	.1585	.2039	.2270	.2225	.1934	.1489	.1009	.0591	.0291	.0115	.0033	.0006	.0000	.0000
	6	.0000	.0005	.0040	.0155	.0401	.0792	.1281	.1766	.2124	.2256	.2124	.1766	.1281	.0792	.0401	.0155	.0040	.0005	.0000
	7	.0000	.0000	.0006	.0033	.0115	.0291	.0591	.1009	.1489	.1934	.2225	.2270	.2039	.1585	.1032	.0532	.0193	.0038	.0002
	8	.0000	.0000	.0001	.0005	.0024	.0078	.0199	.0420	.0762	.1208	.1700	.2128	.2367	.2311	.1936	.1329	.0683	.0213	.0021
	9	.0000	.0000	.0000	.0001	.0004	.0015	.0048	.0125	.0277	.0537	.0923	.1419	.1954	.2397	.2581	.2362	.1720	.0852	.0173
	10	.0000	.0000	.0000	.0000	.0000	.0002	.0008	.0025	.0068	.0161	.0339	.0639	.1088	.1678	.2323	.2835	.2924	.2301	.0988
	11	.0000	.0000	.0000	.0000	.0000	.0000	.0001	.0003	.0010	.0029	.0075	.0174	.0368	.0712	.1267	.2062	.3012	.3766	.3413
	12	.0000	.0000	.0000	.0000	.0000	.0000	.0000	.0000	.0001	.0002	.0008	.0022	.0057	.0138	.0317	.0687	.1422	.2824	.5404
13	0	.5133	.2542	.1209	.0550	.0238	.0097	.0037	.0013	.0004	.0001	.0000	.0000	.0000	.0000	.0000	.0000	.0000	.0000	.0000
	1	.3512	.3672	.2774	.1787	.1029	.0540	.0259	.0113	.0045	.0016	.0005	.0001	.0000	.0000	.0000	.0000	.0000	.0000	.0000
	2	.1109	.2448	.2937	.2680	.2059	.1388	.0836	.0453	.0220	.0095	.0036	.0012	.0003	.0001	.0000	.0000	.0000	.0000	.0000
	3	.0214	.0997	.1900	.2457	.2517	.2181	.1651	.1107	.0660	.0349	.0162	.0065	.0022	.0006	.0001	.0000	.0000	.0000	.0000
	4	.0028	.0277	.0838	.1535	.2097	.2337	.2222	.1845	.1350	.0873	.0495	.0243	.0101	.0034	.0009	.0001	.0000	.0000	.0000
	5	.0003	.0055	.0266	.0691	.1258	.1803	.2154	.2214	.1989	.1571	.1089	.0656	.0336	.0142	.0047	.0011	.0001	.0000	.0000
	6	.0000	.0008	.0063	.0230	.0559	.1030	.1546	.1968	.2169	.2095	.1775	.1312	.0833	.0442	.0186	.0058	.0011	.0001	.0000
	7	.0000	.0001	.0011	.0058	.0186	.0442	.0833	.1312	.1775	.2095	.2169	.1968	.1546	.1030	.0559	.0230	.0063	.0008	.0000
	8	.0000	.0000	.0001	.0011	.0047	.0142	.0336	.0656	.1089	.1571	.1989	.2214	.2154	.1803	.1258	.0691	.0266	.0055	.0003
	9	.0000	.0000	.0000	.0001	.0009	.0034	.0101	.0243	.0495	.0873	.1350	.1845	.2222	.2337	.2097	.1535	.0838	.0277	.0028
	10	.0000	.0000	.0000	.0000	.0001	.0006	.0022	.0065	.0162	.0349	.0660	.1107	.1651	.2181	.2517	.2457	.1900	.0997	.0214
	11	.0000	.0000	.0000	.0000	.0000	.0001	.0003	.0012	.0036	.0095	.0220	.0453	.0836	.1388	.2059	.2680	.2937	.2448	.1109
	12	.0000	.0000	.0000	.0000	.0000	.0000	.0000	.0001	.0005	.0016	.0045	.0113	.0259	.0540	.1029	.1787	.2774	.3672	.3512
	13	.0000	.0000	.0000	.0000	.0000	.0000	.0000	.0000	.0000	.0001	.0004	.0013	.0037	.0097	.0238	.0550	.1209	.2542	.5133
14	0	.4877	.2288	.1028	.0440	.0178	.0068	.0024	.0008	.0002	.0001	.0000	.0000	.0000	.0000	.0000	.0000	.0000	.0000	.0000
	1	.3593	.3559	.2539	.1539	.0832	.0407	.0181	.0073	.0027	.0009	.0002	.0001	.0000	.0000	.0000	.0000	.0000	.0000	.0000
	2	.1229	.2570	.2912	.2501	.1802	.1134	.0634	.0317	.0141	.0056	.0019	.0005	.0001	.0000	.0000	.0000	.0000	.0000	.0000
	3	.0259	.1142	.2056	.2501	.2402	.1943	.1366	.0845	.0462	.0222	.0093	.0033	.0010	.0002	.0000	.0000	.0000	.0000	.0000
	4	.0037	.0349	.0998	.1720	.2202	.2290	.2022	.1549	.1040	.0611	.0312	.0136	.0049	.0014	.0003	.0000	.0000	.0000	.0000
	5	.0004	.0078	.0352	.0860	.1468	.1963	.2178	.2066	.1701	.1222	.0762	.0408	.0183	.0066	.0018	.0003	.0000	.0000	.0000
	6	.0000	.0013	.0093	.0322	.0734	.1262	.1759	.2066	.2088	.1833	.1398	.0918	.0510	.0232	.0082	.0020	.0003	.0000	.0000
	7	.0000	.0002	.0019	.0092	.0280	.0618	.1082	.1574	.1952	.2095	.1952	.1574	.1082	.0618	.0280	.0092	.0019	.0002	.0000
	8	.0000	.0000	.0003	.0020	.0082	.0232	.0510	.0918	.1398	.1833	.2088	.2066	.1759	.1262	.0734	.0322	.0093	.0013	.0000
	9	.0000	.0000	.0000	.0003	.0018	.0066	.0183	.0408	.0762	.1222	.1701	.2066	.2178	.1963	.1468	.0860	.0352	.0078	.0004
	10	.0000	.0000	.0000	.0000	.0003	.0014	.0049	.0136	.0312	.0611	.1040	.1549	.2022	.2290	.2202	.1720	.0998	.0349	.0037
	11	.0000	.0000	.0000	.0000	.0000	.0002	.0010	.0033	.0093	.0222	.0462	.0845	.1366	.1943	.2402	.2501	.2056	.1142	.0259
	12	.0000	.0000	.0000	.0000	.0000	.0000	.0001	.0005	.0019	.0056	.0141	.0317	.0634	.1134	.1802	.2501	.2912	.2570	.1229
	13	.0000	.0000	.0000	.0000	.0000	.0000	.0000	.0001	.0002	.0009	.0027	.0073	.0181	.0407	.0832	.1539	.2539	.3559	.3593
	14	.0000	.0000	.0000	.0000	.0000	.0000	.0000	.0000	.0000	.0001	.0002	.0008	.0024	.0068	.0178	.0440	.1028	.2288	.4877

E. Binomial Distributions (*continued*)

		π																		
n	r	.05	.10	.15	.20	.25	.30	.35	.40	.45	.50	.55	.60	.65	.70	.75	.80	.85	.90	.95
15	0	.4633	.2059	.0874	.0352	.0134	.0047	.0016	.0005	.0001	.0000	.0000	.0000	.0000	.0000	.0000	.0000	.0000	.0000	.0000
	1	.3658	.3432	.2312	.1319	.0668	.0305	.0126	.0047	.0016	.0005	.0001	.0000	.0000	.0000	.0000	.0000	.0000	.0000	.0000
	2	.1348	.2669	.2856	.2309	.1559	.0916	.0476	.0219	.0090	.0032	.0010	.0003	.0001	.0000	.0000	.0000	.0000	.0000	.0000
	3	.0307	.1285	.2184	.2501	.2252	.1700	.1110	.0634	.0318	.0139	.0052	.0016	.0004	.0001	.0000	.0000	.0000	.0000	.0000
	4	.0049	.0428	.1156	.1876	.2252	.2186	.1792	.1268	.0780	.0417	.0191	.0074	.0024	.0006	.0001	.0000	.0000	.0000	.0000
	5	.0006	.0105	.0449	.1032	.1651	.2061	.2123	.1859	.1404	.0916	.0515	.0245	.0096	.0030	.0007	.0001	.0000	.0000	.0000
	6	.0000	.0019	.0132	.0430	.0917	.1472	.1906	.2066	.1914	.1527	.1048	.0612	.0298	.0116	.0034	.0007	.0001	.0000	.0000
	7	.0000	.0003	.0030	.0138	.0393	.0811	.1319	.1771	.2013	.1964	.1647	.1181	.0710	.0348	.0131	.0035	.0005	.0000	.0000
	8	.0000	.0000	.0005	.0035	.0131	.0348	.0710	.1181	.1647	.1964	.2013	.1771	.1319	.0811	.0393	.0138	.0030	.0003	.0000
	9	.0000	.0000	.0001	.0007	.0034	.0116	.0298	.0612	.1048	.1527	.1914	.2066	.1906	.1472	.0917	.0430	.0132	.0019	.0000
	10	.0000	.0000	.0000	.0001	.0007	.0030	.0096	.0245	.0515	.0916	.1404	.1859	.2123	.2061	.1651	.1032	.0449	.0105	.0006
	11	.0000	.0000	.0000	.0000	.0001	.0006	.0024	.0074	.0191	.0417	.0780	.1268	.1792	.2186	.2252	.1876	.1156	.0428	.0049
	12	.0000	.0000	.0000	.0000	.0000	.0001	.0004	.0016	.0052	.0139	.0318	.0634	.1110	.1700	.2252	.2501	.2184	.1285	.0307
	13	.0000	.0000	.0000	.0000	.0000	.0000	.0001	.0003	.0010	.0032	.0090	.0219	.0476	.0916	.1559	.2309	.2856	.2669	.1348
	14	.0000	.0000	.0000	.0000	.0000	.0000	.0000	.0000	.0001	.0005	.0016	.0047	.0126	.0305	.0668	.1319	.2312	.3432	.3658
	15	.0000	.0000	.0000	.0000	.0000	.0000	.0000	.0000	.0000	.0000	.0001	.0005	.0016	.0047	.0134	.0352	.0874	.2059	.4633
16	0	.4401	.1853	.0743	.0281	.0100	.0033	.0010	.0003	.0001	.0000	.0000	.0000	.0000	.0000	.0000	.0000	.0000	.0000	.0000
	1	.3706	.3294	.2097	.1126	.0535	.0228	.0087	.0030	.0009	.0002	.0001	.0000	.0000	.0000	.0000	.0000	.0000	.0000	.0000
	2	.1463	.2745	.2775	.2111	.1336	.0732	.0353	.0150	.0056	.0018	.0005	.0001	.0000	.0000	.0000	.0000	.0000	.0000	.0000
	3	.0359	.1423	.2285	.2463	.2079	.1465	.0888	.0468	.0215	.0085	.0029	.0008	.0002	.0000	.0000	.0000	.0000	.0000	.0000
	4	.0061	.0514	.1311	.2001	.2252	.2040	.1553	.1014	.0572	.0278	.0115	.0040	.0011	.0002	.0000	.0000	.0000	.0000	.0000
	5	.0008	.0137	.0555	.1201	.1802	.2099	.2008	.1623	.1123	.0667	.0337	.0142	.0049	.0013	.0002	.0000	.0000	.0000	.0000
	6	.0001	.0028	.0180	.0550	.1101	.1649	.1982	.1983	.1684	.1222	.0755	.0392	.0167	.0056	.0014	.0002	.0000	.0000	.0000
	7	.0000	.0004	.0045	.0197	.0524	.1010	.1524	.1889	.1969	.1746	.1318	.0840	.0442	.0185	.0058	.0012	.0001	.0000	.0000
	8	.0000	.0001	.0009	.0055	.0197	.0487	.0923	.1417	.1812	.1964	.1812	.1417	.0923	.0487	.0197	.0055	.0009	.0001	.0000
	9	.0000	.0000	.0001	.0012	.0058	.0185	.0442	.0840	.1318	.1746	.1969	.1889	.1524	.1010	.0524	.0197	.0045	.0004	.0000
	10	.0000	.0000	.0000	.0002	.0014	.0056	.0167	.0392	.0755	.1222	.1684	.1983	.1982	.1649	.1101	.0550	.0180	.0028	.0001
	11	.0000	.0000	.0000	.0000	.0002	.0013	.0049	.0142	.0337	.0667	.1123	.1623	.2008	.2099	.1802	.1201	.0555	.0137	.0008
	12	.0000	.0000	.0000	.0000	.0000	.0002	.0011	.0040	.0115	.0278	.0572	.1014	.1553	.2040	.2252	.2001	.1311	.0514	.0061
	13	.0000	.0000	.0000	.0000	.0000	.0000	.0002	.0008	.0029	.0085	.0215	.0468	.0888	.1465	.2079	.2463	.2285	.1423	.0359
	14	.0000	.0000	.0000	.0000	.0000	.0000	.0000	.0001	.0005	.0018	.0056	.0150	.0353	.0732	.1336	.2111	.2775	.2745	.1463
	15	.0000	.0000	.0000	.0000	.0000	.0000	.0000	.0000	.0001	.0002	.0009	.0030	.0087	.0228	.0535	.1126	.2097	.3294	.3706
	16	.0000	.0000	.0000	.0000	.0000	.0000	.0000	.0000	.0000	.0000	.0001	.0003	.0010	.0033	.0100	.0281	.0743	.1853	.4401

F. Random Rectangular Numbers

10 09 73 25 33	76 52 01 35 86	34 67 35 48 76	80 95 90 91 17	39 29 27 49 45
37 54 20 48 05	64 89 47 42 96	24 80 52 40 37	20 63 61 04 02	00 82 29 16 65
08 42 26 89 53	19 64 50 93 03	23 20 90 25 60	15 95 33 47 64	35 08 03 36 06
99 01 90 25 29	09 37 67 07 15	38 31 13 11 65	88 67 67 43 97	04 43 62 76 59
12 80 79 99 70	80 15 73 61 47	64 03 23 66 53	98 95 11 68 77	12 17 17 68 33
66 06 57 47 17	34 07 27 68 50	36 69 73 61 70	65 81 33 98 85	11 19 92 91 70
31 06 01 08 05	45 57 18 24 06	35 30 34 26 14	86 79 90 74 39	23 40 30 97 32
85 26 97 76 02	02 05 16 56 92	68 66 57 48 18	73 05 38 52 47	16 62 38 85 79
63 57 33 21 35	05 32 54 70 48	90 55 35 75 48	28 46 82 87 09	83 49 12 56 24
73 79 64 57 53	03 52 96 47 78	35 80 83 42 82	60 93 52 03 44	35 27 38 84 35
98 52 01 77 67	14 90 56 86 07	22 10 94 05 58	60 97 09 34 33	50 50 07 39 98
11 80 50 54 31	39 80 82 77 32	50 72 56 82 48	29 40 52 42 01	52 77 56 78 51
83 45 29 96 34	06 28 89 80 83	13 74 67 00 78	18 47 54 06 10	68 71 17 78 17
88 68 54 02 00	86 50 75 84 01	36 76 66 79 51	90 36 47 64 93	29 60 91 10 62
99 59 46 73 48	87 51 76 49 69	91 82 60 89 28	93 78 56 13 68	23 47 83 41 13
65 48 11 76 74	17 46 85 09 50	58 04 77 69 74	73 03 95 71 86	40 21 81 65 44
80 12 43 56 35	17 72 70 80 15	45 31 82 23 74	21 11 57 82 53	14 38 55 37 63
74 35 09 98 17	77 40 27 72 14	43 23 60 02 10	45 52 16 42 37	96 28 60 26 55
69 91 62 68 03	66 25 22 91 48	36 93 68 72 03	76 62 11 39 90	94 40 05 64 18
09 89 32 05 05	14 22 56 85 14	46 42 75 67 88	96 29 77 88 22	54 38 21 45 98
91 49 91 45 23	68 47 91 76 86	46 16 28 35 54	94 75 08 99 23	37 08 92 00 48
80 33 69 45 98	26 94 03 68 58	70 29 73 41 35	53 14 03 33 40	42 05 08 23 41
44 10 48 19 49	85 15 74 79 54	32 97 92 65 75	57 60 04 08 81	22 22 20 64 13
12 55 07 37 42	11 10 00 20 40	12 86 07 46 97	96 64 48 94 39	28 70 72 58 15
63 60 64 93 29	16 50 53 44 84	40 21 95 25 63	43 65 17 70 82	07 20 73 17 90
61 19 69 04 46	26 45 74 77 74	51 92 43 37 29	65 39 45 95 93	42 58 26 05 27
15 47 44 52 66	95 27 07 99 53	59 36 78 38 48	82 39 61 01 18	33 21 15 94 66
94 55 72 85 73	67 89 75 43 87	54 62 24 44 31	91 19 04 25 92	92 92 74 59 73
42 48 11 62 13	97 34 40 87 21	16 86 84 87 67	03 07 11 20 59	25 70 14 66 70
23 52 37 83 17	73 20 88 98 37	68 93 59 14 16	26 25 22 96 63	05 52 28 25 62
04 49 35 24 94	75 24 63 38 24	45 86 25 10 25	61 96 27 93 35	65 33 71 24 72
00 54 99 76 54	64 05 18 81 59	96 11 96 38 96	54 69 28 23 91	23 28 72 95 29
35 96 31 53 07	26 89 80 93 54	33 35 13 54 62	77 97 45 00 24	90 10 33 93 33
59 80 80 83 91	45 42 72 68 42	83 60 94 97 00	13 02 12 48 92	78 56 52 01 06
46 05 88 52 36	01 39 09 22 86	77 28 14 40 77	93 91 08 36 47	70 61 74 29 41
32 17 90 05 97	87 37 92 52 41	05 56 70 70 07	86 74 31 71 57	85 39 41 18 38
69 23 46 14 06	20 11 74 52 04	15 95 66 00 00	18 74 39 24 23	07 11 89 63 38
19 56 54 14 30	01 75 87 53 79	40 41 92 15 85	66 67 43 68 06	84 96 28 52 07
45 15 51 49 38	19 47 60 72 46	43 66 79 45 43	59 04 79 00 33	20 82 66 95 41
94 86 43 19 94	36 16 81 08 51	34 88 88 15 53	01 54 03 54 56	05 01 45 11 76
98 08 62 48 26	45 24 02 84 04	44 99 90 88 96	39 09 47 34 07	35 44 13 18 80
33 18 51 62 32	41 94 15 09 49	89 43 54 85 81	88 69 54 19 94	37 54 87 30 43
80 95 10 04 06	96 38 27 07 74	20 15 12 33 87	25 01 62 52 98	94 62 46 11 71
79 75 24 91 40	71 96 12 82 96	69 86 10 25 91	74 85 22 05 39	00 38 75 95 79
18 63 33 25 37	98 14 50 65 71	31 01 02 46 74	05 45 56 14 27	77 93 89 19 36
74 02 94 39 02	77 55 73 22 70	97 79 01 71 19	52 52 75 80 21	80 81 45 17 48
54 17 84 56 11	80 99 33 71 43	05 33 51 29 69	56 12 71 92 55	36 04 09 03 24
11 66 44 98 83	52 07 98 48 27	59 38 17 15 39	09 97 33 34 40	88 46 12 33 56
48 32 47 79 28	31 24 96 47 10	02 29 53 68 70	32 30 75 75 46	15 02 00 99 94
69 07 49 41 38	87 63 79 19 76	35 58 40 44 01	10 51 82 16 15	01 84 87 69 38

F. Random Rectangular Numbers (*continued*)

09 18 82 00 97	32 82 53 95 27	04 22 08 63 04	83 38 98 73 74	64 27 85 80 44
90 04 58 54 97	51 98 15 06 54	94 93 88 19 97	91 87 07 61 50	68 47 66 46 59
73 18 95 02 07	47 67 72 62 69	62 29 06 44 64	27 12 46 70 18	41 36 18 27 60
75 76 87 64 90	20 97 18 17 49	90 42 91 22 72	95 37 50 58 71	93 82 34 31 78
54 01 64 40 56	66 28 13 10 03	00 68 22 73 98	20 71 45 32 95	07 70 61 78 13
08 35 86 99 10	78 54 24 27 85	13 66 15 88 73	04 61 89 75 53	31 22 30 84 20
28 30 60 32 64	81 33 31 05 91	40 51 00 78 93	32 60 46 04 75	94 11 90 18 40
53 84 08 62 33	81 59 41 36 28	51 21 59 02 90	28 46 66 87 95	77 76 22 07 91
91 75 75 37 41	61 61 36 22 69	50 26 39 02 12	55 78 17 65 14	83 48 34 70 55
89 41 59 26 94	00 38 75 83 91	12 60 71 76 46	48 94 97 23 06	94 54 13 74 08
77 51 30 38 20	86 83 42 99 01	68 41 48 27 74	51 90 81 39 80	72 89 35 55 07
19 50 23 71 74	69 97 92 02 88	55 21 02 97 73	74 28 77 52 51	65 34 46 74 15
21 81 85 93 13	93 27 88 17 57	04 68 67 31 56	07 08 28 50 46	31 85 33 84 52
51 47 46 64 99	68 10 72 36 21	94 04 99 13 45	42 83 60 91 91	08 00 74 54 49
99 55 96 83 31	62 53 52 41 70	69 77 71 28 30	74 81 97 81 42	43 86 07 28 34
33 71 34 80 07	93 58 47 28 69	51 92 66 47 21	58 30 32 98 22	93 17 49 39 72
85 27 48 68 93	11 30 32 92 70	28 83 43 41 37	73 51 59 04 00	71 14 84 36 43
84 13 38 96 40	44 03 55 21 66	73 85 27 00 91	61 22 26 05 61	62 32 71 84 23
56 73 21 62 34	17 39 59 61 31	10 12 39 16 22	85 49 65 75 60	81 60 41 88 80
65 13 85 68 06	87 64 88 52 61	34 31 36 58 61	45 87 52 10 69	85 64 44 72 77
38 00 10 21 76	81 71 91 17 11	71 60 29 29 37	74 21 96 40 49	65 58 44 96 98
37 40 29 63 97	01 30 47 75 86	56 27 11 00 86	47 32 46 26 05	40 03 03 74 38
97 12 54 03 48	87 08 33 14 17	21 81 53 92 50	75 23 76 20 47	15 50 12 95 78
21 82 64 11 34	47 14 33 40 72	64 63 88 59 02	49 13 90 64 41	03 85 65 45 52
73 13 54 27 42	95 71 90 90 35	85 79 47 42 96	08 78 98 81 56	64 69 11 92 02
07 63 87 79 29	03 06 11 80 72	96 20 74 41 56	23 82 19 95 38	04 71 36 69 94
60 52 88 34 41	07 95 41 98 14	59 17 52 06 95	05 53 35 21 39	61 21 20 64 55
83 59 63 56 55	06 95 89 29 83	05 12 80 97 19	77 43 35 37 83	92 30 15 04 98
10 85 06 27 46	99 59 91 05 07	13 49 90 63 19	53 07 57 18 39	06 41 01 93 62
39 82 09 89 52	43 62 26 31 47	64 42 18 08 14	43 80 00 93 51	31 02 47 31 67
59 58 00 64 78	75 56 97 88 00	88 83 55 44 86	23 76 80 61 56	04 11 10 84 08
38 50 80 73 41	23 79 34 87 63	90 82 29 70 22	17 71 90 42 07	95 95 44 99 53
30 69 27 06 68	94 68 81 61 27	56 19 68 00 91	82 06 76 34 00	05 46 26 92 00
65 44 39 56 59	18 28 82 74 37	49 63 22 40 41	08 33 76 56 76	96 29 99 08 36
27 26 75 02 64	13 19 27 22 94	07 47 74 46 06	17 98 54 89 11	97 34 13 03 58
91 30 70 69 91	19 07 22 42 10	36 69 95 37 28	28 82 53 57 93	28 97 66 62 52
68 43 49 46 88	84 47 31 36 22	62 12 69 84 08	12 84 38 25 90	09 81 59 31 46
48 90 81 58 77	54 74 52 45 91	35 70 00 47 54	83 82 45 26 92	54 13 05 51 60
06 91 34 51 97	42 67 27 86 01	11 88 30 95 28	63 01 19 89 01	14 97 44 03 44
10 45 51 60 19	14 21 03 37 12	91 34 23 78 21	88 32 58 08 51	43 66 77 08 83
12 88 39 73 43	65 02 76 11 84	04 28 50 13 92	17 97 41 50 77	90 71 22 67 69
21 77 83 09 76	38 80 73 69 61	31 64 94 20 96	63 28 10 20 23	08 81 64 74 49
19 52 35 95 15	65 12 25 96 59	86 28 36 82 58	69 57 21 37 98	16 43 59 15 29
67 24 55 26 70	35 58 31 65 63	79 24 68 66 86	76 46 33 42 22	26 65 59 08 02
60 58 44 73 77	07 50 03 79 92	45 13 42 65 29	26 76 08 36 37	41 32 64 43 44
53 85 34 13 77	36 06 69 48 50	58 83 87 38 59	49 36 47 33 31	96 24 04 36 42
24 63 73 87 36	74 38 48 93 42	52 62 30 79 92	12 36 91 86 01	03 74 28 38 73
83 08 01 24 51	38 99 22 28 15	07 75 95 17 77	97 37 72 75 85	51 97 23 78 67
16 44 42 43 34	36 15 19 90 73	27 49 37 09 39	85 13 03 25 52	54 84 65 47 59
60 79 01 81 57	57 17 86 57 62	11 16 17 85 76	45 81 95 29 79	65 13 00 48 60

G. Wilcoxon's Rank-Sum Test

Critical Lower-Tail Values of W
[Largest value of W' for which $Pr(W \leq W') \leq \alpha$]

$n_1 = 1$

n_2	α .005	.010	.025	.050
3				
4				
5				
6				
7				
8				
9				
10				
11				
12				
13				
14				
15				
16				
17				
18				
19				1
20				1

$n_1 = 2$

n_2	α .005	.010	.025	.050
3				
4				
5				3
6				3
7				3
8			3	4
9			3	4
10			3	4
11			3	4
12			4	5
13		3	4	5
14		3	4	6
15		3	4	6
16		3	4	6
17		3	5	6
18		3	5	7
19	3	4	5	7
20	3	4	5	7

$n_1 = 3$

n_2	α .005	.010	.025	.050
3				6
4				6
5			6	7
6			7	8
7		6	7	8
8		6	8	9
9	6	7	8	10
10	6	7	9	10
11	6	7	9	11
12	7	8	10	11
13	7	8	10	12
14	7	8	11	13
15	8	9	11	13
16	8	9	12	14
17	8	10	12	15
18	8	10	13	15
19	9	10	13	16
20	9	11	14	17

$n_1 = 4$

n_2	α .005	.010	.025	.050
4			10	11
5		10	11	12
6	10	11	12	13
7	10	11	13	14
8	11	12	14	15
9	11	13	14	16
10	12	13	15	17
11	12	14	16	18
12	13	15	17	19
13	13	15	18	20
14	14	16	19	21
15	15	17	20	22
16	15	17	21	24
17	16	18	21	25
18	16	19	22	26
19	17	19	23	27
20	18	20	24	28

$n_1 = 5$

n_2	α .005	.010	.025	.050
5	15	16	17	19
6	16	17	18	20
7	16	18	20	21
8	17	19	21	23
9	18	20	22	24
10	19	21	23	26
11	20	22	24	27
12	21	23	26	28
13	22	24	27	30
14	22	25	28	31
15	23	26	29	33
16	24	27	30	34
17	25	28	32	35
18	26	29	33	37
19	27	30	34	38
20	28	31	35	40

$n_1 = 6$

n_2	α .005	.010	.025	.050
6	23	24	26	28
7	24	25	27	29
8	25	27	29	31
9	26	28	31	33
10	27	29	32	35
11	28	30	34	37
12	30	32	35	38
13	31	33	37	40
14	32	34	38	42
15	33	36	40	44
16	34	37	42	46
17	36	39	43	47
18	37	40	45	49
19	38	41	46	51
20	39	43	48	53

From Table 1, L. R. Verdooren, "Extended Tables of Critical Values for Wilcoxon's Test Statistic," *Biometrika* 50 (1963), 177–186. Used with permission of the author and editor.

G. Wilcoxon's Rank-Sum Test (*continued*)

$n_1 = 7$

n_2	α .005	.010	.025	.050
7	32	34	36	39
8	34	35	38	41
9	35	37	40	43
10	37	39	42	45
11	38	40	44	47
12	40	42	46	49
13	41	44	48	52
14	43	45	50	54
15	44	47	52	56
16	46	49	54	58
17	47	51	56	61
18	49	52	58	63
19	50	54	60	65
20	52	56	62	67
21	53	58	64	69
22	55	59	66	72
23	57	61	68	74
24	58	63	70	76
25	60	64	72	78

$n_1 = 8$

n_2	α .005	.010	.025	.050
8	43	45	49	51
9	45	47	51	54
10	47	49	53	56
11	49	51	55	59
12	51	53	58	62
13	53	56	60	64
14	54	58	62	67
15	56	60	65	69
16	58	62	67	72
17	60	64	70	75
18	62	66	72	77
19	64	68	74	80
20	66	70	77	83
21	68	72	79	85
22	70	74	81	88
23	71	76	84	90
24	73	78	86	93
25	75	81	89	96

$n_1 = 9$

n_2	α .005	.010	.025	.050
9	56	59	62	66
10	58	61	65	69
11	61	63	68	72
12	63	66	71	75
13	65	68	73	78
14	67	71	76	81
15	69	73	79	84
16	72	76	82	87
17	74	78	84	90
18	76	81	87	93
19	78	83	90	96
20	81	85	93	99
21	83	88	95	102
22	85	90	98	105
23	88	93	101	108
24	90	95	104	111
25	92	99	107	114

$n_1 = 10$

n_2	α .005	.010	.025	.050
10	71	74	78	82
11	73	77	81	86
12	76	79	84	89
13	79	82	88	92
14	81	85	91	96
15	84	88	94	99
16	86	91	97	103
17	89	93	100	106
18	92	96	103	110
19	94	99	107	113
20	97	102	110	117
21	99	105	113	120
22	102	108	116	123
23	105	110	119	127
24	107	113	122	130
25	110	116	126	134

$n_1 = 11$

n_2	α .005	.010	.025	.050
11	87	91	96	100
12	90	94	99	104
13	93	97	103	108
14	96	100	106	112
15	99	103	110	116
16	102	107	113	120
17	105	110	117	123
18	108	113	121	127
19	111	116	124	131
20	114	119	128	135
21	117	123	131	139
22	120	126	135	143
23	123	129	139	147
24	126	132	142	151
25	129	136	146	155

$n_1 = 12$

n_2	α .005	.010	.025	.050
12	105	109	115	120
13	109	113	119	125
14	112	116	123	129
15	115	120	127	133
16	119	124	131	138
17	122	127	135	142
18	125	131	139	146
19	129	134	143	150
20	132	138	147	155
21	136	142	151	159
22	139	145	155	163
23	142	149	159	168
24	146	153	163	172
25	149	156	167	176

H. Squares and Square Roots

How to Find Square Roots

1. If the number contains more than three significant digits, round it to just three significant digits. Find the significant digits under column N.
2. Move the decimal point either left or right an *even* number of places until a number from 1 to 100 is found. If the result is less than 10, use the column under $\sqrt{N}$. If the result is greater than 10, use the column under $\sqrt{10N}$.
3. For the appropriate entry under either $\sqrt{N}$ or $\sqrt{10N}$, move the decimal point *half* as many places in the *opposite* direction as you moved it in step 2.

Example A Find $\sqrt{12345}$.

Step 1: Change 12345 to 12300.

Step 2: Change 12300 to 1.23 by moving the decimal point four places left. For the row with digits 1.23 under N in the table, use the entry under $\sqrt{N}$, which is 1.10905.

Step 3: Move the decimal two places *right* in 1.10905 to get 110.905, the square root of 12345 as accurately as is possible from the table.

Example B Find $\sqrt{0.0093}$.

Step 2: Change 0.0093 to 93 by moving the decimal point four places right. For the row with digits 9.30 under N in the table, use the entry under $\sqrt{10N}$, which is 9.64365.

Step 3: Move the decimal two places *left* in 9.64365 to get 0.0964365, the square root of 0.0093 to seven decimal places.

H. Squares and Square Roots (*continued*)

N	N^2	$\sqrt{N}$	$\sqrt{10N}$	N	N^2	$\sqrt{N}$	$\sqrt{10N}$
1.00	1.0000	1.00000	3.16228	1.50	2.2500	1.22474	3.87298
1.01	1.0201	1.00499	3.17805	1.51	2.2801	1.22882	3.88587
1.02	1.0404	1.00995	3.19374	1.52	2.3104	1.23288	3.89872
1.03	1.0609	1.01489	3.20936	1.53	2.3409	1.23693	3.91152
1.04	1.0816	1.01980	3.22490	1.54	2.3716	1.24097	3.92428
1.05	1.1025	1.02470	3.24037	1.55	2.4025	1.24499	3.93700
1.06	1.1236	1.02956	3.25576	1.56	2.4336	1.24900	3.94968
1.07	1.1449	1.03441	3.27109	1.57	2.4649	1.25300	3.96232
1.08	1.1664	1.03923	3.28634	1.58	2.4964	1.25698	3.97492
1.09	1.1881	1.04403	3.30151	1.59	2.5281	1.26095	3.98748
1.10	1.2100	1.04881	3.31662	1.60	2.5600	1.26491	4.00000
1.11	1.2321	1.05357	3.33167	1.61	2.5921	1.26886	4.01248
1.12	1.2544	1.05830	3.34664	1.62	2.6244	1.27279	4.02492
1.13	1.2769	1.06301	3.36155	1.63	2.6569	1.27671	4.03733
1.14	1.2996	1.06771	3.37639	1.64	2.6896	1.28062	4.04969
1.15	1.3225	1.07238	3.39116	1.65	2.7225	1.28452	4.06202
1.16	1.3456	1.07703	3.40588	1.66	2.7556	1.28841	4.07431
1.17	1.3689	1.08167	3.42053	1.67	2.7889	1.29228	4.08656
1.18	1.3924	1.08628	3.43511	1.68	2.8224	1.29615	4.09878
1.19	1.4161	1.09087	3.44964	1.69	2.8561	1.30000	4.11096
1.20	1.4400	1.09545	3.46410	1.70	2.8900	1.30384	4.12311
1.21	1.4641	1.10000	3.47851	1.71	2.9241	1.30767	4.13521
1.22	1.4884	1.10454	3.49285	1.72	2.9584	1.31149	4.14729
1.23	1.5129	1.10905	3.50714	1.73	2.9929	1.31529	4.15933
1.24	1.5376	1.11355	3.52136	1.74	3.0276	1.31909	4.17133
1.25	1.5625	1.11803	3.53553	1.75	3.0625	1.32288	4.18330
1.26	1.5876	1.12250	3.54965	1.76	3.0976	1.32665	4.19524
1.27	1.6129	1.12694	3.56371	1.77	3.1329	1.33041	4.20714
1.28	1.6384	1.13137	3.57771	1.78	3.1684	1.33417	4.21900
1.29	1.6641	1.13578	3.59166	1.79	3.2041	1.33791	4.23084
1.30	1.6900	1.14018	3.60555	1.80	3.2400	1.34164	4.24264
1.31	1.7161	1.14455	3.61939	1.81	3.2761	1.34536	4.25441
1.32	1.7424	1.14891	3.63318	1.82	3.3124	1.34907	4.26615
1.33	1.7689	1.15326	3.64692	1.83	3.3489	1.35277	4.27785
1.34	1.7956	1.15758	3.66060	1.84	3.3856	1.35647	4.28952
1.35	1.8225	1.16190	3.67423	1.85	3.4225	1.36015	4.30116
1.36	1.8496	1.16619	3.68782	1.86	3.4596	1.36382	4.31277
1.37	1.8769	1.17047	3.70135	1.87	3.4969	1.36748	4.32435
1.38	1.9044	1.17473	3.71484	1.88	3.5344	1.37113	4.33590
1.39	1.9321	1.17898	3.72827	1.89	3.5721	1.37477	4.34741
1.40	1.9600	1.18322	3.74166	1.90	3.6100	1.37840	4.35890
1.41	1.9881	1.18743	3.75500	1.91	3.6481	1.38203	4.37035
1.42	2.0164	1.19164	3.76829	1.92	3.6864	1.38564	4.38178
1.43	2.0449	1.19583	3.78153	1.93	3.7249	1.38924	4.39318
1.44	2.0736	1.20000	3.79473	1.94	3.7636	1.39284	4.40454
1.45	2.1025	1.20416	3.80789	1.95	3.8025	1.39642	4.41588
1.46	2.1316	1.20830	3.82099	1.96	3.8416	1.40000	4.42719
1.47	2.1609	1.21244	3.83406	1.97	3.8809	1.40357	4.43847
1.48	2.1904	1.21655	3.84708	1.98	3.9204	1.40712	4.44972
1.49	2.2201	1.22066	3.86005	1.99	3.9601	1.41067	4.46094

H. Squares and Square Roots (*continued*)

N	N^2	$\sqrt{N}$	$\sqrt{10N}$	N	N^2	$\sqrt{N}$	$\sqrt{10N}$
2.00	4.0000	1.41421	4.47214	2.50	6.2500	1.58114	5.00000
2.01	4.0401	1.41774	4.48330	2.51	6.3001	1.58430	5.00999
2.02	4.0804	1.42127	4.49444	2.52	6.3504	1.58745	5.01996
2.03	4.1209	1.42478	4.50555	2.53	6.4009	1.59060	5.02991
2.04	4.1616	1.42829	4.51664	2.54	6.4516	1.59374	5.03984
2.05	4.2025	1.43178	4.52769	2.55	6.5025	1.59687	5.04975
2.06	4.2436	1.43527	4.53872	2.56	6.5536	1.60000	5.05964
2.07	4.2849	1.43875	4.54973	2.57	6.6049	1.60312	5.06952
2.08	4.3264	1.44222	4.56070	2.58	6.6564	1.60624	5.07937
2.09	4.3681	1.44568	4.57165	2.59	6.7081	1.60935	5.08920
2.10	4.4100	1.44914	4.58258	2.60	6.7600	1.61245	5.09902
2.11	4.4521	1.45258	4.59347	2.61	6.8121	1.61555	5.10882
2.12	4.4944	1.45602	4.60435	2.62	6.8644	1.61864	5.11859
2.13	4.5369	1.45945	4.61519	2.63	6.9169	1.62173	5.12835
2.14	4.5796	1.46287	4.62601	2.64	6.9696	1.62481	5.13809
2.15	4.6225	1.46629	4.63681	2.65	7.0225	1.62788	5.14782
2.16	4.6656	1.46969	4.64758	2.66	7.0756	1.63095	5.15752
2.17	4.7089	1.47309	4.65833	2.67	7.1289	1.63401	5.16720
2.18	4.7524	1.47648	4.66905	2.68	7.1824	1.63707	5.17687
2.19	4.7961	1.47986	4.67974	2.69	7.2361	1.64012	5.18652
2.20	4.8400	1.48324	4.69042	2.70	7.2900	1.64317	5.19615
2.21	4.8841	1.48661	4.70106	2.71	7.3441	1.64621	5.20577
2.22	4.9284	1.48997	4.71169	2.72	7.3984	1.64924	5.21536
2.23	4.9729	1.49332	4.72229	2.73	7.4529	1.65227	5.22494
2.24	5.0176	1.49666	4.73286	2.74	7.5076	1.65529	5.23450
2.25	5.0625	1.50000	4.74342	2.75	7.5625	1.65831	5.24404
2.26	5.1076	1.50333	4.75395	2.76	7.6176	1.66132	5.25357
2.27	5.1529	1.50665	4.76445	2.77	7.6729	1.66433	5.26308
2.28	5.1984	1.50997	4.77493	2.78	7.7284	1.66733	5.27257
2.29	5.2441	1.51327	4.78539	2.79	7.7841	1.67033	5.28205
2.30	5.2900	1.51658	4.79583	2.80	7.8400	1.67332	5.29150
2.31	5.3361	1.51987	4.80625	2.81	7.8961	1.67631	5.30094
2.32	5.3824	1.52315	4.81664	2.82	7.9524	1.67929	5.31037
2.33	5.4289	1.52643	4.82701	2.83	8.0089	1.68226	5.31977
2.34	5.4756	1.52971	4.83735	2.84	8.0656	1.68523	5.32917
2.35	5.5225	1.53297	4.84768	2.85	8.1225	1.68819	5.33854
2.36	5.5696	1.53623	4.85798	2.86	8.1796	1.69115	5.34790
2.37	5.6169	1.53948	4.86826	2.87	8.2369	1.69411	5.35724
2.38	5.6644	1.54272	4.87852	2.88	8.2944	1.69706	5.36656
2.39	5.7121	1.54596	4.88876	2.89	8.3521	1.70000	5.37587
2.40	5.7600	1.54919	4.89898	2.90	8.4100	1.70294	5.38516
2.41	5.8081	1.55242	4.90918	2.91	8.4681	1.70587	5.39444
2.42	5.8564	1.55563	4.91935	2.92	8.5264	1.70880	5.40370
2.43	5.9049	1.55885	4.92950	2.93	8.5849	1.71172	5.41295
2.44	5.9536	1.56205	4.93964	2.94	8.6436	1.71464	5.42218
2.45	6.0025	1.56525	4.94975	2.95	8.7025	1.71756	5.43139
2.46	6.0516	1.56844	4.95984	2.96	8.7616	1.72047	5.44059
2.47	6.1009	1.57162	4.96991	2.97	8.8209	1.72337	5.44977
2.48	6.1504	1.57480	4.97996	2.98	8.8804	1.72627	5.45894
2.49	6.2001	1.57797	4.98999	2.99	8.9401	1.72916	5.46809

H. Squares and Square Roots (*continued*)

N	N^2	$\sqrt{N}$	$\sqrt{10N}$	N	N^2	$\sqrt{N}$	$\sqrt{10N}$
3.00	9.0000	1.73205	5.47723	3.50	12.2500	1.87083	5.91608
3.01	9.0601	1.73494	5.48635	3.51	12.3201	1.87350	5.92453
3.02	9.1204	1.73781	5.49545	3.52	12.3904	1.87617	5.93296
3.03	9.1809	1.74069	5.50454	3.53	12.4609	1.87883	5.94138
3.04	9.2416	1.74356	5.51362	3.54	12.5316	1.88149	5.94979
3.05	9.3025	1.74642	5.52268	3.55	12.6025	1.88414	5.95819
3.06	9.3636	1.74929	5.53173	3.56	12.6736	1.88680	5.96657
3.07	9.4249	1.75214	5.54076	3.57	12.7449	1.88944	5.97495
3.08	9.4864	1.75499	5.54977	3.58	12.8164	1.89209	5.98331
3.09	9.5481	1.75784	5.55878	3.59	12.8881	1.89473	5.99166
3.10	9.6100	1.76068	5.56776	3.60	12.9600	1.89737	6.00000
3.11	9.6721	1.76352	5.57674	3.61	13.0321	1.90000	6.00833
3.12	9.7344	1.76635	5.58570	3.62	13.1044	1.90263	6.01664
3.13	9.7969	1.76918	5.59464	3.63	13.1769	1.90526	6.02495
3.14	9.8596	1.77200	5.60357	3.64	13.2496	1.90788	6.03324
3.15	9.9225	1.77482	5.61249	3.65	13.3225	1.91050	6.04152
3.16	9.9856	1.77764	5.62139	3.66	13.3956	1.91311	6.04979
3.17	10.0489	1.78045	5.63028	3.67	13.4689	1.91572	6.05805
3.18	10.1124	1.78326	5.63915	3.68	13.5424	1.91833	6.06630
3.19	10.1761	1.78606	5.64801	3.69	13.6161	1.92094	6.07454
3.20	10.2400	1.78885	5.65685	3.70	13.6900	1.92354	6.08276
3.21	10.3041	1.79165	5.66569	3.71	13.7641	1.92614	6.09098
3.22	10.3684	1.79444	5.67450	3.72	13.8384	1.92873	6.09918
3.23	10.4329	1.79722	5.68331	3.73	13.9129	1.93132	6.10737
3.24	10.4976	1.80000	5.69210	3.74	13.9876	1.93391	6.11555
3.25	10.5625	1.80278	5.70088	3.75	14.0625	1.93649	6.12372
3.26	10.6276	1.80555	5.70964	3.76	14.1376	1.93907	6.13188
3.27	10.6929	1.80831	5.71839	3.77	14.2129	1.94165	6.14003
3.28	10.7584	1.81108	5.72713	3.78	14.2884	1.94422	6.14817
3.29	10.8241	1.81384	5.73585	3.79	14.3641	1.94679	6.15630
3.30	10.8900	1.81659	5.74456	3.80	14.4400	1.94936	6.16441
3.31	10.9561	1.81934	5.75326	3.81	14.5161	1.95192	6.17252
3.32	11.0224	1.82209	5.76194	3.82	14.5924	1.95448	6.18061
3.33	11.0889	1.82483	5.77062	3.83	14.6689	1.95704	6.18870
3.34	11.1556	1.82757	5.77927	3.84	14.7456	1.95959	6.19677
3.35	11.2225	1.83030	5.78792	3.85	14.8225	1.96214	6.20484
3.36	11.2896	1.83303	5.79655	3.86	14.8996	1.96469	6.21289
3.37	11.3569	1.83576	5.80517	3.87	14.9769	1.96723	6.22093
3.38	11.4244	1.83848	5.81378	3.88	15.0544	1.96977	6.22896
3.39	11.4921	1.84120	5.82237	3.89	15.1321	1.97231	6.23699
3.40	11.5600	1.84391	5.83095	3.90	15.2100	1.97484	6.24500
3.41	11.6281	1.84662	5.83952	3.91	15.2881	1.97737	6.25300
3.42	11.6964	1.84932	5.84808	3.92	15.3664	1.97990	6.26099
3.43	11.7649	1 85203	5 85662	3.93	15.4449	1.98242	6.26897
3.44	11.8336	1.85472	5.86515	3.94	15.5236	1.98494	6.27694
3.45	11.9025	1.85742	5.87367	3.95	15.6025	1.98746	6.28490
3.46	11.9716	1.86011	5.88218	3.96	15.6816	1.98997	6.29285
3.47	12.0409	1.86279	5.89067	3.97	15.7609	1.99249	6.30079
3.48	12.1104	1.86548	5.89915	3.98	15.8408	1.99499	6.30872
3.49	12.1801	1.86815	5.90762	3.99	15.9201	1.99750	6.31664

H. Squares and Square Roots (*continued*)

N	N^2	$\sqrt{N}$	$\sqrt{10N}$	N	N^2	$\sqrt{N}$	$\sqrt{10N}$
4.00	16.0000	2.00000	6.32456	4.50	20.2500	2.12132	6.70820
4.01	16.0801	2.00250	6.33246	4.51	20.3401	2.12368	6.71565
4.02	16.1604	2.00499	6.34035	4.52	20.4304	2.12603	6.72309
4.03	16.2409	2.00749	6.34823	4.53	20.5209	2.12838	6.73053
4.04	16.3216	2.00998	6.35610	4.54	20.6116	2.13073	6.73795
4.05	16.4025	2.01246	6.36396	4.55	20.7025	2.13307	6.74537
4.06	16.4836	2.01494	6.37181	4.56	20.7936	2.13542	6.75278
4.07	16.5649	2.01742	6.37966	4.57	20.8849	2.13776	6.76018
4.08	16.6464	2.01990	6.38749	4.58	20.9764	2.14009	6.76757
4.09	16.7281	2.02237	6.39531	4.59	21.0681	2.14243	6.77495
4.10	16.8100	2.02485	6.40312	4.60	21.1600	2.14476	6.78233
4.11	16.8921	2.02731	6.41093	4.61	21.2521	2.14709	6.78970
4.12	16.9744	2.02978	6.41872	4.62	21.3444	2.14942	6.79706
4.13	17.0569	2.03224	6.42651	4.63	21.4369	2.15174	6.80441
4.14	17.1396	2.03470	6.43428	4.64	21.5296	2.15407	6.81175
4.15	17.2225	2.03715	6.44205	4.65	21.6225	2.15639	6.81909
4.16	17.3056	2.03961	6.44981	4.66	21.7156	2.15870	6.82642
4.17	17.3889	2.04206	6.45755	4.67	21.8089	2.16102	6.83374
4.18	17.4724	2.04450	6.46529	4.68	21.9024	2.16333	6.84105
4.19	17.5561	2.04695	6.47302	4.69	21.9961	2.16564	6.84836
4.20	17.6400	2.04939	6.48074	4.70	22.0900	2.16795	6.85565
4.21	17.7241	2.05183	6.48845	4.71	22.1841	2.17025	6.86294
4.22	17.8084	2.05426	6.49615	4.72	22.2784	2.17256	6.87023
4.23	17.8929	2.05670	6.50384	4.73	22.3729	2.17486	6.87750
4.24	17.9776	2.05913	6.51153	4.74	22.4676	2.17715	6.88477
4.25	18.0625	2.06155	6.51920	4.75	22.5625	2.17945	6.89202
4.26	18.1476	2.06398	6.52687	4.76	22.6576	2.18174	6.89928
4.27	18.2329	2.06640	6.53452	4.77	22.7529	2.18403	6.90652
4.28	18.3184	2.06882	6.54217	4.78	22.8484	2.18632	6.91375
4.29	18.4041	2.07123	6.54981	4.79	22.9441	2.18861	6.92098
4.30	18.4900	2.07364	6.55744	4.80	23.0400	2.19089	6.92820
4.31	18.5761	2.07605	6.56506	4.81	23.1361	2.19317	6.93542
4.32	18.6624	2.07846	6.57267	4.82	23.2324	2.19545	6.94262
4.33	18.7489	2.08087	6.58027	4.83	23.3289	2.19773	6.94982
4.34	18.8356	2.08327	6.58787	4.84	23.4256	2.20000	6.95701
4.35	18.9225	2.08567	6.59545	4.85	23.5225	2.20227	6.96419
4.36	19.0096	2.08806	6.60303	4.86	23.6196	2.20454	6.97137
4.37	19.0969	2.09045	6.61060	4.87	23.7169	2.20681	6.97854
4.38	19.1844	2.09284	6.61816	4.88	23.8144	2.20907	6.98570
4.39	19.2721	2.09523	6.62571	4.89	23.9121	2.21133	6.99285
4.40	19.3600	2.09762	6.63325	4.90	24.0100	2.21359	7.00000
4.41	19.4481	2.10000	6.64078	4.91	24.1081	2.21585	7.00714
4.42	19.5364	2.10238	6.64831	4.92	24.2064	2.21811	7.01427
4.43	19.6249	2.10476	6.65582	4.93	24.3049	2.22036	7.02140
4.44	19.7136	2.10713	6.66333	4.94	24.4036	2.22261	7.02851
4.45	19.8025	2.10950	6.67083	4.95	24.5025	2.22486	7.03562
4.46	19.8916	2.11187	6.67832	4.96	24.6016	2.22711	7.04273
4.47	19.9809	2.11424	6.68581	4.97	24.7009	2.22935	7.04982
4.48	20.0704	2.11660	6.69328	4.98	24.8004	2.23159	7.05691
4.49	20.1601	2.11896	6.70075	4.99	24.9001	2.23383	7.06399

H. Squares and Square Roots (*continued*)

N	N^2	$\sqrt{N}$	$\sqrt{10N}$
5.00	25.0000	2.23607	7.07107
5.01	25.1001	2.23830	7.07814
5.02	25.2004	2.24054	7.08520
5.03	25.3009	2.24277	7.09225
5.04	25.4016	2.24499	7.09930
5.05	25.5025	2.24722	7.10634
5.06	25.6036	2.24944	7.11337
5.07	25.7049	2.25167	7.12039
5.08	25.8064	2.25389	7.12741
5.09	25.9081	2.25610	7.13442
5.10	26.0100	2.25832	7.14143
5.11	26.1121	2.26053	7.14843
5.12	26.2144	2.26274	7.15542
5.13	26.3169	2.26495	7.16240
5.14	26.4196	2.26716	7.16938
5.15	26.5225	2.26936	7.17635
5.16	26.6256	2.27156	7.18331
5.17	26.7289	2.27376	7.19027
5.18	26.8324	2.27596	7.19722
5.19	26.9361	2.27816	7.20417
5.20	27.0400	2.28035	7.21110
5.21	27.1441	2.28254	7.21803
5.22	27.2484	2.28473	7.22496
5.23	27.3529	2.28692	7.23187
5.24	27.4576	2.28910	7.23878
5.25	27.5625	2.29129	7.24569
5.26	27.6676	2.29347	7.25259
5.27	27.7729	2.29565	7.25948
5.28	27.8784	2.29783	7.26636
5.29	27.9841	2.30000	7.27324
5.30	28.0900	2.30217	7.28011
5.31	28.1961	2.30434	7.28697
5.32	28.3024	2.30651	7.29383
5.33	28.4089	2.30868	7.30068
5.34	28.5156	2.31084	7.30753
5.35	28.6225	2.31301	7.31437
5.36	28.7296	2.31517	7.32120
5.37	28.8369	2.31733	7.32803
5.38	28.9444	2.31948	7.33485
5.39	29.0521	2.32164	7.34166
5.40	29.1600	2.32379	7.34847
5.41	29.2681	2.32594	7.35527
5.42	29.3764	2.32809	7.36205
5.43	29.4849	2.33024	7.36885
5.44	29.5936	2.33238	7.37564
5.45	29.7025	2.33452	7.38241
5.46	29.8116	2.33666	7.38918
5.47	29.9209	2.33880	7.39594
5.48	30.0304	2.34094	7.40270
5.49	30.1401	2.34307	7.40945

N	N^2	$\sqrt{N}$	$\sqrt{10N}$
5.50	30.2500	2.34521	7.41620
5.51	30.3601	2.34734	7.42294
5.52	30.4704	2.34947	7.42967
5.53	30.5809	2.35160	7.43640
5.54	30.6916	2.35372	7.44312
5.55	30.8025	2.35584	7.44983
5.56	30.9136	2.35797	7.45654
5.57	31.0249	2.36008	7.46324
5.58	31.1364	2.36220	7.46994
5.59	31.2481	2.36432	7.47663
5.60	31.3600	2.36643	7.48331
5.61	31.4721	2.36854	7.48999
5.62	31.5844	2.37065	7.49667
5.63	31.6969	2.37276	7.50333
5.64	31.8096	2.37487	7.50999
5.65	31.9225	2.37697	7.51665
5.66	32.0356	2.37908	7.52330
5.67	32.1489	2.38118	7.52994
5.68	32.2624	2.38328	7.53658
5.69	32.3761	2.38537	7.54321
5.70	32.4900	2.38747	7.54983
5.71	32.6041	2.38956	7.55645
5.72	32.7184	2.39165	7.56307
5.73	32.8329	2.39374	7.56968
5.74	32.9476	2.39583	7.57628
5.75	33.0625	2.39792	7.58288
5.76	33.1776	2.40000	7.58947
5.77	33.2929	2.40208	7.59605
5.78	33.4084	2.40416	7.60263
5.79	33.5241	2.40624	7.60920
5.80	33.6400	2.40832	7.61577
5.81	33.7561	2.41039	7.62234
5.82	33.8724	2.41247	7.62889
5.83	33.9889	2.41454	7.63544
5.84	34.1056	2.41661	7.64199
5.85	34.2225	2.41868	7.64853
5.86	34.3396	2.42074	7.65506
5.87	34.4569	2.42281	7.66159
5.88	34.5744	2.42487	7.66812
5.89	34.6921	2.42693	7.67463
5.90	34.8100	2.42899	7.68115
5.91	34.9281	2.43105	7.68765
5.92	35.0464	2.43311	7.69415
5.93	35.1649	2.43516	7.70065
5.94	35.2836	2.43721	7.70714
5.95	35.4025	2.43926	7.71362
5.96	35.5216	2.44131	7.72010
5.97	35.6409	2.44336	7.72658
5.98	35.7604	2.44540	7.73305
5.99	35.8801	2.44745	7.73951

H. Squares and Square Roots (*continued*)

N	N^2	$\sqrt{N}$	$\sqrt{10N}$	N	N^2	$\sqrt{N}$	$\sqrt{10N}$
6.00	36.0000	2.44949	7.74597	6.50	42.2500	2.54951	8.06226
6.01	36.1201	2.45153	7.75242	6.51	42.3801	2.55147	8.06846
6.02	36.2404	2.45357	7.75887	6.52	42.5104	2.55343	8.07465
6.03	36.3609	2.45561	7.76531	6.53	42.6409	2.55539	8.08084
6.04	36.4816	2.45764	7.77174	6.54	42.7716	2.55734	8.08703
6.05	36.6025	2.45967	7.77817	6.55	42.9025	2.55930	8.09321
6.06	36.7236	2.46171	7.78460	6.56	43.0336	2.56125	8.09938
6.07	36.8449	2.46374	7.79102	6.57	43.1649	2.56320	8.10555
6.08	36.9664	2.46577	7.79744	6.58	43.2964	2.56515	8.11172
6.09	37.0881	2.46779	7.80385	6.59	43.4281	2.56710	8.11788
6.10	37.2100	2.46982	7.81025	6.60	43.5600	2.56905	8.12404
6.11	37.3321	2.47184	7.81665	6.61	43.6921	2.57099	8.13019
6.12	37.4544	2.47386	7.82304	6.62	43.8244	2.57294	8.13634
6.13	37.5769	2.47588	7.82943	6.63	43.9569	2.57488	8.14248
6.14	37.6996	2.47790	7.83582	6.64	44.0896	2.57682	8.14862
6.15	37.8225	2.47992	7.84219	6.65	44.2225	2.57876	8.15475
6.16	37.9456	2.48193	7.84857	6.66	44.3556	2.58070	8.16088
6.17	38.0689	2.48395	7.85493	6.67	44.4889	2.58263	8.16701
6.18	38.1924	2.48596	7.86130	6.68	44.6224	2.58457	8.17313
6.19	38.3161	2.48797	7.86766	6.69	44.7561	2.58650	8.17924
6.20	38.4400	2.48998	7.87401	6.70	44.8900	2.58844	8.18535
6.21	38.5641	2.49199	7.88036	6.71	45.0241	2.59037	8.19146
6.22	38.6884	2.49399	7.88670	6.72	45.1584	2.59230	8.19756
6.23	38.8129	2.49600	7.89303	6.73	45.2929	2.59422	8.20366
6.24	38.9376	2.49800	7.89937	6.74	45.4276	2.59615	8.20975
6.25	39.0625	2.50000	7.90569	6.75	45.5625	2.59808	8.21584
6.26	39.1876	2.50200	7.91202	6.76	45.6976	2.60000	8.22192
6.27	39.3129	2.50400	7.91833	6.77	45.8329	2.60192	8.22800
6.28	39.4384	2.50599	7.92465	6.78	45.9684	2.60384	8.23408
6.29	39.5641	2.50799	7.93095	6.79	46.1041	2.60576	8.24015
6.30	39.6900	2.50998	7.93725	6.80	46.2400	2.60768	8.24621
6.31	39.8161	2.51197	7.94355	6.81	46.3761	2.60960	8.25227
6.32	39.9424	2.51396	7.94984	6.82	46.5124	2.61151	8.25833
6.33	40.0689	2.51595	7.95613	6.83	46.6489	2.61343	8.26438
6.34	40.1956	2.51794	7.96241	6.84	46.7856	2.61534	8.27043
6.35	40.3225	2.51992	7.96869	6.85	46.9225	2.61725	8.27647
6.36	40.4496	2.52190	7.97496	6.86	47.0596	2.61916	8.28251
6.37	40.5769	2.52389	7.98123	6.87	47.1969	2.62107	8.28855
6.38	40.7044	2.52587	7.98749	6.88	47.3344	2.62298	8.29458
6.39	40.8321	2.52784	7.99375	6.89	47.4721	2.62488	8.30060
6.40	40.9600	2.52982	8.00000	6.90	47.6100	2.62679	8.30662
6.41	41.0881	2.53180	8.00625	6.91	47.7481	2.62869	8.31264
6.42	41.2164	2.53377	8.01249	6.92	47.8864	2.63059	8.31865
6.43	41.3449	2.53574	8.01873	6.93	48.0249	2.63249	8.32466
6.44	41.4736	2.53772	8.02496	6.94	48.1636	2.63439	8.33067
6.45	41.6025	2.53969	8.03119	6.95	48.3025	2.63629	8.33667
6.46	41.7316	2.54165	8.03741	6.96	48.4416	2.63818	8.34266
6.47	41.8609	2.54362	8.04363	6.97	48.5809	2.64008	8.34865
6.48	41.9904	2.54558	8.04984	6.98	48.7204	2.64197	8.35464
6.49	42.1201	2.54755	8.05605	6.99	48.8601	2.64386	8.36062

H. Squares and Square Roots (*continued*)

N	N^2	$\sqrt{N}$	$\sqrt{10N}$	N	N^2	$\sqrt{N}$	$\sqrt{10N}$
7.00	49.0000	2.64575	8.36660	7.50	56.2500	2.73861	8.66025
7.01	49.1401	2.64764	8.37257	7.51	56.4001	2.74044	8.66603
7.02	49.2804	2.64953	8.37854	7.52	56.5504	2.74226	8.67179
7.03	49.4209	2.65141	8.38451	7.53	56.7009	2.74408	8.67756
7.04	49.5616	2.65330	8.39047	7.54	56.8516	2.74591	8.68332
7.05	49.7025	2.65518	8.39643	7.55	57.0025	2.74773	8.68907
7.06	49.8436	2.65707	8.40238	7.56	57.1536	2.74955	8.69483
7.07	49.9849	2.65895	8.40833	7.57	57.3049	2.75136	8.70057
7.08	50.1264	2.66083	8.41427	7.58	57.4564	2.75318	8.70632
7.09	50.2681	2.66271	8.42021	7.59	57.6081	2.75500	8.71206
7.10	50.4100	2.66458	8.42615	7.60	57.7600	2.75681	8.71780
7.11	50.5521	2.66646	8.43208	7.61	57.9121	2.75862	8.72353
7.12	50.6944	2.66833	8.43801	7.62	58.0644	2.76043	8.72926
7.13	50.8369	2.67021	8.44393	7.63	58.2169	2.76225	8.73499
7.14	50.9796	2.67208	8.44985	7.64	58.3696	2.76405	8.74071
7.15	51.1225	2.67395	8.45577	7.65	58.5225	2.76586	8.74643
7.16	51.2656	2.67582	8.46168	7.66	58.6756	2.76767	8.75214
7.17	51.4089	2.67769	8.46759	7.67	58.8289	2.76948	8.75785
7.18	51.5524	2.67955	8.47349	7.68	58.9824	2.77128	8.76356
7.19	51.6961	2.68142	8.47939	7.69	59.1361	2.77308	8.76926
7.20	51.8400	2.68328	8.48528	7.70	59.2900	2.77489	8.77496
7.21	51.9841	2.68514	8.49117	7.71	59.4441	2.77669	8.78066
7.22	52.1284	2.68701	8.49706	7.72	59.5984	2.77849	8.78635
7.23	52.2729	2.68887	8.50294	7.73	59.7529	2.78029	8.79204
7.24	52.4176	2.69072	8.50882	7.74	59.9076	2.78209	8.79773
7.25	52.5625	2.69258	8.51469	7.75	60.0625	2.78388	8.80341
7.26	52.7076	2.69444	8.52056	7.76	60.2176	2.78568	8.80909
7.27	52.8529	2.69629	8.52643	7.77	60.3729	2.78747	8.81476
7.28	52.9984	2.69815	8.53229	7.78	60.5284	2.78927	8.82043
7.29	53.1441	2.70000	8.53815	7.79	60.6841	2.79106	8.82610
7.30	53.2900	2.70185	8.54400	7.80	60.8400	2.79285	8.83176
7.31	53.4361	2.70370	8.54985	7.81	60.9961	2.79464	8.83742
7.32	53.5824	2.70555	8.55570	7.82	61.1524	2.79643	8.84308
7.33	53.7289	2.70740	8.56154	7.83	61.3089	2.79821	8.84873
7.34	53.8756	2.70924	8.56738	7.84	61.4656	2.80000	8.85438
7.35	54.0225	2.71109	8.57321	7.85	61.6225	2.80179	8.86002
7.36	54.1696	2.71293	8.57904	7.86	61.7796	2.80357	8.86566
7.37	54.3169	2.71477	8.58487	7.87	61.9369	2.80535	8.87130
7.38	54.4644	2.71662	8.59069	7.88	62.0944	2.80713	8.87694
7.39	54.6121	2.71846	8.59651	7.89	62.2521	2.80891	8.88257
7.40	54.7600	2.72029	8.60233	7.90	62.4100	2.81069	8.88819
7.41	54.9081	2.72213	8.60814	7.91	62.5681	2.81247	8.89382
7.42	55.0564	2.72397	8.61394	7.92	62.7264	2.81425	8.89944
7.43	55.2049	2.72580	8.61974	7.93	62.8849	2.81603	8.90505
7.44	55.3536	2.72764	8.62554	7.94	63.0436	2.81780	8.91067
7.45	55.5025	2.72947	8.63134	7.95	63.2025	2.81957	8.91628
7.46	55.6516	2.73130	8.63713	7.96	63.3616	2.82135	8.92188
7.47	55.8009	2.73313	8.64292	7.97	63.5209	2.82312	8.92749
7.48	55.9504	2.73496	8.64870	7.98	63.6804	2.82489	8.93308
7.49	56.1001	2.73679	8.65448	7.99	63.8401	2.82666	8.93868

H. Squares and Square Roots (*continued*)

N	N^2	$\sqrt{N}$	$\sqrt{10N}$	N	N^2	$\sqrt{N}$	$\sqrt{10N}$
8.00	64.0000	2.82843	8.94427	8.50	72.2500	2.91548	9.21954
8.01	64.1601	2.83019	8.94986	8.51	72.4201	2.91719	9.22497
8.02	64.3204	2.83196	8.95545	8.52	72.5904	2.91890	9.23038
8.03	64.4809	2.83373	8.96103	8.53	72.7609	2.92062	9.23580
8.04	64.6416	2.83549	8.96660	8.54	72.9316	2.92233	9.24121
8.05	64.8025	2.83725	8.97218	8.55	73.1025	2.92404	9.24662
8.06	64.9636	2.83901	8.97775	8.56	73.2736	2.92575	9.25203
8.07	65.1249	2.84077	8.98332	8.57	73.4449	2.92746	9.25743
8.08	65.2864	2.84253	8.98888	8.58	73.6164	2.92916	9.26283
8.09	65.4481	2.84429	8.99444	8.59	73.7881	2.93087	9.26823
8.10	65.6100	2.84605	9.00000	8.60	73.9600	2.93258	9.27362
8.11	65.7721	2.84781	9.00555	8.61	74.1321	2.93428	9.27901
8.12	65.9344	2.84956	9.01110	8.62	74.3044	2.93598	9.28440
8.13	66.0969	2.85132	9.01665	8.63	74.4769	2.93769	9.28978
8.14	66.2596	2.85307	9.02219	8.64	74.6496	2.93939	9.29516
8.15	66.4225	2.85482	9.02774	8.65	74.8225	2.94109	9.30054
8.16	66.5856	2.85657	9.03327	8.66	74.9956	2.94279	9.30591
8.17	66.7489	2.85832	9.03881	8.67	75.1689	2.94449	9.31128
8.18	66.9124	2.86007	9.04434	8.68	75.3424	2.94618	9.31665
8.19	67.0761	2.86182	9.04986	8.69	75.5161	2.94788	9.32202
8.20	67.2400	2.86356	9.05539	8.70	75.6900	2.94958	9.32738
8.21	67.4041	2.86531	9.06091	8.71	75.8641	2.95127	9.33274
8.22	67.5684	2.86705	9.06642	8.72	76.0384	2.95296	9.33809
8.23	67.7329	2.86880	9.07193	8.73	76.2129	2.95466	9.34345
8.24	67.8976	2.87054	9.07744	8.74	76.3876	2.95635	9.34880
8.25	68.0625	2.87228	9.08295	8.75	76.5625	2.95804	9.35414
8.26	68.2276	2.87402	9.08845	8.76	76.7376	2.95973	9.35949
8.27	68.3929	2.87576	9.09395	8.77	76.9129	2.96142	9.36483
8.28	68.5584	2.87750	9.09945	8.78	77.0884	2.96311	9.37017
8.29	68.7241	2.87924	9.10494	8.79	77.2641	2.96479	9.37550
8.30	68.8900	2.88097	9.11043	8.80	77.4400	2.96648	9.38083
8.31	69.0561	2.88271	9.11592	8.81	77.6161	2.96816	9.38616
8.32	69.2224	2.88444	9.12140	8.82	77.7924	2.96985	9.39149
8.33	69.3889	2.88617	9.12688	8.83	77.9689	2.97153	9.39681
8.34	69.5556	2.88791	9.13236	8.84	78.1456	2.97321	9.40213
8.35	69.7225	2.88964	9.13783	8.85	78.3225	2.97489	9.40744
8.36	69.8896	2.89137	9.14330	8.86	78.4996	2.97658	9.41276
8.37	70.0569	2.89310	9.14877	8.87	78.6769	2.97825	9.41807
8.38	70.2244	2.89482	9.15423	8.88	78.8544	2.97993	9.42338
8.39	70.3921	2.89655	9.15969	8.89	79.0321	2.98161	9.42868
8.40	70.5600	2.89828	9.16515	8.90	79.2100	2.98329	9.43398
8.41	70.7281	2.90000	9.17061	8.91	79.3881	2.98496	9.43928
8.42	70.8964	2.90172	9.17606	8.92	79.5664	2.98664	9.44458
8.43	71.0649	2.90345	9.18150	8.93	79.7449	2.98831	9.44987
8.44	71.2336	2.90517	9.18695	8.94	79.9236	2.98998	9.45516
8.45	71.4025	2.90689	9.19239	8.95	80.1025	2.99166	9.46044
8.46	71.5716	2.90861	9.19783	8.96	80.2816	2.99333	9.46573
8.47	71.7409	2.91033	9.20326	8.97	80.4609	2.99500	9.47101
8.48	71.9104	2.91204	9.20869	8.98	80.6404	2.99666	9.47629
8.49	72.0801	2.91376	9.21412	8.99	80.8201	2.99833	9.48156

H. Squares and Square Roots (*continued*)

N	N^2	$\sqrt{N}$	$\sqrt{10N}$	N	N^2	$\sqrt{N}$	$\sqrt{10N}$
9.00	81.0000	3.00000	9.48683	9.50	90.2500	3.08221	9.74679
9.01	81.1801	3.00167	9.49210	9.51	90.4401	3.08383	9.75192
9.02	81.3604	3.00333	9.49737	9.52	90.6304	3.08545	9.75705
9.03	81.5409	3.00500	9.50263	9.53	90.8209	3.08707	9.76217
9.04	81.7216	3.00666	9.50789	9.54	91.0116	3.08869	9.76729
9.05	81.9025	3.00832	9.51315	9.55	91.2025	3.09031	9.77241
9.06	82.0836	3.00998	9.51840	9.56	91.3936	3.09192	9.77753
9.07	82.2649	3.01164	9.52365	9.57	91.5849	3.09354	9.78264
9.08	82.4464	3.01330	9.52890	9.58	91.7764	3.09516	9.78775
9.09	82.6281	3.01496	9.53415	9.59	91.9681	3.09677	9.79285
9.10	82.8100	3.01662	9.53939	9.60	92.1600	3.09839	9.79796
9.11	82.9921	3.01828	9.54463	9.61	92.3521	3.10000	9.80306
9.12	83.1744	3.01993	9.54987	9.62	92.5444	3.10161	9.80816
9.13	83.3569	3.02159	9.55510	9.63	92.7369	3.10322	9.81326
9.14	83.5396	3.02324	9.56033	9.64	92.9296	3.10483	9.81835
9.15	83.7225	3.02490	9.56556	9.65	93.1225	3.10644	9.82344
9.16	83.9056	3.02655	9.57079	9.66	93.3156	3.10805	9.82853
9.17	84.0889	3.02820	9.57601	9.67	93.5089	3.10966	9.83362
9.18	84.2724	3.02985	9.58123	9.68	93.7024	3.11127	9.83870
9.19	84.4561	3.03150	9.58645	9.69	93.8961	3.11288	9.84378
9.20	84.6400	3.03315	9.59166	9.70	94.0900	3.11448	9.84886
9.21	84.8241	3.03480	9.59687	9.71	94.2841	3.11609	9.85393
9.22	85.0084	3.03645	9.60208	9.72	94.4784	3.11769	9.85901
9.23	85.1929	3.03809	9.60729	9.73	94.6729	3.11929	9.86408
9.24	85.3776	3.03974	9.61249	9.74	94.8676	3.12090	9.86914
9.25	85.5625	3.04138	9.61769	9.75	95.0625	3.12250	9.87421
9.26	85.7476	3.04302	9.62289	9.76	95.2576	3.12410	9.87927
9.27	85.9329	3.04467	9.62808	9.77	95.4529	3.12570	9.88433
9.28	86.1184	3.04631	9.63328	9.78	95.6484	3.12730	9.88939
9.29	86.3041	3.04795	9.63846	9.79	95.8441	3.12890	9.89444
9.30	86.4900	3.04959	9.64365	9.80	96.0400	3.13050	9.89949
9.31	86.6761	3.05123	9.64883	9.81	96.2361	3.13209	9.90454
9.32	86.8624	3.05287	9.65401	9.82	96.4324	3.13369	9.90959
9.33	87.0489	3.05450	9.65919	9.83	96.6289	3.13528	9.91464
9.34	87.2356	3.05614	9.66437	9.84	96.8256	3.13688	9.91968
9.35	87.4225	3.05778	9.66954	9.85	97.0225	3.13847	9.92472
9.36	87.6096	3.05941	9.67471	9.86	97.2196	3.14006	9.92975
9.37	87.7969	3.06105	9.67988	9.87	97.4169	3.14166	9.93479
9.38	87.9844	3.06268	9.68504	9.88	97.6144	3.14325	9.93982
9.39	88.1721	3.06431	9.69020	9.89	97.8121	3.14484	9.94485
9.40	88.3600	3.06594	9.69536	9.90	98.0100	3.14643	9.94987
9.41	88.5481	3.06757	9.70052	9.91	98.2081	3.14802	9.95490
9.42	88.7364	3.06920	9.70567	9.92	98.4064	3.14960	9.95992
9.43	88.9249	3.07083	9.71082	9.93	98.6049	3.15119	9.96494
9.44	89.1136	3.07246	9.71597	9.94	98.8036	3.15278	9.96995
9.45	89.3025	3.07409	9.72111	9.95	99.0025	3.15436	9.97497
9.46	89.4916	3.07571	9.72625	9.96	99.2016	3.15595	9.97998
9.47	89.6809	3.07734	9.73139	9.97	99.4009	3.15753	9.98499
9.48	89.8704	3.07896	9.73653	9.98	99.6004	3.15911	9.98999
9.49	90.0601	3.08058	9.74166	9.99	99.8001	3.16070	9.99500

Students: There is *extra help* available . . .

This text has been written to make your study of statistics as simple and rewarding as possible. But even so there may be concepts or techniques that will be difficult for you. Or you may want a little "insurance" for success as mid-terms and finals come up. If so, you will want to buy a copy of the *Student Supplement for Basic Statistics: An Introduction.* It's described in the preface of this book. Your campus bookstore either has it in stock or will order it for you.

Answers to Selected Exercises

Chapter 1: Set A

1. (a) Discrete; (b) shipment of bumpers; (c) nominal.
3. (a) Nominal; (c) ratio; (e) ratio; (g) nominal; (i) ratio; (k) interval; (m) ratio.
5. (a) (i) 3.25; (ii) 3.246.
   (c) (i) 1.02; (ii) 1.020.
   (e) (i) 3.56; (ii) 3.555.
6. (c) (i) 16; (iii) 7; (iv) 2; (vii) 74; (ix) 20.
7. (a) $\sum_{i=1}^{5} x_i$ (e) $\sum_{i=1}^{3} (x_i - i)^2$

   (c) $\sum_{i=1}^{4} x_{2i}$ (g) $\sum_{i=1}^{3} 5x_i$

Chapter 1: Set B

1. (a) $\mu = 3.23$, $\tilde{\mu} = 3$; (b) $\mu = 3$, $\tilde{\mu} = 3$; (c) $\mu = 3$, $\tilde{\mu} = 3$; (d) $\mu = 1.5$, $\tilde{\mu} = 1$.
3. (a) The 16.5th (midway between 16th and 17th); (b) 44th.
5. (a) \$20; (b) \$21.

7. 6.1 years.

9. (a) 50; (b) 50; $\sigma_1^2 = 2500$; $\sigma_2^2 = 1250$; $\sigma_3^2 = 0$.

11. (a) (i) -1.5; (iii) 1; (v) -7.
(b) 4 and 16.

Chapter 1: Overview

1. (a) $\sum_{i=1}^{N} (aw_i + b)$.

(b) $\sum_{i=1}^{4} (100i)$.

3. (a) 170.6; (c) 25,100; (e) 1.234.

5. (a) $\mu = 2.5$; (b) $\tilde{\mu} = 2$.

7. 1 week.

9. 1416 pounds and 1584 pounds.

11. (a) $+0.71$ and -0.71; (b) 100 percent;
(c) no conflict, because 100 percent is certainly "at least 90 percent."

Chapter 2: Set A

1. 41.97.

3.

x:	32	33	34	35	36	37	38	39	40	
f:	1	0	2	1	1	2	4	1	1	
x:	41	42	43	44	45	46	47	48	49	
f:	2	3	2	1	0	1	3	0	2	
x:	50	51	52	53	54	55	56	57	58	Total
f:	1	0	1	0	0	0	0	0	1	30

(The distribution has been placed horizontally to save space.)

6. (a) 0.13; (b) 0.03; (c) 0.03.

9. 27.

11. 5.

13. No.

15.

x	f
30–34	3
35–39	9
40–44	9
45–49	6
50–54	2
55–59	1
	30

19.

x	f
32–45	21
46–59	9

Not effective.

21. Too few classes summarize the pattern too much; too many classes overemphasize the detail.

23. A bar correctly conveys the impression that all observations represented have the same value; a rectangle, that observations have different values.

Chapter 2: Set B

1. (d) 10.245 rounds to 10.24.
   (e) 10.255 rounds to 10.26.
   (f) 10.025 rounds to 10.02.
   (g) 10.2555 rounds to 10.26.
   (i) 10.258 rounds to 10.26.
   All others round to 10.25.

3. 8.2500005 is one of infinite possible numbers of such values. All such values round to 8.3.

5. Continuous.

7. Eight classes are recommended, but seven are acceptable.

9. (a) 16.5°.
   (b) The upper class boundary of the third class, 28.5, is equal to the lower class boundary of the fourth class.

11. (a) Six classes.
    (b) 2.01
    (c) Stating the interval in hundredths of a degree will make it possible for class boundaries to end in 0.005.
    (d) 0.34.

13. 0.795–1.135
 1.135–1.475
 1.475–1.815
 1.815–2.155
 2.155–2.495
 2.495–2.835

17.

x	F
0.80–1.13	9
1.14–1.47	15
1.48–1.81	20
1.82–2.15	24
2.16–2.49	26
2.50–2.83	30

19.

x	F'
0.80–1.13	0.30
1.14–1.47	0.50
1.48–1.81	0.67
1.82–2.15	0.80
2.16–2.49	0.87
2.50–2.83	1.00

21. Nominal scale; cross-section data.

Chapter 2: Overview

1. Lower class boundary, lower class limit, upper class limit, upper class boundary.

2.

x:	32	33	34	35	36	37	38	39	40
F:	1	1	3	4	5	7	11	12	13
x:	41	42	43	44	45	46	47	48	49
F:	15	18	20	21	21	22	25	25	27
x:	50	51	52	53	54	55	56	57	58
F:	28	28	29	29	29	29	29	29	30

(The distribution has been placed horizontally to save space.)

5.

x	F'
81–85	0.13
86–90	0.30
91–95	0.60
96–100	0.82
101–105	0.90
106–110	0.97
111–115	1.00

7. More than one value occurs in each class.

9. (a) Frequency distribution or relative frequency distribution for many-value classes, continuous or discrete variables.

   (b) Cumulative frequency distribution or cumulative relative frequency distribution for many-value classes, continuous or discrete variables.

11. Upper end. There are a few widely scattered high incomes, but many tightly packed low incomes.

Chapter 3: Set A

2. (a) The actual values of the observations in each class; (b) the class midvalues.

3. (a) $f_1 = 9$; $f_2 = 12$.
   $\sum f = 41$.
   $N = 41$.
   $f_1 x_1 = 27$; $f_2 x_2 = 48$.

   (b) $\sum fx = 190$.

   (c) $\mu = 4.63$.

   (d) 4.

   (e) 4.

5. $^8/_{13}$.

7. 35.

9. (a) 1.608; (b) 1.48; (c) 1.00.

11. 10.15.

13. Mean: 104.81; median: 105; mode: 105.5.

Chapter 3: Set B

1. (a) 1; (b) 2.

3. Distribution A would not have as many frequencies near the center as B, and A would have more frequencies in the tails than B. A would tend to be flatter and more scattered than B.

5. Distribution (a) has the greatest dispersion; (c) has the least. The variance is greatest for (a) and least for (c).

7. (a) 1.73; (b) 1.73.

9. $\sigma^2 = 19.64$; $\sigma = 4.4317$, or 4.43.

11. The answers are and should be the same. Equation 3.2(a) was "messier" because deviations had to be found and squared.

13. $\sigma = i\sqrt{\dfrac{\sum fd^2}{N} - \left(\dfrac{\sum fd}{N}\right)^2}$.

15. $\mu = 5.1$, $\sigma = 3.36$.
    The fourth class interval is 4 while the others are 3.

17. (a) No skewness. Symmetrical around 5.
    (b) Positive skewness. Long tail to the right.

19. (a) Positive; (b) none.

Chapter 3: Overview

1. (a) Mean: $4\frac{1}{12}$; median: 3.5; mode: 3.
   (b) Mean: $4\frac{5}{13}$; median: 4; mode: 3.

3. (a) 9.25.
   (b) 9.
   (c) 9.
   (d) 0.4875.
   (e) 0.6982.
   (f) Positive.

5. (a) All equal; (b) none; (c) either.

7. (a) 1.95, meaningless; (b) 2 (meaningless); (c) 1 (black).

9. (a) Meaningless (see Exercise 8); (b) meaningless (see Exercise 8); (c) Boozo.

Chapter 4: Set A

1. $S \cup L = \{1,3,7,8,9,10,11,12\}$; $B \cap U = \{1,4\}$; $U \cup S = \{1,3,4,7,8,9,10\}$;
   $U \cap S = \{1,7\}$; $B \cap M = \{\varnothing\}$; $L \cap M' = \{10,11,12\}$; $U' \cap S' = \{2,5,6,11,12\}$.

3. $1 = P(\mathrm{U}) = P(A \cup A') = P(A) + P(A')$; hence, $1 - P(A) = P(A')$.

5. (a) 52.
   (b) 4,0,1,48,12,3,0,52,26,0.

7. (b), (c), (e).

9. (a) $\frac{1}{13}$; (b) $\frac{12}{13}$; (c) 0; (d) $\frac{1}{2}$; (e) $\frac{10}{13}$; (f) $\frac{3}{13}$; (g) $\frac{1}{52}$; (h) $\frac{1}{52}$;
   (i) $\frac{1}{2}$; (j) $\frac{24}{52}$.

11. Parts (b) and (c) are not valid because $P(A \cup B) \neq 1.0$. Also in (c), $P(A \cap B) > P(A)$.

Chapter 4: Set B

1. (a) $C \cap S$ = event that a serving contains cream and sugar.
   (b) C' = event that a serving does not contain cream.

3. (a) If the events are independent; (b) yes.

4. (a) Dependent; (b) (i) 0, (ii) 0.

6. (a) 0.15; (b) 0.75; (c) 0.25; (d) 0.45.

8. (a) 0; (b) 0; (c) 0.65; (d) 1.0.

13. (a) 25; (b) 14; (c) 3; (d) 5; (e) 0.

14. (a) $\frac{14}{25}$; (b) $\frac{3}{25}$; (c) $\frac{3}{25}$; (d) 0.

17. (a) 1.0; (b) 0; (c) $\frac{3}{14}$; (d) 0.

20. (a) 0.60; (b) 0.55; (c) $\frac{11}{12}$; (d) $\frac{11}{14}$; (e) $\frac{5}{8}$.

21. (a) Independent; (b) $(\frac{1}{6})^6$; (c) $1 - (\frac{1}{6})^6$; (d) $(\frac{1}{6})^6$;
    (e) $2(\frac{1}{6})^6$; (f) 0.

Chapter 4: Overview

1. $N(R) = 8$; $N(R') = 4$; $N(W) = 7$; $N(W') = 5$.

2. $N(R \cup W) = 8$; $N(R' \cup W') = 6$.

3. $N(C \cap R \cap W) = 3$; $N(C \cap R' \cap W) = 4$; $N(C \cap R \cap W') = 5$;
   $N(C' \cap R \cap W) = 2$.

7. If they are mutually exclusive; otherwise $P(A \cup B) = P(A) + P(B) - P(A \cap B)$.

8. $P(B \cap C) = 0.3$.

10. (a) Independent; (b), (c), (d) not independent.

14. (a) Dependent.
    (b) (i) Both balls drawn are red.

(ii) At least one red ball is drawn.
(iii) Both balls drawn are even-numbered red ones.
(iv) First ball drawn is white or black or odd-numbered.
(v) First ball drawn is a white or black odd-numbered one.
(vi) First ball drawn is red or black.

(c) (i) R_2'; (ii) $(E_1' \cap E_2')'$; (iii) $R_2 \cup W_2$.

(d) (i) $5/39$; (ii) $10/39$; (iii) $29/39$; (iv) $1/39$; (v) $14/39$; (vi) $1/26$.

(e) $(R_1 \cap E_1) \cap (W_2 \cap E_2')$.

Chapter 5: Set A

2. Yes; $n = 4$.

4. The distributions differ in dispersion, with the dispersion of the sampling distribution of the mean becoming smaller (in relation to the original distribution) as sample size increases.

6. 36.

8. (a)

6	7	8	9	10	11	12	13	14	15	16	17
0	0	0	0	0	0	0	0.02	0.02	0.03	0.01	0

(b) $\mu = 0.0067$; $\sigma^2 = 0.00011$.

x	$P(x)$
0.00	$8/12$
0.01	$1/12$
0.02	$2/12$
0.03	$1/12$

(c) (i) $8/12$; (ii) $4/12$; (iii) 0; (iv) $4/12$; (v) $4/12$.

11. (a) By relative frequency for a large number of nights; (b) 2.07; (c) 756 approximately.

12. (a) 0.01; (b) 0.1104; (c) 0.0625.

Chapter 5: Set B

2. (a) 0.0504; (b) 0.1225.

3. (a) 0.3674; (b) 0.4850.

6. Its expected value is zero.

7.

x	6	11	15	21	25	30
$P(x)$	$1/6$	$1/6$	$1/6$	$1/6$	$1/6$	$1/6$

9.

x	0	1	2	3	4	5	6
$P(x)$	$1/27$	$3/27$	$6/27$	$7/27$	$6/27$	$3/27$	$1/27$

11. (a) 0.36; (b) 0.60.

13. 0.364.

16. (a) (i) 343/1728; (ii) 232/1728; (iii) 12/1728.
(b) (i) 343/1728; (ii) 0; (iii) 232/1728; (iv) 0.

17. (a) 6; (b) 5040; (c) 3,628,800; (d) 10; (e) 7; (f) 5040; (g) 3,628,800; (h) 10; (i) $N(N-1)(N-2)/6$; (j) $N(N-1)(N-2)(N-3)(N-4)$.

19. 50,965,320.

21. (a) 1000; (b) 720.

Chapter 5: Overview

4. (a) 5.0.

(b)

$\bar{x}$	2.0	2.5	3.0	4.5	5.0	5.5	7.0	7.5	8.0
$f(\bar{x})$	1	2	1	2	4	2	1	2	1

$\mu_{\bar{X}} = 5.0$; $\sigma_{\bar{X}} = 1.803$.

(c) Yes.

(d)

$\bar{x}$	2.5	4.5	5.0	5.5	7.5
$f(\bar{x})$	2	2	4	2	2

$\mu_{\bar{X}} = 5.0$; $\sigma_{\bar{X}} = 1.472$.

9. 0.88; if six calls are made, the probability of a sale is 0.93952.

10. 120.

12. (a) $1/6$; (b) $3/10$.

15. (a) (i) 18,564; (ii) 125,970; (iii) 924; (iv) (i) × (ii) × (iii).
(b) Twice.

16. The answers are numerically the same.

Chapter 6: Set A

3. (a) ${}_7C_0(0.23)^0(0.77)^7$.

(b) ${}_7C_3(0.23)^3(0.77)^4 + {}_7C_4(0.23)^4(0.77)^3 + {}_7C_5(0.23)^5(0.77)^2 + {}_7C_6(0.23)^6(0.77) + {}_7C_7(0.23)^7$.

(c) ${}_7C_0(0.23)^0(0.77)^7 + {}_7C_1(0.23)^1(0.77)^6$.

(d) ${}_7C_3(0.23)^3(0.77)^4$.

(e) Same as part (b).

(f) ${}_7C_2(0.23)^2(0.77)^5 + {}_7C_1(0.23)(0.77)^6 + {}_7C_0(0.23)^0(0.77)^7$.

7. (a) 0.9437; (b) 0.1493; (c) 0.9165.

8. Separate probabilities from Equation 6.3 have to be calculated and summed in each case. Appendix E does not include probabilities for $\pi = 0.71$.

11. (a) $\frac{1}{16}$; (b) $\frac{1}{4}$, binomial applies here.

14. 0.2211.

Chapter 6: Set B

2. 0.0010; 0.9990; 0.5; 500; 500.5; 0.10.

4. (a) (i) 0.2937; (ii) 0.2680; (iii) 0.2448.
   (b) Ms. Fitch; no.

6. (a) 6, 2.19; (b) 8.0, 2.71; (c) 1.0, 0.707; (d) 13.3, 0.815.

8. The mean increases; the standard deviation increases at first and then decreases.

Chapter 6: Overview

2. (a) 625/1296; (b) 25/216; (c) 171/1296; (d) 0; (e) 1/16; (f) 16/81; (g) 1/81; (h) 1/1296.

4. $\frac{1}{2}$; sampling is without replacement.

5. (a) 0.2123; (b) 0.0515.

7. (a) 0.0116; (b) 0.1672.

8. Yes; a maximum acceptance number of $r = 3$.

Chapter 7: Set A

1. (a) 1.0; (b) $\frac{2}{9}$; (c) Yes; $\mu = 9.45$.

3. 8.30 inches.

5. (a) 9.5; (b) 6.5; (c) 12.1.
8. (a) 1.0; (b) 0.14; (c) -1.08; (d) -3.90.
9. (a) 0.8413; (b) 0.4443; (c) 0.1401; (d) 0.14 approximately.
10. (a) and (h) are false.
12. The mean and standard deviation of X.
15. (a) No general conditions; (b) if $x < \mu$; (c) never; (d) never.
16. The probability density is less in the second interval.
18. (b), (c), (e), and (f).
19. (a) 1.28; (b) -1.28; (c) -2.33; (d) 0.00; (e) -0.01; (f) -0.42; (g) 1.96.
22. $z < -1.64$ or $z > 1.64$.
23. $x < 5.44$ or > 18.56; 0.05, 0.95.
27. 0.21864 gallon.
28. 3.2 percent.

Chapter 7: Set B

2. 0.9082.
4. 0.1587; 0.0062.
6. (a) \$3492.72; (b) \$3450.97, \$3534.77.
8. (a) 0.5517; (b) 0.2514; (c) 0.3936; (d) 0.5000.
11. (a) 0.3413; (b) 0.3153.
13. (a) 0.6179; (b) 0.7257; (c) 0.9332; (d) 0.9986.
16. (a) 0.1140; (b) 0.0456.
17. (a) 104.8, 104.8, 104.8, 104.8; (b) 105.47, 105.14, 104.93, 104.87;
(c) 102.47, 103.64, 104.33, 104.57.

Chapter 7: Overview

3. (a) -1.28; (b) 0.25; (c) 1.28; (d) 2.33; (e) 2.58.
4. $1/16$; $3/16$; $5/16$; $7/16$.

6. (a) 10.0; (b) −8.0; (c) 11.20; (d) −23.72; (e) 4.0; (f) 46.

8. (a) 0.2266; (b) 0.3085; (c) 0.8849; (d) 0.3830; (e) 0.8904; (f) 0.7066.

11. It is more concentrated about the population mean–in accordance with the smaller standard errors for increasing sample size.

15. Virtually 1.0.

16. (a) (i) 0; (ii) $\frac{1}{16}$; (iii) $\frac{3}{32}$; (iv) $\frac{1}{4}$.
    (b) 1.0.
    (c) (i) $\frac{1}{16}$; (ii) $\frac{15}{16}$; (iii) 1.0; (iv) 0.0; (v) $\frac{5}{16}$; (vi) 0.0.
    (d) The area under the entire function would not equal 1.0.

Chapter 8: Set A

1. Absence of bias and the existence of a confidence coefficient.

3. Accompanied by a measure of confidence in the estimate.

5. (a) 0.9876; (b) $\mu_L = 2.33$, $\mu_U = 15.67$.

8. (a) 25 to 35; (c) 28.59 to 31.41.

11. (a) −0.658 to +0.658; (c) −0.932 to +0.932.

13. The parent population must be normally distributed to use Equation 8.2 with a sample of only 16. We can take a sample of at least 30 and use it.

15. By choosing a sufficiently large sample. In practice, we stop when the cost of another observation is greater than its value in reducing error.

17. (a) 58; (b) 230; (c) 918.

19. $773.20 to $890.80.

Chapter 8: Set B

1. (a) If H_0 were rejected, an upper one-tailed test would allow them to say that the evidence shows the deodorant protects for at least 25 hours.
   (b) A two-tailed test would help to protect users from both overdosage and underdosage.

3. (a) 1.28; (b) 1.645.

5. (b) 10.92; (e) yes, $8 < 10.92$.

6. (a) H_0: $z = 0$; H_A: $z \neq 0$.
   (b) Acceptance region: $-1.96 < z < 1.96$.
   (c) $z = -2.93$ for $\bar{x} = 8$; reject H_0.

9. (a) 0.02 for $\mu = 12$; (c) 0.3632; (f) meaningless.

11. Alpha is smaller and β is larger in each case.

13. For 9.80, z is -5. The critical value of $z = -2.33$. Conclude that the lot does not exceed 10 grains.

Chapter 8: Overview

1. Estimation.

3. (a) Type I; (b) Type II.

5. μ_0 is the hypothesized value of the population mean. It usually differs somewhat from μ, the actual value. The sample mean, $\bar{x}$, is compared with the critical value of $\bar{x}$, or it is used to find a value of z to compare with the critical value of z.

7. $14.97 < \mu < 25.29$.

9. $128.08 < \mu < 135.92$.

Chapter 9: Set A

1. The population variance σ^2; yes; the population mean μ; the sample mean $\bar{x}$.

3.

$S^2 = (n - 1)s^2/n$	$P(S^2)$
0	4/16
1	6/16
4	4/16
9	2/16

Expected value of $S^2 = 2.5$, which is smaller than σ^2.

5. (a) 38/9; (b) 38/9.

7. There are fewer than 30 observations in each sample. No, even if the population is normal there must be at least 30 observations.

10. When σ is known we use $\sigma/\sqrt{n}$, the actual standard error of the mean. When σ is unknown we use $s/\sqrt{n}$, the estimated standard error of the mean.

12. (a) 62; (b) 16.

14. A sample of slightly over 30 should be enough. Hence 62 is more than adequate.

16. (a) 195, 984; (b) 195, 984.

Chapter 9: Set B

2. (a) None, the sample size must be at least 50; (b) none, $n \geq 50$;
   (c) none, the sample size must be at least 50; (d) z.
4. (a) -3.182, -2.131, -2.052; (c) 3.355, 2.797.
6. (a) 3.059 to 8.941; (b) 0.324 to 11.676; (c) -1.301 to 13.301.
8. (a) 2.16; (b) 2.14.
10. z.
12. (a) σ^2, yes; (b) σ, no.
14. 11.

Chapter 9: Overview

1. 1172.36 to 1207.64.
3. 227.4 to 244.6, using z.
5. (a) $\bar{x} = 15$, $s = 2$; (b) 13.25 and 18.75; (c) no, 15 is in the acceptance region.
7. (a) Lower limit: $80 - 1.2z_{\alpha/2}$; large sample ($n > 50$).
   (b) Reject H_0: $\mu = 80$.
9. (a) Type II; $\beta = 0.386$ ($z = -2/7$).
   (b) Type II; $\beta = 0.92$ ($z = -10/7$).
11. Critical value $= -1.860$; since $\bar{x}$ is 1.8 and t is -2.143, the null hypothesis ($\mu \geq 2.3$) should be rejected.
13. (a) None ($n < 50$); (b) none ($n < 50$);
    (c) lower critical value $= 11.846$; reject null hypothesis that μ is 12.

Chapter 10: Set A

1. (a) $r_L = 1$, $r_U = 10$; (b) no lower limit, $r_U = 6$.
4. (a) 0.0192; (b) 0.0; both are less than 0.02.
6. (a) (i) None; (ii) 0.10 to 0.90; (iii) 0.05 to 0.95.
   (b) (i) 13; (ii) 100; (iii) 167.

9. (a) $26.5 \leq x \leq 27.5$; (b) $1.5 \leq x \leq 5.5$; (c) $0.5 \leq x \leq 6.5$; (d) $x \geq 69.5$.

12. (a) 0.006; (b) 0.3812; (c) 0.6141 approximately; (d) 0.8980.

14. $z = -0.22$; accept H_0.

Chapter 10: Set B

2. 14.07; 2.17.

3. (a) 0.2107; (b) 12.44; (c) 100.62.

5. The distribution of discrepancies under the null hypothesis is shifting to the right as the degrees of freedom (opportunities for discrepancy) increase.

8. $\chi^2_{\text{obs}} = 1.03$; accept H_0.

11. City A: $\chi^2 = 0.7$; City B: $\chi^2 = 3.7$; City C: $\chi^2 = 3.9$. Accept H_0 in each case for any $\alpha = 0.10$ or less.

Chapter 10: Overview

4. (a) 0.0401, 0.0383; (b) 0.1492, 0.1501; (c) 0.1210, 0.1222.

5. (a) 0.0150; (b) 0.1052; (c) 0.0668.

6. $z = 2.7$; reject H_0.

11. 0.4019 – 0.6781; yes.

12. 543; no; the confidence level and the error limit.

14. $\chi^2_{\text{obs}} = 1.334$, $\chi^2_{.90,m=1} = 2.71$; no.

16. $\chi^2_{\text{obs}} = 13.440$, $\chi^2_{.99,m=3} = 9.21$; reject H_0.

Chapter 11: Set A

2. The standard error of estimate is smaller for the least-squares regression line than for any other.

4. $y = 2.829 + 1.686x$.

6. $s_{y \cdot x} = 0.653$; see the answer to Exercise 2.

8. Zero; zero.

Chapter 11: Set B

1. (a), (c), and (f) are parameters.

3. (a) The square of a number (r^2) can never be negative;
   (b) r^2 is the proportion of y variance explained by regression and a proportion cannot exceed 1.

7. (a) The one for $x = 9.2$; (b) the one for $x = 12$;
   (c) the interval closest to $\bar{x}$ will be smaller; the interval farthest from $\bar{x}$ will be largest.

10. $t = 1.79$; critical value: $t = \pm 2.365$. Accept the null hypothesis that β is -0.35.

11. (a) 2.69 ± 1.24; (b) 2.21 ± 1.25; (c) 3.88 ± 1.48.
    Because the values of $(x' - \bar{x})$ differ.

13. 0.67; 0.33.

Chapter 11: Overview

1. (a) Regression; (c) correlation; (e) correlation; (g) correlation.

3. (a) $y_c = 6.37 - 0.22x$; (b) 4.61, or 5.

6. 0.8958.

7. 0.8906.

Chapter 12: Set A

1. (a) They must be normally distributed for small samples.
   (b) 0.1702.

(c) H_0: $\mu_1 - \mu_2 = 0$; two-tailed: $z = -4.70$; criterion: -1.96; conclusion: means significantly different.

3. (a) σ_1 and σ_2 are unknown so Equation 12.3 applies. Minimums are 30. Populations must be normal.

   (b) 1.387.

   (c) $z = -0.50$; criterion: -2.58; conclusion: means not significantly different.

5. H_0: $\mu_2 - \mu_1 \leq 0$; $t = 0.77$; criterion: 1.895; conclusion: μ_2 not significantly greater than μ_1.

Chapter 12: Set B

1. 2.84; 2.09.

3. (a) Row 3, column 5; (c) row 2, column j; (e) row i, column j.

5. SST = 52; SSA = 30; $F = 3.64$; criterion: 4.07; conclusion: no significant difference in means.

Chapter 12: Overview

1. 2.30.

3. $\bar{d} = 0.25$; $s_{\bar{d}} = 1.238$; $t = 0.202$; criterion: 2.201; conclusion: no significant difference.

5. (a) SSA = 35.37; SSE = 31.55; $F = 5.04$; criterion: 4.26; conclusion: means significantly different at the 0.05 level.

   (b) The three means are 7, 8.2, and 11.25. Choose A.

Chapter 13: Set A

1. (a) \$11.31; (b) \$9.82.

7. 2.99 – 4.51.

8. 3.00 – 4.50

14. (a) 2.00, 4.75, 1.75, 3.50, 2.00; 10.00, 2.75, 8.75, 1.00, 2.00; (b) 1.307; (c) 1.232; (d) 2.80, 0.307; (e) 1.64 – 3.96.

15. (b) 1.633; (c) 1.261; (e) 1.59 – 4.01.

Chapter 13: Set B

3. (a) 39.57–52.43; (b) 38.84–53.16.
4. Proportions (of this order) in samples of size 50 will be normally distributed.
6. (a) Definition of full-time; (b) meaning is vague; (c) wording–better to ask what would attract respondent to lecture series; (d) probably OK; (e) leads respondent to answer yes; (f) is this the most understandable classification? In any event, survey should provide for other answers.
7. The phone survey would get a more complete response from intended respondents, leading to less opportunity for selection bias.

Chapter 13: Overview

1. (a) Stratified is more work; (b) stratified is more accurate.
2. (a) Cluster is less work; (b) cluster is less accurate.
3. (a) Strata would be effective in reducing sampling error in determining mean age; (b) among-cluster variance would add to the sampling error in determining mean age.
4. The numerical order of dwelling units is associated with neighborhoods, so that the systematic samples would possess some of the benefits of stratification.

Chapter 14: Set A

1. (a) $n_1 = 5; n_2 = 6$.
   (b) 1; 11.
   (c) 11.

(d) 26.

(e) 20; 18; 17.

(f) Does the open classroom produce significantly higher reading comprehension scores?

3. (a) H_0: $W \leq S$; H_A: $W > S$. Or H_0: $\pi \geq 0.5$; H_A: $\pi < 0.5$ where π is the probability of a plus sign.

(b)

Pair:	1	2	3	4	5	6	7	8	9	10	11	12	13	14	15	16
Sign:	−	−	+	+	−	−	−	−	−	+	−	−	0	−	+	−

(c) Zero; this pair is dropped to make 15 pairs in all.

(d) $n = 15$.

(e) 0.5.

(f) $P(r \leq 4 \mid \pi = 0.5, n = 15) = 0.0593$; criterion: 0.05; conclusion: accept H_0.

5. The direction of the differences must conform to the null hypothesis.

7. $P(r \leq 2 \mid \pi = 0.5, n = 10) = 0.0547$. "After" not significantly less than "before" weights.

9. $P(r \geq 32 \mid \pi = 0.5, n = 40)$ is zero to three places by normal approximation. $z = 3.64$; criterion: 2.58. Brand B got significantly better mileage.

Chapter 14: Set B

1.

A	B	C	D
9	1	24	23(22)
12	8	22(23)	5
13	11	6	19
16	14	3	17
15	18	10	7
4		21	2
			20

Subtract numbers shown from 25 if you ranked from largest to smallest.

3. (a) $k - 1 = 3$; (b) 9.35.

5. The outcome is not affected.

9. (a) 16.

(b) Either.

(c) -0.4059.

(d) $t = -1.66$; df = 14; criterion: ± 2.624; conclusion: r_s not significant.

(e) IQ and competency are not related. Don't use IQ as a guide to competency based on these findings.

Chapter 14: Overview

1. Chi square for independence; rank-sum; sign; Kruskal-Wallis; rank-difference correlation.
5. Beginning with 0.79 as rank 1, the sum of ranks for B is 41. The criterion is 45. Hence B is significantly weaker or, alternatively, A is significantly stronger.
7. $z = -3.037$; criterion: ± 2.58. Productivities significantly different. A appears better.
9. The test should be run twice, once with rank 14 in column 1 and rank 15 in column 2 and once with the ranks reversed. If the same conclusion results, there is no problem. If different conclusions result, the test is inconclusive.

Chapter 15: Set A

3. 88.6; 90.7; 88.9; 100.0; 85.9; 103.9.
5. 116.0.
8. 128.0.
9. The quantity index measures the effect of quantity changes on total value. In the total carats produced, gems increased and industrial diamonds decreased. The effect on total value, since gems are more valued in price, was a substantial increase.
13. No; only that prices in the U.K. have increased more since 1970 than prices in the U.S.

Chapter 15: Set B

3. $468.52 million; $360.56 million.
4. 100.9.
5. $464.34 million. The change in imports valued at 1965 prices is $+28.8$ percent, as compared with $+28.0$ percent from the quantity index in Exercise 8 of Set A. They should agree, since value $\div$ price $=$ quantity.
8. 133.4; 132.5.
9. No; pork consumption has stayed about the same and beef consumption has increased. The Paasche index weights beef more heavily, and it has gone up less in price.
13. $2158, $2547, $3046; the figures measure change in real per capita personal income (or income adjusted for price change) in terms of 1958 price levels.

Chapter 15: Overview

4. Clams; clams have the smallest relative price increases and are given much greater relative weight by the 1970 quantities than by the 1950 quantities.
5. 150.14; the shellfish catch, valued at 1965 prices, increased by 50.14 percent from 1950 to 1970.
6. $81.20 million, $101.50 million, $125.52 million; the 1970 shellfish catch, valued at 1965 prices, was $125.52 million.
7. 111.32.

Index

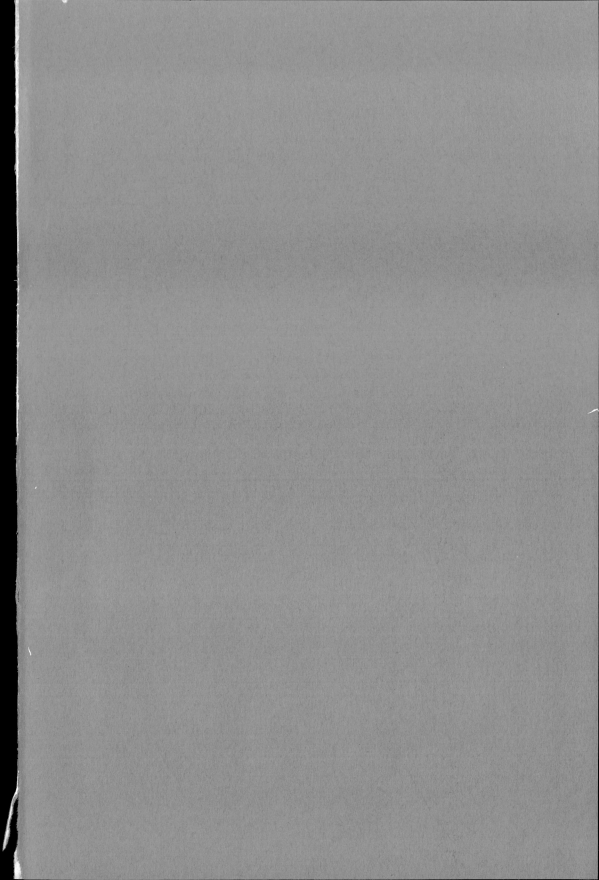